Pravin Kumar Singh and Dilip Kumar Dwivedi (Eds.)
Perovskite Solar Cells

Also of interest

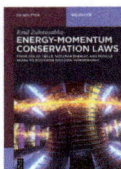

Energy-Momentum Conservation Laws
From Solar Cells, Nuclear Energy, and Muscle Work to Positron Emission
Tomography
Emil Zolotoyabko, 2024
ISBN 978-3-11-134345-7, e-ISBN (PDF) 978-3-11-134354-9
e-ISBN (EPUB) 978-3-11-134392-1

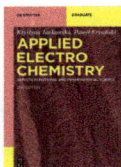

Applied Electrochemistry
Aspects in Material and Environmental Science
Krystyna Jackowska, Paweł Krysiński, 2nd Edition 2024
ISBN 978-3-11-116034-4, e-ISBN (PDF) 978-3-11-116098-6,
e-ISBN (EPUB) 978-3-11-116344-4

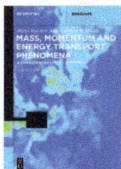

Mass, Momentum and Energy Transport Phenomena
A Consistent Balances Approach
Harry Van den Akker, Robert F. Mudde, 2023
ISBN 978-3-11-124623-9, e-ISBN (PDF) 978-3-11-124657-4
e-ISBN (EPUB) 978-3-11-124715-1

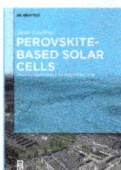

Perovskite-Based Solar Cells
From Fundamentals to Tandem Devices
Saida Laalioui, 2022
ISBN 978-3-11-076060-6, e-ISBN (PDF) 978-3-11-076061-3,
e-ISBN (EPUB) 978-3-11-076065-1

Organic and Hybrid Solar Cells
Lukas Schmidt-Mende, Stefan Kraner, Azhar Fakharuddin, 2nd Edition
2024
ISBN 978-3-11-073692-2, e-ISBN (PDF) 978-3-11-073693-9,
e-ISBN (EPUB) 978-3-11-073152-1

Perovskite Solar Cells

Sustainable Materials, Device Design, Renewable Energy Applications

Edited by
Pravin Kumar Singh and Dilip Kumar Dwivedi

DE GRUYTER

Editors

Dr. Pravin Kumar Singh
Institute of Advanced Materials
IAAM
Gammalkilsvägen 18
590 53 Ulrika
Sweden
Email: pravin.singh@iaam.se

Prof. Dilip Kumar Dwivedi
Madan Mohan Malaviya
University of Technology
273016 Gorakhpur
Uttar Pradesh
India
Email: todkdwivedi@gmail.com

ISBN 978-3-11-172682-3
e-ISBN (PDF) 978-3-11-172684-7
e-ISBN (EPUB) 978-3-11-172716-5

Library of Congress Control Number: 2025942102

Bibliographic information published by the Deutsche Nationalbibliothek
The Deutsche Nationalbibliothek lists this publication in the Deutsche Nationalbibliografie;
detailed bibliographic data are available on the Internet at http://dnb.dnb.de.

© 2025 Walter de Gruyter GmbH, Berlin/Boston, Genthiner Straße 13, 10785 Berlin
Cover image: sankai/E+/Getty Images
Typesetting: Integra Software Services Pvt. Ltd.

www.degruyterbrill.com
Questions about General Product Safety Regulation:
productsafety@degruyterbrill.com

Preface

The field of perovskite solar cells (PSCs) has witnessed unprecedented progress over the past decade, transforming from a scientific curiosity to a leading contender in next-generation photovoltaics. With record efficiencies now surpassing 26% in lab-scale devices and rapid advancements in stability and scalability, perovskite technology is poised to revolutionize solar energy harvesting.

This book provides a comprehensive exploration of the science, engineering, and commercialization pathways of this groundbreaking technology. Each chapter systematically covers key aspects from material synthesis and device physics to advanced fabrication techniques and emerging applications. Whether you are a researcher, engineer, or industry professional, this book is an authoritative reference and a forward-looking guide to the challenges and opportunities in perovskite photovoltaics.

We have carefully structured the content to balance fundamental principles with cutting-edge innovations. The opening chapters delve into PSCs' historical evolution, material properties, and interfacial engineering. Subsequent sections explore advanced device architectures, tandem solar cells, and hybrid energy systems. Finally, the book examines real-world applications, including building-integrated photovoltaics, wearable electronics, and sustainable energy solutions.

We aim to equip readers with a holistic understanding of perovskite solar cells while inspiring further research and development in this dynamic field. We extend our deepest gratitude to the contributing authors whose expertise has made this book possible.

We hope this work will serve as a valuable resource for the scientific community and accelerate the transition toward efficient, low-cost, and sustainable solar energy technologies.

https://doi.org/10.1515/9783111726847-202

Preface

Contents

About the editors

Dr. Pravin Kumar Singh is a scientist at the Institute of Advanced Materials, IAAM, Sweden. He has been awarded Marie Skłodowska-Curie Individual Fellowships (MSCA IF) funded by the European Research Grant Agency, European Commission. Before this fellowship, he worked as a Prof. Kobayashi Shunsuke Postdoctoral Fellow at KTH Royal Institute of Technology, Sweden, and served as a group leader at VBRI Innovation Centre, New Delhi, India. He was an assistant professor in physics at KIPM College of Engineering and Technology in Gorakhpur, India. Before that, he worked as a guest faculty in Madan Mohan Malaviya University of Technology, Gorakhpur, India, for 2 years. He has a Ph.D. in physics and two master's degrees, an M.Sc. in electronics and communication, and an M.Tech. in optoelectronics and laser technology. As per his research contributions, he has written a book, and more than 65 papers published in well-known international journals. He is the managing editor of *Advanced Materials Proceedings*, associate editor of *Science Advances*, and reviewer of many high-impact reputed journals.

Prof. D. K. Dwivedi is currently professor and head, Department of Physics and Material Science, Madan Mohan Malaviya University of Technology, Gorakhpur. He obtained his M.Sc. and Ph.D. from D.D.U. Gorakhpur University, Gorakhpur. He started his carrier as scientific officer in Bhabha Atomic Research Centre, Mumbai, in 2001. He has served as lecturer in physics at D.D.U. Gorakhpur University for nearly 8 years. He joined as reader in Madan Mohan Malaviya University of Technology, Gorakhpur, in 2009. He is a globally recognized scientist (top 2% worldwide, 2024). He has more than 300 publications in SCI indexed international journals, 42 papers in national journals, and 206 conference papers to his credit. He has delivered 53 invited lectures. He has authored five books. He has supervised 15 doctoral thesis and 8 are in progress. He has supervised 42 M.Sc./M.Tech. students in dissertation project. He has been awarded three major research projects. He has organized 11 national level conferences/workshops/short-term courses. His areas of interest are amorphous semiconductors, nanostructured materials, energy storage devices (Li-ion batteries), solar cell devices, and photonic crystal fiber sensors. He has served at almost all the administrative positions such as head of physics and material science department, dean of undergraduate studies and entrepreneurship, dean of postgraduate studies and research and development, dean of faculty affairs, chairman of the administrative committee, chairman recruitment cell, chairman of board of studies, chairman of departmental purchase committee, member of board of management, member secretary of IQAC, member of apace advisory committee, member of examination committee, member of university admission committee, member of university student grievance redressal committee, QIP co-ordinator, chairman of the council of student activities, member of project monitoring unit RUSA, member of research and consultancy management committee, member of flexible cadre structure, member of board of studies of different universities, officiating vice chancellor, and many more. He is editor in 12 reputed journals including *Nature Scientific Reports* and referee in more than 60 international journals. He has three national patents. He is member of six academic societies/professional body and associations of national as well as international level. He is life fellow of Optical Society of India and senior member of IEEE society. He has been awarded by the Har Gobind Khorana Diamond Jubilee Medal and BEST RESEARCHER AWARD (2023) on SENSING TECHNOLOGY Sensors 2023.

https://doi.org/10.1515/9783111726847-204

Anjali Gupta, Anchal Srivastava, and R. K. Shukla*

Chapter 1
History and development of perovskite solar cells

Abstract: Perovskite solar cells (PSCs) have emerged as a ground breaking advancement in the field of photovoltaics due to their remarkable power conversion efficiencies and cost-effective fabrication processes. Initially discovered in the early nineteenth century as a calcium titanium oxide mineral (CaTiO$_3$), perovskite materials have undergone extensive structural and functional evolution, leading to their adoption in solar energy applications. The development of hybrid organic–inorganic halide perovskites, especially methylammonium lead halides, has revolutionized solar cell technology, achieving efficiencies surpassing 25% in under two decades. This review outlines the historical milestones, structural transformations, material compositions, and architectural optimizations that have shaped the growth of PSCs. Special attention is given to the transition from dye-sensitized solar cells to solid-state PSCs, the introduction of tandem structures, and the on-going efforts to overcome challenges related to stability, toxicity, and scalability. The document also explores the impact of cation/anion engineering, interfacial modifications, and the pursuit of lead-free alternatives, positioning perovskites as a viable future technology in global energy solutions.

Keywords: Crystal, conductivity, inorganic, photovoltaics and halides

1.1 Introduction

Over the past two centuries, few minerals have witnessed as profound a shift in scientific perception as perovskite; initially regarded as a minor constituent of rare silica-under saturated rocks, it has since emerged as a focal point of extensive research and global scientific interest [1]. Perovskite, a mineral composed of calcium titanium oxide (CaTiO$_3$) was named after the Russian mineralogist Lev Perovski (1792–1856), when it was found in the Ural Mountains of Russia by Gustav Rose in 1839, who further contributed to its study [2].

Following Gustav Rose's first Latin description of perovskite, research for seven decades mostly concentrated on an imprecise crystallographic inconsistency: despite

*Corresponding author: R. K. Shukla,** Department of Physics, University of Lucknow, Lucknow 226007, Uttar Pradesh, India, e-mail: rajeshkumarshukla00@gmail.com
Anjali Gupta, Anchal Srivastava, Department of Physics, University of Lucknow, Lucknow 226007, Uttar Pradesh, India

https://doi.org/10.1515/9783111726847-001

exhibiting an isometric external morphology – particularly in specimens from the Urals and Alps – perovskite crystals demonstrated biaxial optical properties along with characteristic lamellar twinning, as briefly reviewed by Bowman (1908) [3]. This apparent contradiction, deeply rooted in the subtleties of crystal structure, remained unresolved before the development and application of X-ray diffraction techniques nearly fifty years later (Kay and Bailey, 1957) [4]. Despite being artificial, calcium titanate and structurally connected sodium–niobium oxides were initially synthesized in the late nineteenth century – concurrent with the creation of Verneuil's synthetic ruby and Spezia's quartz (Holmquist, 1898) [5] – a significant acceleration in perovskite research occurred silently in the 1920s. This recovery was marked by the issuance of the first industrial patent for perovskite-related applications (US1436164) to Victor Moritz Goldschmidt, whose influence extended to comprehensive structural investigations of $CaTiO_3$ and related compounds (verwandten Verbindungen), conducted by his students Thomas F.W. Barth (1925) [6] and Frederik W.H. Zachariasen (1928) [7] in Oslo. Their pioneering contributions, involving the introduction of the tolerance factor, laid the groundwork for modern crystallographic understanding of the perovskite structure. Table 1.1 shows the entire discovery related to perovskite with their lead researchers.

The unique crystal structure of perovskite was first thoroughly described by Victor Goldschmidt in 1926, when he introduced the concept of the tolerance factor – a geometric parameter used to forecast the stability and formability of perovskite-type compounds [8]. This work not only provided a quantitative framework for understanding the structural versatility of perovskites but also laid the foundation for subsequent studies in crystallography, solid-state chemistry, and materials science. Later, in 1945, Helen Dick Megaw reported the crystal structure based on X-ray diffraction data on barium titanate [9]. Due to its abundance in nature, perovskite has attracted a lot of attention from material scientists. Furthermore, it is at the forefront of research on super conductivity, magneto resistance, ionic conductivity, and the collection of dielectric properties – which are all crucial for microelectronics and telephony. The first published research was Thomas Barth's (1975) analysis of the fundamental structural arrangement of the atoms in the perovskite mineral, as shown in Figure 1.1.

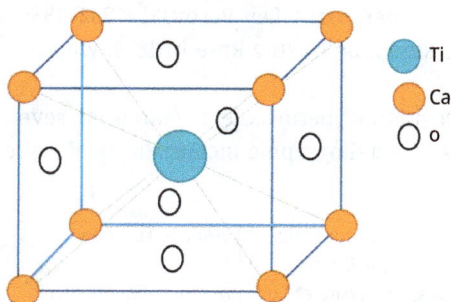

Figure 1.1: The fundamental atomic structure of the perovskite material.

Table 1.1: Timeline of perovskites: significant scientific and technological advancements [10].

Year	A brief outline	Lead researcher
1839	Perovskite ($CaTiO_3$) was discovered and named in a skarn sample from the Ural Mountains in Russia (Schisto chloritico)	Gustav Rose (Prussia)
1851	The first $CaTiO_3$ synthesis ever recorded (flux growth)	Jacques-Joseph Ebelmen (France)
1876	Perovskite from igneous rock (melilitic rocks at Deṭvín and other places in Bohemia) was first reported	Emanuel Borˇický (Bohemia)
1877	The first report of widespread element changes in Kaiserstuhl carbonatites' perovskite ("Dysanalyt")	Adolph Knop (Germany)
1898	The first synthetic perovskite-type chemical ($NaNbO_3$)	Per Johan Holmquist (Sweden)
1912	Verification of $CaTiO_3$ orthorhombic symmetry	O. Balthasar Bøggild (Denmark)
1922	The first commercial patent for pigment $CaTiO_3$	Victor M. Goldschmidt (Norway)
1925	The crystal structure of perovskites was first described using metamorphic rocks from Zermatt, Switzerland	Thomas F.W. Barth (Norway)
1940	Ferroelectric ceramics were found to have a high dielectric constant ($BaTiO_3$)	B.M. Vul and I.M. Gol'dman (Physical Institute, USSR); A. von Hippel and co-workers (MIT, USA)
1945	Ferroelectric $BaTiO_3$ tetragonal structure was identified	Helen D. Megaw (Philips Lamps, England); Harold Percy Rooksby (GE, England)
1946	Essential thermodynamic characteristics of $CaTiO_3$ were measured	C. Howard Shomate (Bureau of Mines, USA)
1949	$Ca_2Nb_2Bi_2O_9$ was the first "layered perovskites" (Aurivillius phases) to be discovered	B. Aurivillius (Stockholm University, Sweden)
1950	The discovery of magnetoresistance and ferromagnetism in $La_{1-x}(Ca,Sr,Ba)_xMnO_3$ mixed-valence manganites	G.H. Jonker and J.H. Van Santen (Philips, Netherlands)
1970	Perovskite was found in crude meteorites (Lancé carbonaceous chondrite)	M.J. Frost and R.F. Symes (NHM, England)
1970	Perovskite-type cobaltite catalysts are discovered	D.B. Meadowcroft (CERL, England)
1974	Synthesis of high-pressure silicate perovskite $MgSiO_3$	Lin-gun Liu (ANU, Australia)

Table 1.1 (continued)

Year	A brief outline	Lead researcher
1975	The first low carrier density superconductor was discovered in perovskites (BaPb$_{1-x}$Bi$_x$O$_3$)	Arthur W. Sleight (DuPont, USA) and co-workers
1978	The use of perovskite-bearing ceramics, or "SYNROC," to immobilize high-level radioactive wastes	A.E. (Ted) Ringwood (ANU, Australia) and co-workers
1986	Perovskite-type cuprate ceramics have been found to exhibit high-T superconductivity	J.G. Bednorz and K.A. Müller (IBM, Switzerland)
1994	Creation of perovskites with hybrid organic–inorganic halides for use in thin-film transistors	D. B. Mitzi (IBM, USA) and co-workers
2006	Creation of a solar cell that is sensitive to perovskites (based on the hybrid halide perovskites CH$_3$NH$_3$PbBr$_3$)	A. Kojima, T. Miyasaka et al. (Tokyo Polytech and Toin Universities, Japan)

1.2 History of perovskite solar cell

Perovskite minerals possess exceptional light-absorbing capabilities, requiring less than 1 μm of material to capture a comparable amount of sunlight as conventional solar cells. As semiconductors, perovskites efficiently transport electric charge when exposed to light. In the United Kingdom, researchers at the University of Oxford have demonstrated the potential of perovskites as promising alternatives to traditional thin-film solar cells. Perovskite solar cells (PSCs) typically feature crystal structure, incorporating an organic cation, a halide anion, and a metal cation – commonly lead or tin – forming a highly efficient and tunable photovoltaic material. One of the most exciting developments in photovoltaics is the use of PSCs, showing rapid development since their inception in the early 2000s. A class of materials known as "perovskite" typically refers to organometal halide perovskites, like methylammonium lead iodide (CH$_3$NH$_3$PbI$_3$), which have remarkable capacity to absorb light and carry charges.

Researchers started investigating hybrid perovskites of organic and inorganic materials for optoelectronic utilization in 2006, marking the beginning of the PSCs adventure. However, Tsutomu Miyasaka and his group in Japan reported using perovskite materials in solar cells, initially in 2009. They employed dye-sensitized solar cells (DSSCs), with methylammonium lead halide perovskites as light absorbers. Due to the perovskite dissolving in the liquid electrolyte employed in the cells, these early devices exhibited poor power conversion efficiency (PCE) and stability of only roughly 3.8%.

In 2012, researchers made a major advancement by replacing the liquid electrolyte with a solid-state hole transport medium (HTM), which greatly increased stability and efficiency. This breakthrough, which was partially attributed to Henry Snaith's team at the University of Oxford, helped perovskite materials move from being a specialized topic to a viable option for photovoltaics. Efficiency spiked to over 10% that year, attracting international scientific attention.

1.2.1 Timeline of perovskite solar cell development

1839 – Discovery of the mineral perovskite ($CaTiO_3$) in the Ural Mountains, named after the Russian mineralogist Lev Perovski [10].

1892 – First synthesis of cesium lead halide perovskite material, laying the foundation for modern PSCs [10].

1926 – Theoretical description of the perovskite crystal structure [10].

1945 – Confirmation of the perovskite crystal structure via X-ray crystallography of barium titanate [10].

2009 – Discovery of perovskite's photovoltaic properties:

The first usage of organic-inorganic lead halide perovskites ($CH_3NH_3PbBr_3$ and $CH_3NH_3PbI_3$) as light absorbers in dye-sensitized solar cells was reported by Tsutomu Miyasaka and associates in 2009, marking the beginning of the PSC adventure. Although the perovskite material was unstable in liquid electrolyte, they managed to attain moderate power conversion efficiency (PCE) of around 3.8%.

2011 – Nam-Gyu Park improves the efficiency to 6.5% using the same dye-sensitized concept.

2012 – Solid-state breakthrough:

In 2012, researchers at the Oxford University, led by Henry Snaith, replaced the liquid electrolyte with a solid hole-transport material, Spiro-OMeTAD [11]. This important development raised the efficiency to more than 10% and greatly enhanced cell stability, positioning perovskites as serious contenders in the solar market.

2013–2014 – Rapid efficiency gains:

Within just a few years, research accelerated rapidly. By 2013, efficiencies exceeded 15%, and in 2014, perovskite cells surpassed 20%. During this time, new fabrication techniques, improved materials, and better interfaces helped reduce defects and enhance performance. Burschka et al. used a two-step solution processing method to develop a deposition methodology for the sensitized architecture that exceeded 15% efficiency [12]. During this period, Olga Malinkiewicz et al. and Liu et al. demonstrated that planar solar cells could be created via thermal co-evaporation, attaining over 12% and 15% efficiencies in p-i-n and n-i-p designs, respectively [13, 14]. Researchers from KRICT set a record in November 2014 when they certified a device with a non-stabilized efficiency of 20.1% [15].

2015 – Tandem cell integration begins:

In order to create tandem solar cells, perovskites began to be mixed with silicon and other well-known PV technologies. This made it possible to exceed single-junction silicon cells' Shockley–Queisser limit. Perovskite/silicon tandem cells, with PCEs higher than 21%, were shown by researchers.

2016–2018 – Addressing stability and toxicity:

The main challenges shifted to long-term stability and lead toxicity. Efforts focused on compositional engineering (e.g., mixing formamidinium, cesium, and methylammonium) to stabilize perovskite phases. Encapsulation methods also improved, prolonging device lifetimes. Meanwhile, lead-free alternatives like tin-based perovskites were explored, though with limited efficiency.

2019 – Commercial interest intensifies:

Startups and large corporations began investing in perovskite technologies. Companies like Oxford PV and Saule Technologies started pilot production, focusing on tandem cells and flexible, printed solar panels.

2020 – Tandem cell efficiency milestones:

Perovskite/silicon tandem cells exceeded 25% efficiency in lab settings. Min, H. et al. concludes that an interlayer with atomic coherence between SnO_2 and halide perovskite improves charge extraction and transport, leading to increased stability and 25.8% power conversion efficiency, while also offering design guidance for defects, reducing interfaces in PSCs [16]. Oxford PV set a new world record of 29.52% in 2020, verified by the National Renewable Energy Laboratory (NREL). Researchers also reported over 1,000 hours of operational stability under continuous illumination.

2021–2023 – Scaling and stability progress:

Large-area and roll-to-roll fabrication methods advanced significantly. Stability standards improved, with devices lasting thousands of hours under real-world conditions. Encapsulation and interface engineering minimized degradation, while scalable production techniques pushed perovskites closer to commercialization. In 2022, German team achieves 32.5% efficiency in silicon–perovskite tandem solar cells [17].

2024 – Near-commercial readiness:

By 2024, several companies entered pre-commercial production stages. Tandem cells, with efficiencies above 30%, were demonstrated in labs. Research into recycling, toxicity mitigation, and durability continued to align perovskites with international PV standards.

2025 – Japan invests $1.5 billion in commercializing ultra-thin, flexible PSCs, aiming to diversify energy sources and improve energy security.

1.3 Perovskite structures

Any substance with the same crystal structure as calcium titanate ($CaTiO_3$) is termed as "perovskite." The crystalline structure of perovskite is composed by the formula ABX_3, where X is frequently an oxygen or halogen anion, while A and B are cations. Six anions surround each B-cation, forming $[BX_6]$ [4] – octahedra. Through corner-sharing, these are joined to form the ReO_3 structure [18]. The distinction is that in perovskite structures, the A-cation is found in the cuboctahedral location [19]. Adding elements with varying valences to the structure is one major benefit of the perovskite structure. When X is an anion of oxygen (O_2^-), for instance, cations A and B would be trivalent and divalent, respectively., but if X is a halogen anion (Cl^-, Br^-, I^-), then A and B would be monovalent and divalent, respectively [20]. This provides you with a great opportunity to easily alter the material's bandgap and other characteristics. Many chemists and physicists worldwide are interested in the relevant research being done on the creation of the crystal structure (ABX_3) and the advancements in device construction. PSC efficiencies have thus rapidly grown to 25.5%, which is comparable to the 27.6% efficiency of commercially available single-crystalline silicon solar cells.

PSCs are still laboratory-based technologies that have not yet been commercialized, despite their exceptional effectiveness, because of the use of harmful inorganic lead (Pb) element and device instabilities. Variations in oxygen, UV radiation, temperature, and humidity are some environmental factors that impact the stability of the PSC. Therefore, producing a gadget with great performance, environmental friendliness, and a 25-year lifespan is the main manufacturing difficulty.

As shown in Figure 1.2, a typical PSC structure consists of a light absorber (perovskite) layer, also known as the active layer, and an electron transport layer (ETL), favored by a hole transport layer (HTL). Glass/transparent conductive oxide (TCO)/ETL/perovskite/HTL/metal electrode is the standard device layout. The dye-sensitized solar cells (DSSCs) served as a model for this original n-i-p PSC construction. Halide perovskite specimens such as $MAPbI_3$ and $MAPbBr_3$ are used in PSC in place of the dye ingredient [21]. These devices' active layer, or perovskite layer, can be deposited in two steps using the sequential deposition approach or in one step using the single-step method. The $CH_3NH_3PbI_3$ perovskite layer is deposited in a single step using a solvent solution of lead iodide (PbI_2) and methylammonium iodide (CH_3NH_3I, MAI). In con-

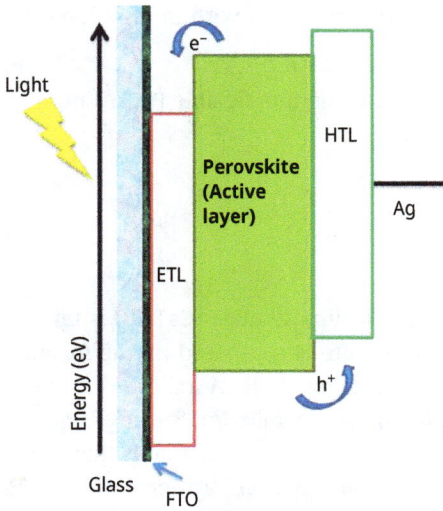

Figure 1.2: A typical perovskite solar cell consists of a layered structure built on an FTO (fluorine-doped tin oxide)-coated glass substrate. The perovskite absorber layer is sandwiched between an electron transport layer (ETL) and a hole transport layer (HTL), with a silver (Ag) layer serving as the top metal electrode.

trast, the two-step process creates $CH_3NH_3PbI_3$ perovskite, by first depositing PbI_2 and then a layer of CH_3NH_3I[22].

Perovskite photovoltaic devices' initial architecture, shown in Figure 1.3, is the n-i-p mesoscopic structure, which continues to be often utilized to create high-performance devices. As shown in Figure 1.3(a), the mesoscopic n-i-p structure device configuration consists of glass/FTO/cTiO$_2$/mp-TiO$_2$ (ETL)/perovskite/HTL/metal electrode. The process of creating the device begins with the deposition of ETL [such as compact-titanium dioxide (cTiO$_2$), mesoporous-titanium dioxide (mp-TiO$_2$), or mesoporous-aluminum oxide (mp-Al$_2$O$_3$)] in a glass substrate, coated with transparent conductive oxides (FTO, fluorine-doped tin oxide, or ITO, indium-doped tin oxide), and then annealing. Next, either a one-step or two-step procedure is used to deposit the perovskite layer on top of the ETL layer. A delicate layer of HTL [such as spiro-MeOTAD (2,2',7,7'-tetrakis(*N*,*N*-dipmethoxyphenylamine)-9,9'-spirobifluorene)] is coated on the layer of perovskite, following annealing. Lastly, the device is finished by evaporating a metal electrode (gold or silver) on HTL [22, 23]. A mesoporous TiO$_2$ layer is absent from the n-i-p planar structure, as shown in Figure 1.3(b), which is nearly identical to the n-i-p mesoscopic structure. Soon after it was discovered that the mesoporous layer of TiO$_2$ wasn't necessary in perovskite based solar cells to work, without the mesoporous TiO$_2$ layer, the device structure was reduced to a planar n-i-p structure [24]. The planar structure devices demonstrated a high PCE of 19.3%, with appropriate optimization, which was equivalent to mesoscopic devices [23]. As a result, the planar configuration was modified for perovskite-based solar cells due to its lower manufacturing costs.

The structure of the inverted p-i-n is derived from the organic solar cell. Numerous materials, both inorganic and organic, have been studied for charge transport layers because of this arrangement. In comparison to n-i-p structures, p-i-n planar

(a) n-i-p mesoscopic	(b) n-i-p planar	(c) p-i-n planar	(d) p-i-n mesoscopic
Metal anode (Au)	Metal anode (Au)	Metal Cathode (Ai)	Metal Cathode (Ai)
HTM(Spiro-MeOTAD)	HTM(Spiro-MeOTAD)	ETM(PCBM)	ETM(PCBM)
Perovskite (MaPbI₃)	Perovskite (MaPbI₃)	Perovskite (MaPbI₃)	Perovskite (MaPbI₃)
ETM (TiO₂)	ETM (TiO₂)	HTM (PEDOT:PSS)	HTM (PEDOT:PSS)
Transparent cathode (FTO)	Transparent cathode (FTO)	Transparent anode (ITO)	Transparent anode (ITO)
Glass	Glass	Glass	Glass

Sunlight

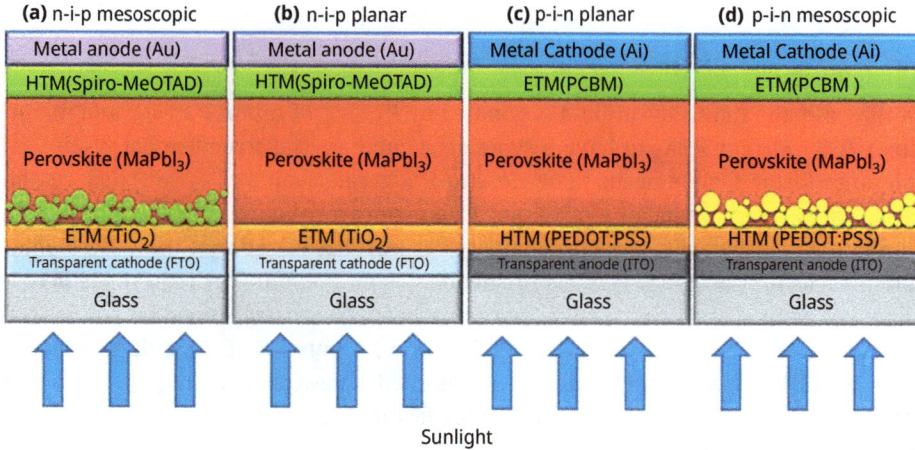

Figure 1.3: Simplified pictures of a typical PSC device with n-i-p configuration are shown in (a, b). The PSC device with inverted p-i-n configuration is represented by (c, d).

PSC provides low-temperature processing and little hysteresis in the current–voltage curve, with similar PCE [25]. Figure 1.3(c) illustrates the FTO/HTL/perovskite/ETL/metal electrode layout of the inverted planar configuration. Here, the device design reverses the n-i-p structure. The structures shown in Figure 1.3(d) are reversed versions of the typical (n-i-p) configuration, with the deposition order reversed. The selection of HTL and ETL components utilized in the manufacture of solar cells affects the device's structure. An inverted structure is not feasible when creating a configuration that calls for TiO₂ film, which should be annealed at a high temperature of around 500 ˚C, since this would damage the active (perovskite) layer that was previously deposited, as well as HTL [26]. To increase stability and efficiency significantly, more interfacial layers are added between the charge transport and perovskite layers. These interfacial layers help inject carriers into the carrier transport layers, prevent reverse flow of carriers, minimize recombination, and preserve high carrier extraction at the electrodes.

1.4 Simulating the crystal structure of perovskites

The device structure of PSCs, which are third-generation photovoltaic cells, are derived from the design of second-generation dye-sensitized solar cells (DSSCs) [27]. The overall structure and stability of perovskites are greatly impacted by the ion sizes utilized. Perovskites' crystal structure is determined by the Goldschmidt tolerance factor (t), as shown in the following equation:

$$t = \frac{R_A + R_X}{\sqrt{2(R_B + R_X)}} \tag{1.1}$$

In this case, the radius of cation A is denoted by R_A, that of cation B by R_B, and that of anion X by R_X. For ABX$_3$ to have a symmetric cubic crystal structure perovskite, its tolerance factor t value needs to be close to 1. A greater value of t ($t > 1$) distorts the structure, whereas a smaller t value may lead to tetragonal (β phase) or orthorhombic (γ phase) structure with reduced symmetry. It was discovered that the Goldschmidt tolerance factor's accuracy (~74%) is inadequate to differentiate between stable perovskite and non-perovskite structures, particularly for perovskites based on halides (Cl, Br, and I), which have the lowest rate of accuracy – only 33% for iodide perovskite. Thus, in 2019, Scheffler and colleagues presented a new tolerance factor (τ) that is more accurate than the Goldsmith tolerance factor [28]. Equation (1.2) provides the τ relation, and a stable perovskite structure is indicated by $\tau < 4.18$:

$$\tau = \frac{r_X}{r_B} - n_A \left(n_A - \frac{r_A/r_B}{\ln(r_A/r_B)} \right) \tag{1.2}$$

where r is ionic radius of the ion, n_A is oxidation state of A, and $r_A > r_B$, by definition. A 92% overall precision was attained, utilizing τ for a variety of compounds, including perovskites based on halides. The accuracy has improved and τ is now a dependable instrument for practically predicting the likelihood of obtaining a stable perovskite structure, due to the significant drop in the wrong prediction rate (from 51% to 11%). In addition to this condition, variations in temperature also cause structural alterations. The crystal orientation of perovskite is likewise intimately related to the PCE of PSC, as understood from the previously published investigations. It has been demonstrated that CH$_3$NH$_3$ aids in the structural cohesion of MAPbI$_3$ perovskites but does not add to their optical or electronic response [29]. Because the valence band and the conduction band are created by the combination of the metal and halide orbitals, the metal–halide primarily controls the optoelectronic characteristics of ABX$_3$ perovskite. Lattice contraction and octahedral tilting can have an indirect impact on the band locations, even if the band edge energy levels are not directly influenced by the A-site cation [30]. The bandgap is reduced as a result, and in order to raise the bands to a lower energy, the lattice contraction increases the metal–halide orbital overlap. Conversely, the octahedral tilting raises the bandgap and pushes bands to deeper energy levels by decreasing the metal–halide orbital overlap. When the perovskite layer is properly orientated at the interface with the right charge transport layers, lower PCE and short circuit current density result [31].

1.5 Material selection for perovskite absorbers

As previously stated, single-halide perovskites such as MAPbI$_3$ and MAPbBr$_3$ are often utilized as absorber materials in PSC. Blended-halide perovskite, such as MAPb(I$_{1-x}$Br$_x$)$_3$ and MAPb(I$_{1-x}$Cl$_x$)$_3$ (0 ≤ x ≤ 1), is created when halides (Cl, Br, I) are blended in the perovskite composition. By altering the ratio of two distinct halides in the perovskite composition, the adjustable bandgap – a unique feature of mixed-halide perovskite – can be changed across a wide range. For instance, MAPb(I$_{1-x}$Cl$_x$)$_3$ bandgap is likely adjusted between 1.6 and around 3 eV based on material's halide (I and Cl) ratio [32]. For MAPb(I$_{1-x}$Br$_x$)$_3$ and MAPb (Br$_{1-x}$Cl$_x$)$_3$, the bandgap may be changed from 1.6 to 2.3 eV and 2.42 to 3.16 eV, respectively. Mixed-halide perovskites have altered structural and optical characteristics that result in fluctuating photovoltaic performance and stability according to the halide ratio. The electron-hole diffusion lengths of the MAPbI$_3$ perovskite are around 100 nm, but the mixed-halide MAPbI$_{3-x}$Cl$_x$ perovskite exhibits diffusion lengths longer than 1 μm [33]. As a result, the configuration made using MAPbI$_{3-x}$Cl$_x$ as an active layer demonstrated a greater PCE than the PSC based on MAPbI$_3$. Furthermore, the high bandgap perovskites might be used to produce top cells in tandem solar cells and blue LEDs.

The widespread use of lead halide perovskite in photovoltaics is problematic, as the main perovskite component, Pb, is hazardous to humans and poisonous in nature. The substitution of an appropriate substitute for lead in the perovskite compound has been a primary focus of study in order to overcome the toxicity concern. Nontoxic tin (Sn) and bismuth (Bi) have been used to create PSC in place of lead. Additionally, the bandgap decreases from 1.6 to 1.23 eV when Pb is substituted with Sn. This is near the optimal bandgap energy of 1.34 eV for reaching the highest PCE. According to the detailed balancing limit, also called the Shockley-Queisser efficiency limit, the optimal bandgap for single-junction solar cells is 1.34 eV, in order to get the maximum PCE of 33.16% (for AM1.5 G illumination) [34]. The highest efficiency limit of MAPbI$_3$ perovskite is 31%, and its bandgap (1.6 eV) is larger than the ideal bandgap (1.34 eV). Materials with a lower bandgap can absorb more low-energy photons outside of the visible spectrum than those with a greater bandgap. Therefore, even if the V_{oc} may be somewhat compromised, lowering the bandgap is preferred in order to increase photon absorption and enhance PCE. However, due to rapid oxidation of Sn^{2+} to Sn^{4+}, Sn-based perovskites are less stable. Additionally, inadequate homogeneity of the film and subpar device performance are caused by poor crystallization of Sn-based perovskite [35]. Perovskites based on lead exhibit greater stability compared to those based on bismuth, although Pb-based perovskite devices have higher cell efficiencies than bismuth-based PSC.

A novel concept of mixed-cation perovskites are suggested to improve PSC efficiency because single cation halide perovskite has been found to have a number of problems. The most often employed cations at the moment to create perovskite structures with lead halide are CH$_3$NH$_3^+$ (MA, methylammonium), CH(NH$_2$)$_2^+$ (FA, formamidinium), and Cs^+ (cesium). Cell efficiency may rise from 11.4% to 12.9% if the Jsc value is increased by the proper ratio of FA and MA cations, since $FAPbI_3$ has a larger absorption range of up to 840 nm. Similarly, Singh et al. [36] reported high PCE of up to 20.8% for perovskite devices based on cesium-formamidinium-methylammonium triple cations, with a J_{sc} of 23.6 mA cm^{-2} and a V_{oc} of 1.14 V. Even under ambient circumstances (RH = 20–25%), these configurations maintained the increased PCE of 19.5% for up to 18 weeks. Thus, by adjusting the perovskite content, it is feasible to obtain an optimal composition for high-efficiency PSC with acceptable stability. Table 1.2 lists the different PSC conditions and the performance metrics they have acquired.

The ETL and HTL layers of PSC undergo the majority of the structural changes. The most often used ETL has been the TiO_2 layer since PSC started in 2009. Mesoporous TiO_2 layers are also used as ETL in the fabrication of PSC, despite their claimed high efficiency (PCE-20%, as reported in 2016) [37]. Due to its impact on photovoltaic characteristics in PSC, the TiO_2 ETL has been the subject of much research. Due to the necessity of annealing TiO_2 at 500 °C, it restricts the use of substrates with low deformation temperatures and consumes a lot of energy; this is a significant disadvantage of employing this material. Additionally, it is generally known that UV radiation causes TiO_2-based solar cells to degrade, which lowers PSC stability. The surface of the TiO_2 film has a large number of oxygen vacancies, which are basically deep electron-donating sites, at a density of around 1 eV below the conduction band. At these locations, oxygen is adsorbed as O^{2-}, as a result of the electrons there reacting with atmospheric oxygen molecules. UV photons cause electron–hole pairs to form during the exposure of TiO_2 layer to light [38]. Desorption of oxygen is caused by the recombination of O^{2-} with the holes in the TiO_2 valence band. Deep surface trap sites become vacant as a result of this desorption. These sites subsequently trap the electrons from the TiO_2 conduction band. At the same time, either directly or through the conduction band, in the perovskite layer, the photo-generated electrons leap to the deep surface traps states in TiO_2 that are vacant. Because these captured electrons are immobile, they combine again with the holes in HTL, which impairs solar cell efficiency. The practical use of these solar cells is hindered by this uncertainty. It has been noted that the mesoporous TiO_2-free solar cells exhibit less instability. Ball et al. [39] employed Al_2O_3 nanoparticles that were annealed at a low temperature of around 150 °C only as an alternative to TiO_2, avoiding the requirement of high-temperature processing of PSC. PCE was present in the device up to 12.3%. PSC employing ZnO (zinc oxide), PCBM ((6, 6)-phenyl C61-butyric acid methyl ester), and SnO_2 (tin oxide) as ETL, in place of TiO_2, was reported by several studies shortly after. PCBM (fullerene derivative) was employed as an ETL in the production of PSC in 2013. Since then, TiO_2 has been replaced by ETL of some fullerene derivatives, such as PC61BM and PC71BM. The use of alternative ETM, such as PCBM and ZnO, has the major benefit of processing at low temperatures (~100 °C) [40].

Lack of stability is the main obstacle to PSCs development. These problems arise from chemical interactions inside the perovskite structure. Most of the interactions

Table 1.2: List of the different PSC conditions and the performance year.

Device structure	J_{sc} (mA cm^{-2})	V_{oc} (V)	FF	PCE (%)	Year	References
ITO/TiO$_2$/MAPbBr$_3$/Au	6.96	1.36	0.69	6.53	2016	[41]
FTO/TiO$_2$/MAPbBr$_3$/spiro-OMeTAD/Au	8.77	1.31	0.62	7.11	2017	[42]
FTO/TiO$_2$/MAPbI$_3$/spiro-OMeTAD/Ag	22.28	0.668	0.59	8.78	2017	[43]
ITO/PEDOT:PSS/MAPbI$_3$/PCBM/Ag	22.15	0.75	0.27	4.4	2018	[44]
ITO/PTAA/MAPbI$_3$/PCBM/C60/BCP/Cu	21	1.08	0.786	17.8	2017	[45]
ITO/PTAA/MAPbI$_3$/C60/BCP/Cu	23.46	1.076	0.835	21.09	2019	[46]
ITO/NiO$_x$/(FAPbI$_3$)$_{0.85}$(MAPbBr$_3$)$_{0.15}$/TiO$_2$/Ag	23.14	1.03	0.51	12.18	2017	[47]
FTO/Cu/Au/Ag/LiF/PCBM/MAPbI$_3$/spiro-MeOTAD/ITO/Au	13.5	1	0.296	4	2017	[48]
ITO/PEDOT:PSS/GA$_x$FA$_{0.98-x}$SnI$_3$–1% EDAI2/C60 (20 nm)/BCP/Ag	21.2	0.61	0.72	9.2	2018	[49]
FTO/PEDOT PSS/EA$_{0.98}$EDA$_{0.01}$SnI$_3$/C60BCP/Au	20.32	0.84	0.78	13.24	2020	[50]
ITO/PTAA/Cs$_{0.05}$(FA$_{0.92}$MA$_{0.08}$)$_{0.95}$Pb(I$_{0.92}$Br $_{0.08}$)$_3$/C60/BCP/Cu	24.1	1.71	0.81	22.3	2020	[51]
Glass/ITO/PTAA/(Cs0.05(FA$_5$/MAI)0.95Pb(I0.9 Br0.1)$_3$)/PCBM/BCP/Ag	24	1.16	0.82	23	2021	[52]
Glass/ITO/PTAA/PEAI/(Cs0.05(FA$_5$/MAI)0.95Pb(I0.9 Br0.1)$_3$)/PEAI/PCBM/BCP/Ag	24.16	1.16	0.84	23.7	2021	[53]
FTO/SnO$_2$/MAPbBr$_3$/HTL/back contact	25.09	1.19	0.84	25.4	2021	[54]
FTO/SnO$_2$-Cl/FAPbI$_3$/Spiro-OMETAD/Au	25.74	1.18	0.83	25.2	2021	[55]

are weak ionic bonds, such as the Pb–I connection, which has a bonding energy of 142 kJ mol^{-1}. Van der Waals forces and hydrogen bonds are additional secondary interactions that contribute to the material's softness [25]. Due to its intrinsic sensitivity to heat, light, humidity, electric fields, and oxygen, PSCs are susceptible to degradation. The largest PV market has been dominated by silicon PV technology for more than 50 years. According to studies, using a broad bandgap perovskite material with a bandgap of order 1.7 eV and a thickness of order 1 μm is necessary to maximize tandem cell efficiency. The synthesis of such perovskite material is one of the most prominent research issues within the PV field[33]. If a perovskite material, with a bandgap of 1.7 eV is utilized, the extracted voltage will be around 1.3 V, resulting in a total voltage of 2.0 V for the device. To extract equivalent currents from both cells, the perovskite material's thickness has to be around 1 μm, with a bandgap of 1.7 eV. The usage of lead (Pb^{2+}) as the B-cation site is still necessary for modern PSCs. Because lead is poisonous, using it might cause issues if it were to leak into the surroundings and ultimately make their way into the human meal. As a result, several studies to substitute lead-free perovskite materials have been carried out. PSCs based on a variety of elements, such as bismuth, copper, germanium, antimony, and others, have undergone testing.

The first suggestion is to investigate lead-free perovskites due to the negative consequences of lead intake. Tin-based perovskites have shown the most promise, with PCEs ranging from 10 to 12%. Since lead-based perovskites, which achieved a PCE of 25%, unfortunately cannot be competed with, more research is necessary. Tandem and multi-junction cells, which absorb a wider spectrum of radiation, is the last subject discussed. Tandem cells not only increase the PCE but also address the stability issues with perovskite cells. A perovskite–silicon tandem cell was synthesized, resulting in a high PCE of 29.15%. Tandem cells that employ 2D-perovskites as the high bandgap absorber can achieve high PCE and exceptional stability.

1.6 Conclusion

The use of PSCs has grown quickly and has seen a transformative journey from a laboratory curiosity to a frontrunner in next-generation photovoltaic technologies. Their exceptional optoelectronic properties, tunable bandgap, and compatibility with various device architectures have made them a subject of intense global research. The ability to tailor the crystal structure through compositional engineering has led to continuous improvements in performance and stability. Notably, the integration of mixed halide and mixed cation strategies has proven effective in enhancing efficiency and environmental resilience. The existence of hazardous lead, poor long-term operating stability, and deterioration under environmental stresses, however, continue to be major obstacles to commercialization. The evolution of tandem cell configurations and the emergence of low-temperature processing techniques have brought PSCs closer to industrial scalability. Further advancements in lead-free alternatives and encapsulation methods will be essential to realize sustainable, efficient, and commercially viable PSC technologies. As the world seeks cleaner and more efficient energy sources, the ongoing innovation in perovskite materials and device design holds the potential to revolutionize solar energy harvesting.

References

[1] Chakhmouradian AR, Woodward PM. Celebrating 175 years of perovskite research: a tribute to Roger H. Mitchell, Physics and Chemistry of Minerals. 2014; 41(6): 387–391 doi: 10.1007/s00269-014-0678-9

[2] Sahoo SK, Manoharan B, Sivakumar N. Introduction, Perovskite Photovoltaics. 2018; 1–24, doi: 10.1016/b978-0-12-812915-9.00001-0.

[3]   Bowman HL. On the structure of perovskite from the Burgumer Alp, Pfitschthal, Tyrol, Mineralogical Magazine. 1908; 15: 156–176.

[4]   Kay HF, Bailey PC. Structure and properties of $CaTiO_3$, Acta Crystallographica. 1957; 10: 219–226.

[5]   Holmquist PJ. Synthetische Studien über die Perowskit- und Pyrochlormineralen, Bulletin of the Geological Institutions of the University of Uppsala. 3(5): 88, 1898.

[6]   Barth T. Die Kristallstruktur von Perowskit und verwandten Verbindungen, Norsk Geol Tidsskr. 1925; 8: 201–216.

[7]   Zachariasen WH. Untersuchungen über die Kristallstruktur von Sesquioxyden und verbindungen $ABO3$, Norsk Vidensk Akad Oslo, Mat Naturv Klasse. 1928; 4: 1–165.

[8]   Goldschmidt VM. Die gesetze der krystallochemie, Naturwissenschaften. 14(21): 477–485, 1926.

[9]   Megaw HD. Crystal structure of barium titanate, Nature. 155(3938): 484–485, 1945.

[10] Chakhmouradian AR, Woodward PM. Celebrating 175 years of perovskite research: a tribute to Roger H. Mitchell, Physics and Chemistry of Minerals. 2014; 41(6): 387–391 doi: 10.1007/s00269-014-0678-9

[11] Im JH, Lee CR, Lee JW, Park SW, Park NG. 6.5% efficient perovskite quantum-dot-sensitized solar cell, Nanoscale. 3(10): 4088–4093, 2011.

[12] Burschka J, Pellet N, Moon SJ, Humphry-Baker R, Gao P, Nazeeruddin MK, Grätzel M. Sequential deposition as a route to high-performance perovskite-sensitized solar cells, Nature. 499(7458): 316–319, 2013.

[13] Malinkiewicz O, Yella A, Lee YH, Espallargas GM, Graetzel M, Nazeeruddin MK, Bolink HJ. Perovskite solar cells employing organic charge-transport layers, Nature photonics. 8(2): 128–132, 2014.

[14] Liu M, Johnston MB, Snaith HJ. Efficient planar heterojunction perovskite solar cells by vapour deposition, Nature. 501(7467): 395–398, 2013.

[15] "Best Research-Cell Efficiencies" (PDF). National Renewable Energy Laboratory. 2022-06-30. Archived from the original (PDF) on 2022-08-03. Retrieved 2022-07-12.

[16] Min H, Lee DY, Kim J, Kim G, Lee KS, Kim J, Paik MJ, Kim YK, Kim KS, Kim MG, Shin TJ. Perovskite solar cells with atomically coherent interlayers on SnO2 electrodes, Nature. 598(7881): 444–450, 2021.

[17] Allen TG, Aydin E, Subbiah AS, De Bastiani M, De Wolf S. Perovskite/silicon tandem photovoltaics, Photovoltaic Solar Energy: From Fundamentals to Applications. 2024; 2: 157–177.

[18] Evans HA, Wu Y, Seshadri R, Cheetham A. Perovskite-related ReO3-type structures, Nature Reviews Materials. 2020; 5: 196–213.

[19] Roy A, Ghosh A, Bhandari S, Sundaram S, Mallick T. Perovskite Solar Cells for BIPV Application: A Review, Buildings. 2020; 10: 129.

[20] Park NG. Halide perovskite photovoltaics: History, progress, and perspectives, MRS Bull. 2018; 43: 527–533.

[21] Li D, Cui J, Li H, Huang D, Wang M, Shen Y. Graphene oxide modified hole transport layer for CH3NH3PbI3 planar heterojunction solar cells, Solar Energy. 2016; 131: 176–182.

[22] Im JH, Kim H-S, Park N-G. Morphology-photovoltaic property correlation in perovskite solar cells: One-step versus two-step deposition of CH3NH3PbI3, APL Materials. 2014; 2: 081510.

[23] Zhou H, Chen Q, Li G, Luo S, Song T-B, Duan H-S. Interface engineering of highly efficient perovskite solar cells, Science. 2014; 345: 542–546.

[24] Lee MM, Teuscher J, Miyasaka T, Murakami TN, Snaith HJ. Efficient hybrid solar cells based on meso-superstructured organometal halide perovskites, Science. 2012; 338: 643.

[25] Chiang C-H, Tseng Z-L, Wu C-G. Planar heterojunction perovskite/PC71BM solar cells with enhanced open-circuit voltage via a (2/1)-step spin-coating process, Journal of Materials Chemistry A. 2014; 2: 15897–15903.

[26] Wu Y, Yang X, Chen H, Zhang K, Qin C, Liu J. Highly compact TiO2 layer for efficient hole-blocking in perovskite solar cells, Applied Physics Express. 2014; 7: 052301.

[27]   Ng C, Lim HN, Hayase S, Zainal Z, Huang N. Photovoltaic performances of mono- and mixed-halide structures for perovskite solar cell: A review, Renewable and Sustainable Energy Reviews. 2018; 90: 248–274.

[28]   Bartel CJ, Sutton C, Goldsmith BR, Ouyang R, Musgrave CB, Ghiringhelli LM. New tolerance factor to predict the stability of perovskite oxides and halides, Science Advances. 2019; 5: eaav0693.

[29]   Motta C, El-Mellouhi F, Kais S, Tabet N, Alharbi F, Sanvito S. Revealing the role of organic cations in hybrid halide perovskite CH3NH3PbI3, Nature Communications. 2015; 6: 7026.

[30]   Prasanna R, Gold-Parker A, Leijtens T, Conings B, Babayigit A, Boyen HG. Band gap tuning via lattice contraction and octahedral tilting in perovskite materials for photovoltaics, Journal of the American Chemical Society. 2017; 139: 11117–11124.

[31]   Sfyri G, Kumar CV, Raptis D, Dracopoulos V, Lianos P. Study of perovskite solar cells synthesized under ambient conditions and of the performance of small cell modules, Sol, Solar Energy Materials and Solar Cells. 2015; 134: 60–63.

[32]   Cheng X, Jing L, Zhao Y, Du S, Ding J, Zhou T. Crystal orientation-dependent optoelectronic properties of MAPbCl3 single crystals, Journal of Materials Chemistry C. 2018; 6: 1579–1586.

[33]   Stranks D, Eperon GE, Grancini G, Menelaou C, Alcocer MJ, Leijtens T, Electron-hole diffusion lengths exceeding 1 micrometer in an organometal trihalide perovskite absorber, Science. 2013; 342: 341–344.

[34]   Rühle S. Tabulated values of the Shockley–Queisser limit for single junction solar cells, Solar Powered. 2016; 130: 139–147.

[35]   Wu T, Liu X, Luo X, Lin X, Cui D, Wang Y. Lead-free tin perovskite solar cells, Joule. 2021; 5: 863–886.

[36]   Singh T, Miyasaka T. Stabilizing the efficiency beyond 20% with a mixed cation perovskite solar cell fabricated in ambient air under controlled humidity, Advanced Energy Materials. 2018; 8: 1700677.

[37]   Xiao Y, Han G, Chang Y, Zhang Y, Li Y, Li M. Investigation of perovskite-sensitized nanoporous titanium dioxide photoanodes with different thicknesses in perovskite solar cells, Journal of Power Sources. 2015; 286: 118–123.

[38]   Lu G, Linsebigler A, Yates JT Jr. The adsorption and photodesorption of oxygen on the TiO$_2$ (110) surface, The Journal of chemical physics, 1995; 102: 4657–4662.

[39]   Ball JM, Lee MM, Hey A, Snaith HJ. Low-temperature processed meso-superstructured to thin-film perovskite solar cells, Energ, Environmental Science. 2013; 6: 1739.

[40]   Liu, D., Kelly, T. Perovskite solar cells with a planar heterojunction structure prepared using room-temperature solution processing techniques. *Nature Photon* 8, 133–138 (2014).

[41]   Peng, W., Wang, L., Murali, B., Ho, K.T., Bera, A., Cho, N., Kang, C.F., Burlakov, V.M., Pan, J., Sinatra, L. and Ma, C., 2016. Solution-grown monocrystalline hybrid perovskite films for hole-transporter-free solar cells. *Advanced Materials*, 28(17), pp.3383–3390.

[42]   Rao, H.S., Chen, B.X., Wang, X.D., Kuang, D.B. and Su, C.Y., 2017. A micron-scale laminar MAPbBr 3 single crystal for an efficient and stable perovskite solar cell. Chemical communications, 53(37), pp.5163–5166.

[43]   Zhao, J., Kong, G., Chen, S., Li, Q., Huang, B., Liu, Z., San, X., Wang, Y., Wang, C., Zhen, Y. and Wen, H., 2017. Single crystalline CH3NH3PbI3 self-grown on FTO/TiO2 substrate for high efficiency perovskite solar cells. Science Bulletin, 62(17), pp.1173–1176.

[44]   Yue, H.L., Sung, H.H. and Chen, F.C., 2018. Seeded space-limited crystallization of CH3NH3PbI3 single-crystal plates for perovskite solar cells. *Advanced Electronic Materials*, 4(7), p.1700655.

[45]   Chen, Z., Dong, Q., Liu, Y., Bao, C., Fang, Y., Lin, Y., Tang, S., Wang, Q., Xiao, X., Bai, Y. and Deng, Y., 2017. Thin single crystal perovskite solar cells to harvest below-bandgap light absorption. *Nature communications*, 8(1), p.1890.

[46]   Chen, Z., Turedi, B., Alsalloum, A.Y., Yang, C., Zheng, X., Gereige, I., AlSaggaf, A., Mohammed, O.F. and Bakr, O.M., 2019. Single-crystal MAPbI3 perovskite solar cells exceeding 21% power conversion efficiency. *ACS Energy Letters*, 4(6), pp.1258–1259.

[47] Huang, Y., Zhang, Y., Sun, J., Wang, X., Sun, J., Chen, Q., Pan, C. and Zhou, H., 2018. The exploration of carrier behavior in the inverted mixed perovskite single-crystal solar cells. *Advanced Materials Interfaces*, 5(14), p.1800224.

[48] Liu, Y., Ren, X., Zhang, J., Yang, Z., Yang, D., Yu, F., Sun, J., Zhao, C., Yao, Z., Wang, B. and Wei, Q., 2017. 120 mm single-crystalline perovskite and wafers: towards viable applications. *Science China Chemistry*, 60(10), pp.1367–1376.

[49] Jokar E, Chien C, Tsai C, Fathi A, Diau EW. Robust Tin-Based Perovskite Solar Cells with Hybrid Organic Cations to Attain Efficiency Approaching 10%, Advanced Materials. 2018; 31: 1804835.

[50] Mohanty I, Mangal S, Udai PS. Performance optimization of lead free-MASnI3/CIGS heterojunction solar cell with 28.7% efficiency: A numerical approach, Optical Materials. 2021; 122: 111812.

[51] Zheng X, Hou Y, Bao C, Yin J, Yuan F, Huang Z, Song K, Liu J, Troughton J, Gasparini N, et al. Managing grains and interfaces via ligand anchoring enables 22.3%-efficiency inverted perovskite solar cells. Nature Energy, 5: 131–140, 2020.

[52] Cacovich S, Vidon G, Degani M, Legrand M, Gouda L, Puel J, Vaynzof Y, Guillemoles J, Ory D, Grancini G. Imaging and quantifying non-radiative losses at 23% efficient inverted perovskite solar cells interfaces, Nature Communication. 2022; 13: 2868.

[53] Degani M, An Q, Albaladejo-Siguan M, Hofstetter YJ, Cho C, Paulus F, Grancini G, Vaynzof Y. 23.7% Efficient inverted perovskite solar cells by dual interfacial modification, Science Advances. 2021; 7: eabj7930.

[54] Feng X, Guo Q, Xiu J, Ying Z, Ng KW, Huang L, Wang S, Pan H, Tang Z, He Z. Close-loop recycling of perovskite solar cells through dissolution-recrystallization of perovskite by butylamine, Cell Reports Physical Science. 2021; 2: 100341.

[55] Min H, Lee DY, Kim J, Kim G, Lee KS, Kim J, Paik MJ, Kim YK, Kim KS, Kim MG, et al. Perovskite solar cells with atomically coherent interlayers on SnO2 electrodes. Nature, 598: 444–450, 2021.

Tushar A. Limbani*, A. Mahesh*, and Shivani R. Bharucha

Chapter 2
Synthesis and characterization of perovskite materials

Abstract: Perovskite materials have attracted significant interest in photovoltaics due to their outstanding optoelectronic properties, including high absorption coefficients ($>10^5$ cm^{-1}), tunable bandgaps (1.5–2.3 eV), long carrier diffusion lengths (>1 μm), and excellent defect tolerance. This chapter explores the synthesis and characterization of perovskite materials across various structural dimensions – 0D (nanoparticles and quantum dots), 1D (nanowires and nanorods), 2D (thin films and layered perovskites), and 3D (bulk single crystals and polycrystalline films) – for applications in perovskite solar cells (PSCs). The primary objective is to provide a comprehensive analysis of synthesis methodologies and their impact on material characteristics, charge transport mechanisms, and stability in PSCs.

The chapter systematically categorizes perovskite synthesis methods based on structural dimensionality. Techniques such as hot injection (120–180 °C) and ligand-assisted reprecipitation facilitate the synthesis of quantum dots (2–10 nm) with precise emission control. One-dimensional perovskites, including nanowires and nanorods, are fabricated via template-assisted growth using anodic aluminum oxide templates, solvo-thermal synthesis (100–150 °C), and vapor-phase deposition, enhancing charge transport properties. Two-dimensional perovskite thin films, recognized for their superior moisture resistance, are synthesized using solution-based deposition methods like spin coating (2,000–5,000 rpm), doctor blade coating (200–500 nm thickness), and spray coating, as well as a vapor-assisted deposition to improve crystallinity. Bulk single-crystal 3D perovskites, featuring long carrier lifetimes (>1 μs) and low defect densities ($\sim10^9$ cm^{-3}), are synthesized through inverse temperature crystallization (ITC), slow solvent evaporation, and vapor-phase growth, yielding high-quality crystals (up to $5 \times 5 \times 3$ mm^3).

A structured characterization framework is employed to analyze the structural, optical, and electrical properties of synthesized perovskite materials. X-ray diffraction is utilized for phase identification and crystallinity assessment, while scanning electron microscopy and transmission electron microscopy reveal morphological details

*Corresponding author: Tushar A. Limbani, C.L. Patel Institute of Studies and Research in Renewable Energy (ISRRE), The Charutar Vidya Mandal (CVM) University, New Vallabh Vidyanagar 388121, Gujarat, India, email: tusharlimbani97@gmail.com
*Corresponding author: A. Mahesh, C.L. Patel Institute of Studies and Research in Renewable Energy (ISRRE), The Charutar Vidya Mandal (CVM) University, New Vallabh Vidyanagar 388121, Gujarat, India, email: maheshiit10@gmail.com
Shivani R. Bharucha, N. V. Patel College of Pure & Applied Science, The Charutar Vidya Mandal (CVM) University, Vallabh Vidyanagar 388120, Gujarat, India

https://doi.org/10.1515/9783111726847-002

and grain structures. Ultraviolet-visible spectroscopy measures absorption coefficients and bandgap energies, and photoluminescence spectroscopy evaluates charge carrier recombination kinetics. Ultraviolet photoelectron spectroscopy and X-ray photoelectron spectroscopy investigate energy band alignment with electron transport layers and hole transport layers for optimal charge transfer. Furthermore, electrochemical impedance spectroscopy examines charge transport kinetics and interfacial resistances in perovskite materials.

Beyond synthesis and characterization, the chapter emphasizes the crucial role of surface passivation in reducing defects and improving stability across all perovskite dimensionalities. Strategies such as ligand engineering for quantum dots, defect suppression in nanowires, organic–inorganic interface engineering in 2D perovskites, and grain boundary passivation in 3D perovskites are explored to enhance resilience against environmental factors like moisture, oxygen, and thermal degradation.

By integrating precise synthesis techniques with advanced characterization approaches, this chapter provides a fundamental understanding of perovskite material engineering for high-efficiency and stable photovoltaic applications. The insights presented here contribute to the advancement of scalable and commercially viable perovskite photovoltaics, addressing critical challenges such as defect passivation, stability, and reproducibility in large-scale production.

Keywords: Charge transport, nanocrystals, perovskite nanoparticles, single crystals, spectroscopy, thin film

2.1 Introduction

2.1.1 Importance of dimensionality in perovskite properties

The dimensionality of perovskite materials significantly influences their electrical structure, charge transport characteristics, and stability [1–3]. Researchers may optimize the performance of perovskites for particular uses in solar cells (SCs) by developing them in 0D, 1D, 2D, and 3Dconfigurations [4–6].

2.1.1.1 0D perovskites

Quantum dots (QDs) and nanoparticles (NPs), exhibit quantum confinement effects, allowing precise bandgap tuning within 1.5–2.3 eV [7]. These materials are synthesized via hot injection (120–180 °C) or ligand-assisted reprecipitation (LARP), yielding highly monodisperse QDs (2–10 nm) with high photoluminescence quantum yield (PLQY > 90%) [8]. Their tunable absorption and emission properties make them highly useful in multi-junction SCs and light-emitting applications [9].

2.1.1.2 1D perovskite

Nanowires (NWs) and nanorods (NRs) facilitate anisotropic charge transport, minimizing recombination losses and extending carrier lifetimes [10]. They are synthesized through solvothermal growth (100–150 °C) and template-assisted methods using anodic aluminum oxide (AAO) templates [11]. These structures are promising for high-performance photodetectors and vertically oriented SCs due to their directed charge transport properties.

2.1.1.3 2D perovskites

Layered perovskites and thin films exhibit alternating organic and inorganic layers, enhancing stability against moisture and ion migration [12]. Synthesis methods include spin coating (2,000–5,000 rpm), dip-pen nanolithography, and sequential vapor deposition [13, 14]. Vapor-assisted deposition significantly improves crystallinity and minimizes grain boundary defects, making 2D perovskites ideal for long-lifetime SCs [15].

2.1.1.4 3D perovskites

Bulk single crystals and polycrystalline films possess high charge carrier mobility (>10 cm^2 V^{-1} s^{-1}) and extended carrier lifetimes (>1 μs) [16]. Techniques such as inverse temperature crystallization (ITC) and slow solvent evaporation produce defect-free single crystals (up to $5 \times 5 \times 3$ mm^3) with minimal defect densities ($\sim 10^9$ cm^{-3}) [17–19]. These materials are widely employed in tandem and multi-junction SCs due to their superior optoelectronic properties [20].

2.1.2 Scope of synthesis and characterization

The development and commercialization of perovskite solar cells (PSCs) necessitate precise synthesis methods and robust characterization techniques [21]. Structural analysis employs X-ray diffraction (XRD) for crystallinity assessment, while Fourier-transform infrared spectroscopy identifies functional groups and chemical bonding [22]. Morphological investigations utilize scanning electron microscopy (SEM) and transmission electron microscopy (TEM), supplemented by atomic force microscopy for topographical evaluation [23]. Optical characterization includes ultraviolet-visible (UV-Vis) spectroscopy for absorption analysis and time-resolved photoluminescence (TRPL) for carrier recombination studies [24]. Electrical properties are evaluated through current–voltage (I–V) measurements and electrochemical impedance spectroscopy (EIS), while Kelvin probe force microscopy (KPFM) examines surface potential fluctuations [25]. Energy band alignment and electronic structure are examined by ultraviolet photoelectron spectroscopy (UPS) and X-ray photoelectron spectroscopy (XPS), which ascertain energy level placement and chemical states [26].

Despite their remarkable efficiencies, perovskite materials still face challenges regarding stability, large-scale production, and defect passivation [27]. Strategies such as ligand engineering, compositional tuning, and interfacial modifications are crucial for long-term durability. Additionally, the advancement of scalable deposition techniques, including slot-die coating and roll-to-roll processing, will be essential for commercialization [28].This chapter provides a fundamental understanding of perovskite materials for SCs, emphasizing their dimensional versatility, synthesis methodologies, and characterization techniques. Addressing challenges in stability, charge transport, and scalability will enable the development of next-generation, high-efficiency, and commercially viable solar solutions.

2.2 Synthesis techniques for perovskite materials

The synthesis techniques of perovskite materials significantly influence their optoelectronic properties, stability, and scalability in photovoltaic applications [29]. The choice of synthesis method affects the crystallinity, morphology, and defect density, all of which determine the efficiency of PSCs [30]. This section elaborates on various synthesis techniques categorized based on the dimensionality of perovskite materials, their process optimizations, and associated scalability challenges.

2.2.1 Synthesis of 0D perovskite nanoparticles and quantum dots

Zero-dimensional perovskite materials, such as nanoparticles and QDs, are widely synthesized using solution-based techniques to achieve precise control over particle size, composition, and optical properties.

2.2.1.1 Hot-injection method

The hot-injection technique is widely used for synthesizing 0D halide perovskite (HP) nanoparticles and QDs [31]. This method involves injecting a perovskite precursor into heated solvent containing ligands like oleic acid (OA) and oleylamine (OAm) [32]. It enables the synthesis of highly monodisperse QDs (2–10 nm) with tunable bandgaps (1.5–2.3 eV) and high photoluminescence quantum yield (PLQY > 90%) [33]. In this method, a cesium precursor is rapidly injected into a PbX_2-octadecene (ODE) solution containing OA and OAm at high temperatures [34]. OA plays a crucial role in breaking the precursor into monomers and coordinating Pb^{2+} ions. OAm regulates monomer concentration and influences the reactivity of Pb^{2+} after nucleation [35].

By adjusting the OA-to-OAm ratio, researchers can control the morphology of perovskites, transforming them from 3D $CsPbBr_3$ nanocubes to 1D NRs and eventually to 2D nanoplatelets [36]. Furthermore, consistent morphology in $CsPb(Br/I)_3$ NRs can be achieved using $PbBr_2$ and PbI_2 precursors. The monomer reservoir method enables the production of homogeneous, low-defect $CsPbBr_3$ NRs [37].

In lead-free perovskite synthesis, tin-based perovskite QDs exhibit tunable photoluminescence between 625 and 709 nm [38]. Stable Cs_2SnI_6 nanowires and nanoribbons can be synthesized by reacting tetravalent tin (IV) iodide with cesium oleate at 220 °C in ODE, using OA and OAm as capping ligands [39]. The interaction between OA and OAm influences the selective growth of nanostructures, directing perovskites into different morphologies [40]. Figure 2.1 provides a fundamental structural view of ABX_3 perovskites, illustrates the hot-injection technique for synthesizing colloidal $CsPbX_3$ nanocrystals (NCs), and shows their corresponding photoluminescence (PL) spectra. This highlights the importance of ligand chemistry in controlling the final structure of 0D perovskite materials.

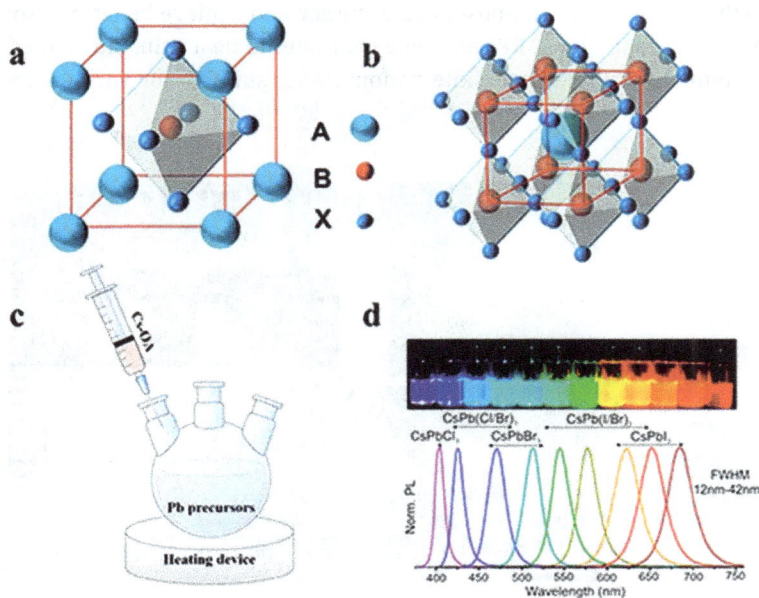

Figure 2.1: (a) Structural depiction of an individual unit cell of ABX_3 perovskites. **(b)** Extended crystal lattice of ABX_3 perovskites, illustrating corner-sharing octahedral connectivity. **(c)** Schematic representation of the hot-injection (HI) technique for the synthesis of colloidal $CsPbX_3$ nanocrystals (NCs). **(d)** Diverse colloidal dispersions of $CsPbX_3$ nanocrystals in toluene subjected to UV light ($\lambda = 365$ nm) together with their respective PL spectra [41].

2.2.1.2 Ligand-assisted reprecipitation (LARP)

LARP method is an efficient approach for synthesizing perovskite QDs and nanoparticles. This method involves rapidly mixing perovskite precursors in a polar solvent with an anti-solvent, resulting in the nucleation of nanoparticles [42]. It enables room-temperature synthesis but requires careful ligand engineering to control size and stability. This technique involves dissolving lead halide (PbX$_2$) and organic halide (CH$_3$NH$_3$X) salts in a polar solvent such as dimethylformamide (DMF) [43]. The solution is then introduced into a nonpolar solvent like toluene, which contains stabilizing ligands. Due to the immiscibility between polar and nonpolar solvents, a rapid decrease in solubility occurs, leading to fast nucleation and crystallization of perovskite nanocrystals [44]. Figure 2.2 presents the LARP synthesis process, morphology of CH$_3$NH$_3$PbBr$_3$ QDs, and their optical properties under ambient and UV light, along with the corresponding PL spectra.

Ligands such as *n*-octylamine and OA play a crucial role in stabilizing the nanoparticles, resulting in highly luminescent MAPbBr$_3$ QDs with photoluminescence quantum yields (PLQYs) reaching up to 70%. Despite its efficiency, a key limitation of the LARP method is its reliance on PbX$_2$ salts as both lead and halide precursors [45]. This restricts precise control over reaction species. Researchers have developed modified synthesis methods to improve compositional accuracy and achieve better control over nanocrystal growth [46]. The LARP technique continues to be a valuable method for producing high-quality, stable perovskite nanoparticles suitable for various optoelectronic applications.

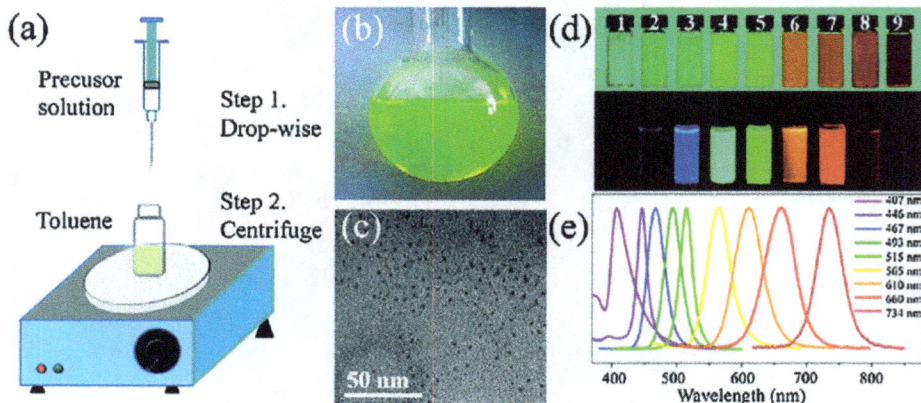

Figure 2.2: (a) Schematic depiction of the LARP process. **(b)** Image of colloidal CH$_3$NH$_3$PbBr$_3$ QDs. **(c)** TEM picture depicting the morphology of CH$_3$NH$_3$PbBr$_3$ QDs. **(d)** Photographic visuals of CH$_3$NH$_3$PbX$_3$ QDs under ambient light and ultraviolet irradiation (λ = 365 nm). **(e)** PL spectra of CH$_3$NH$_3$PbX$_3$ QDs [47].

2.2.2 Synthesis of 1D perovskite nanowires and nanorods

One-dimensional perovskite structures offer anisotropic charge transport and reduced recombination losses. These are synthesized using the following methods.

2.2.2.1 Solvothermal growth

Conducted in a sealed autoclave at 100–150 °C, this method allows the growth of high-
aspect-ratio NWs with improved carrier mobility (>1 cm^2 V^{-1} s^{-1}). It is an effective technique for synthesizing 1D perovskite NWs and NRs [48]. In this process, perovskite
precursors are dissolved in a solvent with high-temperature and high-pressure conditions inside a sealed autoclave. The reaction environment promotes controlled crystal
growth, leading to high-quality, uniform NWs and NRs [49]. The choice of solvents,
temperature, and reaction time plays a crucial role in determining the morphology
and crystallinity of the final perovskite nanostructures.

2.2.2.2 Template-assisted growth

This approach uses anodic aluminum oxide (AAO) or polymer templates to guide the
controlled growth of NRs and NWs, ensuring uniformity and enhanced electrical properties [50]. The template-assisted approach utilizes predefined templates to guide the
formation of 1D perovskite nanostructures [51]. Common templates include AAO, elec-

Figure 2.3: (A) Schematic representation of the template-assisted synthesis of perovskite NCs inside
mesoporous silica. The procedure entails saturating a mesoporous template (with pore diameters of 2.5,
4.0, or 7 nm) with a precursor solution, followed by the elimination of the excess solution and the
subsequent crystallization of APbX$_3$ NCs by a drying process. **(B)** Photograph of mesoporous silica
containing CsPbBr$_3$ (left) and CsPb(Br$_{0.25}$I$_{0.75}$)$_3$ nanocrystals (right) under daylight and ultraviolet lighting
[56].

trospun polymer fibers, and carbon nanotubes [52]. Perovskite precursors are introduced into the template pores, followed by controlled crystallization to form uniform NWs or NRs [53]. After synthesis, the template is removed to obtain free-standing 1D perovskite nanostructures [54]. This method provides excellent control over NW diameter, length, and orientation [55]. Figure 2.3 illustrates the template-assisted synthesis of perovskite NCs within mesoporous silica and shows their optical appearance under daylight and UV illumination.

2.2.2.3 Vapor-phase deposition

In this technique, perovskite precursors are vaporized and recondensed onto a substrate, producing highly crystalline NWs with fewer grain boundaries and enhanced charge transport properties [57]. Vapor-phase deposition techniques, such as chemical vapor deposition (CVD) and physical vapor deposition (PVD), enable the synthesis of high-quality 1D perovskite nanostructures. In CVD, gaseous perovskite precursors are introduced into a reaction chamber, where they react and condense to form NWs or NRs on a substrate [58]. The reaction temperature, precursor concentration, and substrate properties significantly influence the growth process [57].

PVD involves the physical evaporation of perovskite precursors onto a substrate, facilitating controlled growth of 1D nanostructure. Vapor-phase deposition techniques offer precise control over crystallinity, composition, and morphology, making them suitable for high-performance optoelectronic applications [58, 59]. Figure 2.4 shows the dual-source thermal evaporation method for perovskite deposition, compares XRD patterns of solution-processed and vapor-deposited films, and presents the p-i-n device architecture along with the crystal structure of ABX_3 perovskite.

2.2.3 Synthesis of 2D perovskite thin films

2D perovskites exhibit layered structures that enhance moisture resistance and stability. Various deposition techniques are employed for thin-film synthesis [61, 62].

2.2.3.1 Spin coating

Spin coating is a prevalent method for depositing perovskite thin films, owing to its simplicity, cost-efficiency, and capacity to generate homogeneous films with precise thickness control [63, 64]. This approach involves depositing a perovskite precursor solution, usually dissolved in a polar solvent like DMF or dimethyl sulfoxide (DMSO), onto a substrate. The substrate is then spun at high velocities (1,000–5,000 rpm), re-

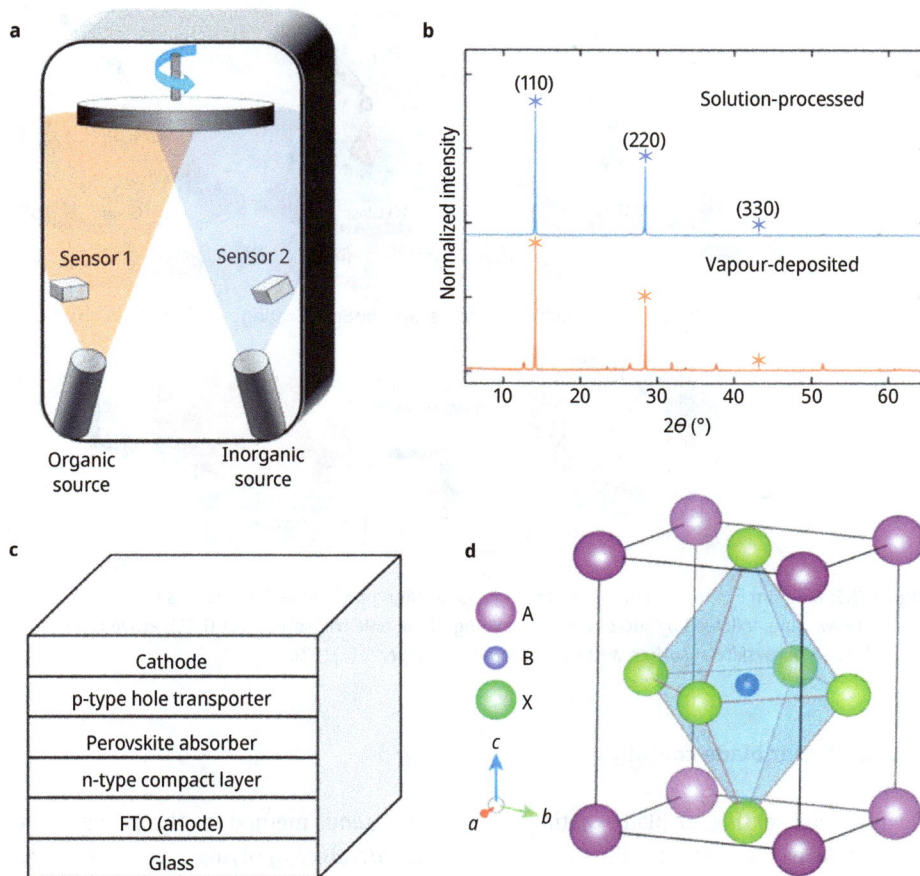

Figure 2.4: (a) Dual-source thermal evaporation technique for the deposition of perovskite absorbers using methylammonium iodide and lead(II) chloride. **(b)** XRD patterns of solution-processed (blue) and vapor-deposited (red) perovskite films. **(c)** Diagram of a planar p-i-n PSC. **(d)** The crystal structure of ABX_3 perovskite, with A as methylammonium, B as lead (Pb), and X as iodine (I) or chlorine (Cl) [60].

sulting in the uniform distribution of the solution throughout the surface as the solvent swiftly evaporates. This procedure yields a uniform and dense perovskite layer [65]. The film thickness is regulated by modifying the spin speed, solution concentration, and spin time. Post-deposition annealing is often necessary to facilitate crystallization and enhance film shape. Notwithstanding its benefits, spin coating is not readily scalable for large-area manufacturing and is prone to material loss, making it less appropriate for industrial-scale production [66]. Figure 2.5 illustrates the PSC fabrication process via spin coating, detailing sequential deposition of the HTL, active layer, perovskite absorber, and ETL.

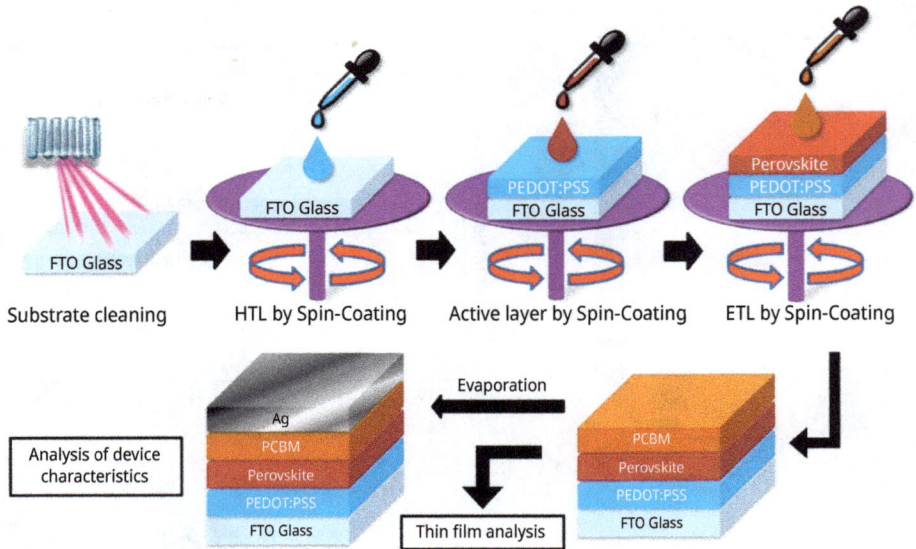

Figure 2.5: Diagram of the PSC fabrication through spin-coating technique. The process includes substrate washing, followed by successive spin coating of the hole transport layer (HTL), active layer (PEDOT: PSS), perovskite absorber, and electron transport layer (ETL) [67].

2.2.3.2 Doctor blade coating

Doctor blade coating, or blade coating, is an advantageous method for the extensive deposition of perovskite thin films [65]. This process involves applying a perovskite precursor solution onto a substrate using a blade, ensuring uniform layer deposition. The film thickness is regulated by modifying the blade speed, solution viscosity, and the gap distance between the blade and the substrate [68]. A primary feature of this approach is its compatibility with roll-to-roll processing, making it appropriate for large-scale production of PSCs [28]. Moreover, it reduces material waste relative to spin coating, enhancing cost efficiency. The resultant films demonstrate commendable uniformity and coverage; nevertheless, attaining high crystallinity and defect-free films requires meticulous optimization of processing parameters, including drying duration and post-deposition annealing. Doctor blade coating is especially beneficial for flexible and printable PSCs, where scalability and material efficiency are paramount [69, 70]. Figure 2.6 shows the doctor blade coating process and optical images of perovskite films deposited at different temperatures, highlighting their morphological and crystallization behavior during annealing.

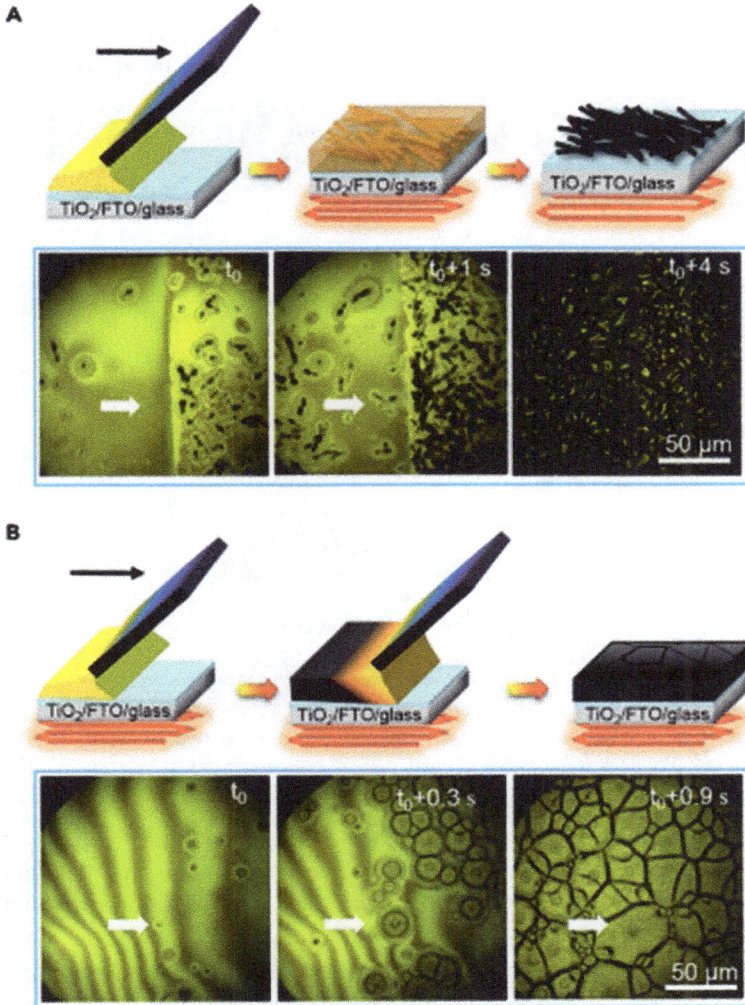

Figure 2.6: Schematic depiction of the doctor blade coating procedure and real-time optical microscope images. **(A)** Perovskite film coated with a DMSO: GBL solvent system at 25 °C during thermal annealing. **(B)** Perovskite film deposited at 150 °C, exhibiting morphological development and crystallization characteristics under regulated annealing conditions [71].

2.2.3.3 Spray coating

Spray coating is a scalable deposition method that facilitates high-throughput production of perovskite thin films [72]. A perovskite precursor solution is atomized into small droplets and applied to a heated substrate. The solvent swiftly evaporates upon interaction with the substrate, resulting in the creation of a thin perovskite layer. This ap-

a

b   **c**

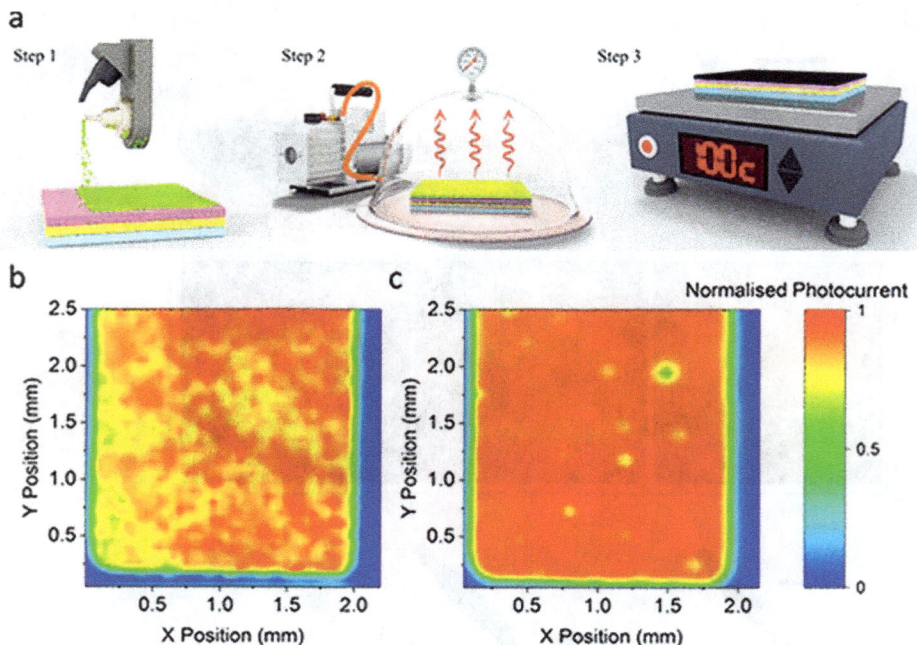

Figure 2.7: Diagram illustrating the spray coating and vacuum-assisted solvent processing for the production of high-quality perovskite films. **(a)** Precursor ink is applied as a spray, creating a wet layer, which is then subjected to vacuum exposure to eliminate DMF, resulting in partial crystallization. Subsequent annealing at 100 °C finalizes perovskite formation. **(b, c)** Laser-beam-induced current maps of films with and without vacuum treatment [74].

proach enables extensive coating, consistent film generation, and precise control of film thickness by modifying spray parameters like droplet size, solution concentration, and substrate temperature [73]. Spray coating is suitable for roll-to-roll manufacturing, making it appealing for large-scale production of PSCs. Achieving high crystallinity and removing flaws necessitates the optimization of deposition circumstances and post-treatment methods. Moreover, ensuring film uniformity over extensive regions might be difficult owing to discrepancies in droplet dispersion and drying kinetics [74]. Figure 2.7 illustrates the spray coating with vacuum-assisted solvent removal for perovskite film formation, along with current maps showing the impact of vacuum treatment.

2.2.3.4 Vapor-assisted deposition

Vapor-assisted deposition (VAD) is a hybrid method that improves film uniformity and crystallinity by subjecting a pre-deposited perovskite precursor layer to halide vapors [75]. This process involves the first deposition of an organic–inorganic perovskite precursor layer using a standard solution-based technique, such as spin coating or blade coating. The film is then positioned in a regulated setting where it is subjected to a halide vapor source, such as MAI or formamidinium bromide (FABr). The vapor interacts with the precursor film, transforming it into a highly crystalline perovskite phase [76]. VAD provides substantial benefits for superior grain development, increased film homogeneity, and less defect density. This technology is particularly successful for producing high-performance PSCs with improved stability and efficiency. Nonetheless, VAD requires meticulous regulation of reaction parameters, such as temperature, vapor pressure, and exposure duration, which limit its scalability [77]. Figure 2.8 depicts the vapor-assisted solution process for controlled growth and crystallization of perovskite films. Where Figure 2.9 illustrates the VAD setup with controlled nitrogen flow and base pressure for optimized film formation.

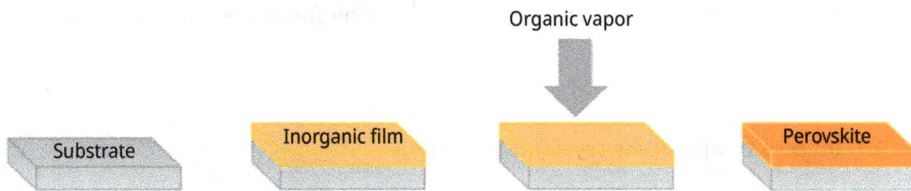

Figure 2.8: Schematic illustration of the vapor-assisted solution process used for the regulated production and crystallization of perovskite films [78].

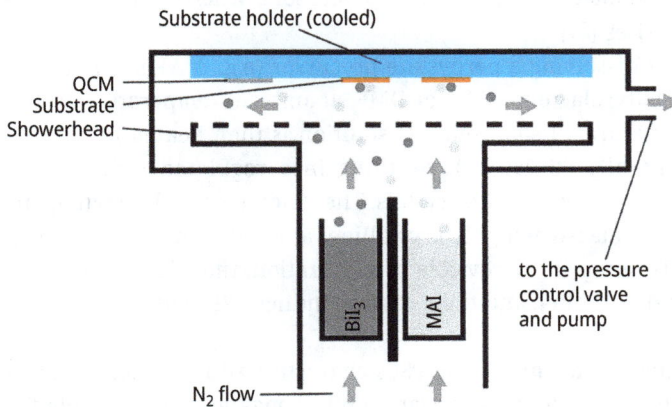

Figure 2.9: Schematic illustration of the VAD apparatus, wherein the nitrogen gas flow is sustained at 500 SCCM, and the base pressure is regulated at 10 hPa to promote film formation [77].

Solution-based and VAD methods are essential for producing high-quality perovskite thin films for optoelectronic applications. Spin coating is a prevalent technique for research-scale device fabrication, valued for its simplicity and homogeneous film quality, while doctor blade coating and spray coating provide scaled alternatives appropriate for industrial manufacturing [79]. VAD promotes layer crystallinity and uniformity, thus enhancing the performance of PSCs. Future developments in these deposition processes will be crucial for enhancing film characteristics, minimizing fault density, and facilitating the large-scale commercialization of perovskite-based photovoltaic devices.

2.2.4 Synthesis of 3D bulk perovskite crystals

Bulk single crystals and polycrystalline films are essential for high-performance PSCs due to their superior charge transport properties. The following techniques are employed:

2.2.4.1 Inverse temperature crystallization (ITC)

ITC is a prevalent method for producing high-quality perovskite SCs characterized by superior structural and optoelectronic characteristics [80]. In contrast to traditional crystallization techniques that depend on solvent evaporation or cooling, ITC utilizes the inverse solubility phenomenon of perovskite precursors, whereby solubility diminishes as temperature rises [81].

This process involves dissolving a perovskite precursor (e.g., $MAPbX_3$) in an organic solvent, such as γ-butyrolactone (GBL) or DMF, at ambient temperature to create a saturated or supersaturated solution. The solution is then heated to a designated temperature (generally 80–150 °C), resulting in a regulated reduction in solubility and the nucleation of perovskite crystals. The gradual crystal growth guarantees the development of massive, highly crystalline, and defect-free SCs [82, 83]. Figure 2.10 shows the ITC setup for perovskite crystallization, time-dependent crystal growth of $MAPbI_3$ and $MAPbBr_3$, and their corresponding XRD patterns with images of the bulk crystals.

ITC is especially beneficial for producing PSCs characterized by negligible grain boundaries, low defect density, and high carrier mobility, making them suitable for optoelectronic applications including photodetectors and single crystals. Nonetheless, careful regulation of temperature, precursor concentration, and solvent choice are crucial for enhancing crystal quality and consistency [84]. The method's scalability and enhanced purity make it a favored option for foundational research and device manufacturing.

a

Heating to a fixed temperature
Fast crystallization

Solution

Oil bath
Room temperature

Solution +
MAPbX crystals

Oil bath
Hot plate

b

15 min   30 min   1 h   2 h   3 h

c

15 min   30 min   1 h   2 h   3 h

d

Powder XRD of cubic MAPbBr$_3$

Intensity (a.u.)

2θ (degree)

e

Powder XRD of tetragonal MAPbI$_3$

Intensity (a.u.)

2θ (degree)

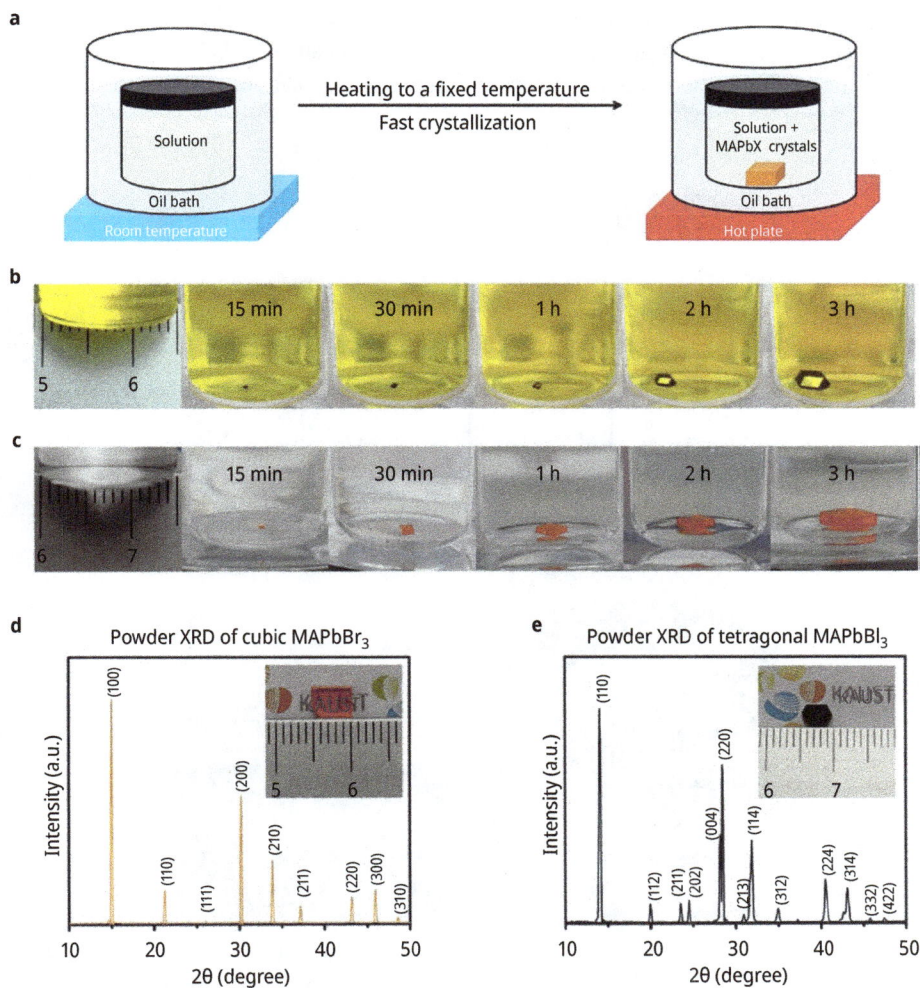

Figure 2.10: (a) Schematic representation of the ITC apparatus, whereby a vial holding the precursor solution is submerged in a heated bath (80 °C for MAPbBr$_3$ and 110 °C for MAPbI$_3$) to facilitate crystallization. **(b, c)** Time-dependent crystallization of MAPbI$_3$ and MAPbBr$_3$. **(d, e)** Powder XRD patterns of pulverized crystals. The insets provide photos of the respective bulk crystals cultivated in an unrestricted environment [82].

2.2.4.2 Slow solvent evaporation method

The slow solvent evaporation method is a recognized approach for producing high-quality perovskite SCs with superior structural and optoelectronic characteristics. This technique depends on the systematic elimination of solvent under regulated circumstances to facilitate nucleation and decelerate crystal development, yielding massive, defect-reduced SCs appropriate for diverse optoelectronic applications [85]. Figure 2.11 illustrates the solvent evaporation crystallization method and shows BA_2PbCl_4 single crystals with simulated morphology and temperature-dependent growth of $MAPbCl_3$ crystals.

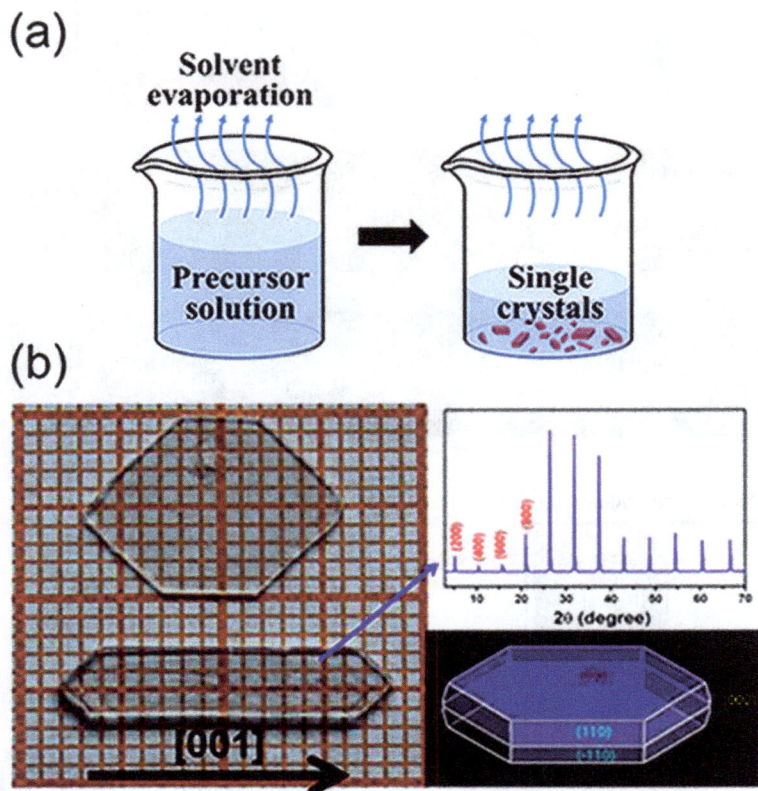

Figure 2.11: (a) Schematic representation of the solvent evaporation crystallization technique. **(b)** Photographs of substantial BA_2PbCl_4 SCs accompanied with their simulated morphology. Images depict the dimensions of the $MAPbCl_3$ SC and its [100] crystallographic plane, with crystals cultivated at different temperatures [86].

This method involves dissolving a perovskite precursor, such as $MAPbX_3$, in an appropriate organic solvent, such as GBL, DMF, or DMSO, to create a saturated or supersaturated solution. The solution is thereafter maintained at a regulated temperature, allowing the solvent to evaporate gradually over hours or days. With a reduction in solvent concentration, perovskite crystallization transpires, resulting in the emergence of substantial, highly crystalline SCs characterized by negligible grain borders and imperfections [87].

This technique is beneficial for generating high-purity SCs with exceptional charge transport characteristics, rendering them suitable for perovskite-based photodetectors, SCs, and light-emitting devices. Nonetheless, meticulous temperature, humidity, and evaporation rate regulation are required to guarantee consistent crystal formation. This approach, while scalable, is rather time-intensive when juxtaposed with other quick crystallization techniques such as ITC [88].

The synthesis techniques for perovskite materials considerably influence their crystallinity, shape, and optoelectronic characteristics. ITC produces defect-free SCs with enhanced charge transfer, while slow solvent evaporation facilitates the formation of massive, high-purity crystals but requires prolonged processing durations. The HI approach yields monodisperse QDs with adjustable optical characteristics, whereas the LARP methodology enables synthesis at ambient temperature but creates polydisperse particles. Solvothermal and vapor phase growth techniques provide meticulous regulation of one-dimensional nanostructures, although they require elevated temperatures and pressures. Template-assisted growth enables the production of homogeneous NWs, although it constrains scalability. Every method necessitates optimization for high-performance perovskite devices.

2.2.5 Scalability challenges and process optimizations

Despite the rapid advancements in perovskite synthesis, challenges remain regarding large-scale production, stability, and reproducibility. Key considerations for scalability include:
–   **Solvent engineering**: Reducing toxic solvents like dimethylformamide (DMF) in favor of eco-friendly alternatives to meet environmental regulations.
–   **Deposition control**: Enhancing film uniformity using roll-to-roll or slot-die coating methods for industrial scalability.
–   **Defect minimization**: Implementing surface passivation strategies and compositional engineering to improve long-term stability and performance.

Addressing these challenges will be crucial for the commercial viability of perovskite photovoltaics. The next section will explore characterization techniques essential for assessing the quality and performance of synthesized perovskite materials.

2.3 Characterization techniques for perovskite materials

Characterization techniques play a crucial role in understanding the structural, optical, and electronic properties of perovskite materials [89, 90]. These techniques help assess crystallinity, charge transport mechanisms, defect densities, and overall material stability, all of which are essential for optimizing PSCs and other optoelectronic devices [91]. The following sections elaborate on the major characterization methods used for perovskite materials.

2.3.1 Structural characterization

2.3.1.1 X-ray diffraction (XRD)

XRD is a fundamental technique for analyzing the crystalline structure and phase purity of perovskites. Using Bragg's law ($n\lambda = 2d \sin \theta$), XRD determines lattice parameters, symmetry, and crystallite size [92, 93]. It is also instrumental in identifying phase transitions, strain effects, and structural degradation due to environmental exposure [94]. Peak shifts indicate changes in crystal structure, while grazing-incidence XRD (GIXRD) is specifically employed for studying thin films [95]. Encapsulation of three-dimensional (3D) perovskites with two-dimensional (2D) layers enhances stability. When combined with Raman spectroscopy and neutron diffraction, XRD provides critical insights into the durability and optimization of perovskites for optoelectronic applications [96].

Methylammonium lead iodide ($MAPbI_3$), a widely studied lead halide perovskite (LHP), undergoes a reversible phase transition from tetragonal to cubic symmetry at temperatures near typical SC operating conditions [97]. This transition negatively impacts solar performance, necessitating the development of thermally stable perovskites. One stabilization strategy involves cation substitution, where formamidinium lead iodide ($FAPbI_3$) is partially substituted with $MAPbI_3$. The introduction of smaller methylammonium (MA) cations stabilizes the trigonal phase of $FAPbI_3$ without inducing lattice contraction or altering its optical properties. This effect arises from enhanced I–H hydrogen bonding and increased Madelung energy, which collectively improve structural integrity [98]. Figure 2.12 presents XRD spectra of $(FAPbI_3)_{1-x}(MAPbBr_3)_x$ perovskites, highlighting diffraction patterns over a broad and zoomed 2θ range.

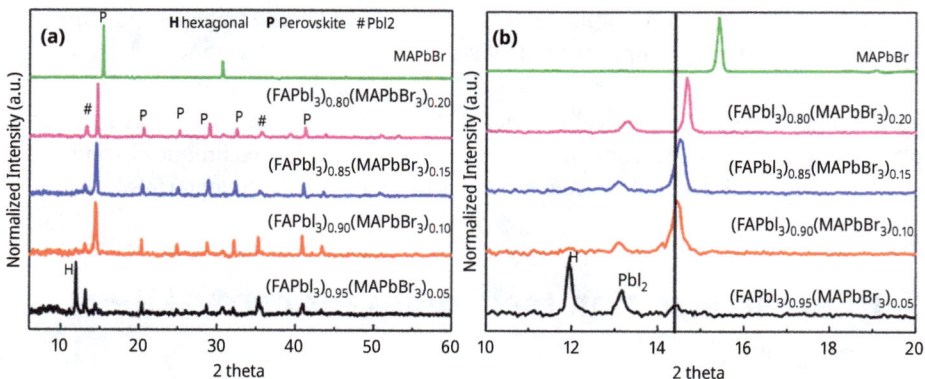

Figure 2.12: XRD spectra of $(FAPbI_3)_{1-x}(MAPbBr_3)_x$ perovskite compositions with x values of 0.05, 0.1, 0.15, 0.2, and 1. **(a)** Diffraction patterns obtained across the range of $2\theta = 5$–$60°$. **(b)** Enlarged perspective of the 2θ interval from $10°$ to $20°$ [98].

(a)　　　　　　　　　　　　　　　　　　　　　　　**(b)**

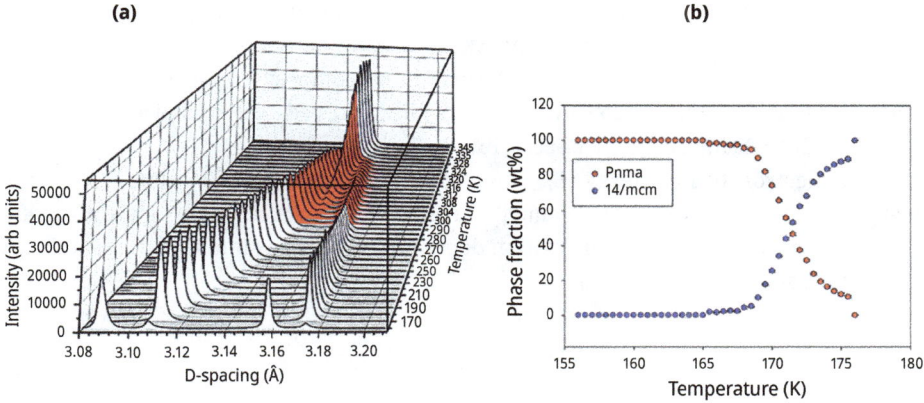

Figure 2.13: (a) Synchrotron X-ray powder diffraction patterns exhibiting Bragg peaks during the cubic-to-tetragonal phase transition. The (200) cubic peak and (220)/(004) tetragonal peaks fluctuate with temperature, indicating phase coexistence between 300 and 330 K. **(b)** Phase fractions of tetragonal and orthorhombic phases, ascertained using neutron diffraction, during the tetragonal-to-orthorhombic transition in d_6-MAPbI$_3$ [99].

2.3.1.2 Temperature-dependent XRD analysis

Temperature-dependent XRD is crucial for studying these phase transitions. XRD patterns reveal peak shifts, peak merging, and the disappearance of characteristic reflections as MAPbI$_3$ transitions from tetragonal to cubic at approximately 54 °C. Contour plots tracking the evolution of XRD peaks across a temperature range of 35–95 °C demonstrate a fully reversible transition within a 1 °C margin of error, reflecting the thermal fluctuations encountered in operational SCs [99]. Figure 2.13 shows synchrotron XRD and neutron diffraction data revealing temperature-induced phase transitions in d_6-MAPbI$_3$, highlighting phase coexistence and evolution from cubic to tetragonal and orthorhombic structures.

Neutron powder diffraction and high-resolution synchrotron XRD further confirm that the cubic and tetragonal phases coexist over a broad temperature range, indicating a first-order phase transition. The orthorhombic-to-tetragonal transition also follows a first-order mechanism, contradicting Landau's criteria for a second-order transition due to distinct irreducible representations. This highlights the complex structural dynamics governing perovskite phase behavior [99].

2.3.1.3 Scanning electron microscopy (SEM) for morphology

SEM is a vital technique for imaging thin-film SCs, providing high-resolution insights into layer thickness, surface topography, and microstructural characteristics, with resolutions below 1 nm [100]. Modern SEM systems, integrated with energy-dispersive X-ray (EDX) detectors, enable elemental composition analysis across 0D to 3D structures. Beyond imaging, SEM provides crucial information on composition, microstructure, and surface potential through secondary electrons (SEs) and backscattered electrons (BSEs), which produce distinct contrast effects [101].

The electron beam energy (E_b), controlled by the acceleration voltage, and the beam current (I_B), adjusted via filament current or aperture settings, significantly impact SEM analysis [102]. The penetration depth (R) of electrons into the sample follows the empirical relationship:

$$R = 4.28 \times 10^2 \frac{\mathrm{mm} \times r}{\mathrm{g/cm^3}} \times \left(\frac{E_b}{\mathrm{keV}} \right)^{1.75} \tag{2.1}$$

where r is the density of the specimen.

In SEM, electromagnetic lenses focus the electron beam to achieve probe sizes below 1 nm, while scanning coils direct the beam across the surface. SEs (0–50 eV) and BSEs (50 eV–E_b) provide structural and compositional contrast. In-column BSE detectors capture Rutherford-type BSEs, enabling phase identification through atomic number variations [103]. Gas injection, such as nitrogen (N_2), during imaging mitigates surface charging effects in perovskite materials spanning 0D to 3D structures. Figure 2.14 presents EDX spectra and elemental mapping of perovskite films spin-coated on SnO_2 nanocomposites, showing elemental distribution at the interfaces.

2.3.1.4 Transmission electron microscopy (TEM) for nanoscale structure

TEM is a powerful tool for investigating the nanoscale architecture of perovskite materials, providing insights into interfaces, grain boundaries, defects, and phase distributions. By employing electron beam energies ranging from 80 to 1,200 keV, TEM enables high-resolution imaging and diffraction analysis, essential for understanding perovskite crystallinity and degradation mechanisms [105].

Conventional TEM (CTEM) and scanning TEM (STEM) are widely utilized. CTEM offers bright-field (BF) and dark-field (DF) imaging to reveal grain orientations and defects, while high-resolution TEM (HR-TEM) provides atomic-scale structural visualization. Selected-area electron diffraction and convergent beam electron diffraction (CBED) facilitate phase identification and strain analysis. STEM, equipped with annular DF (ADF) detectors, enables atomic-resolution imaging with Z-contrast, allowing differentiation of elemental distributions in perovskites [106].

Analytical techniques such as electron energy-loss spectroscopy and energy-filtered TEM provide crucial information on composition and bonding, which is essential for investigating ion migration and degradation pathways. Off-axis electron holography further enhances the understanding of electrostatic potential variations, contributing to

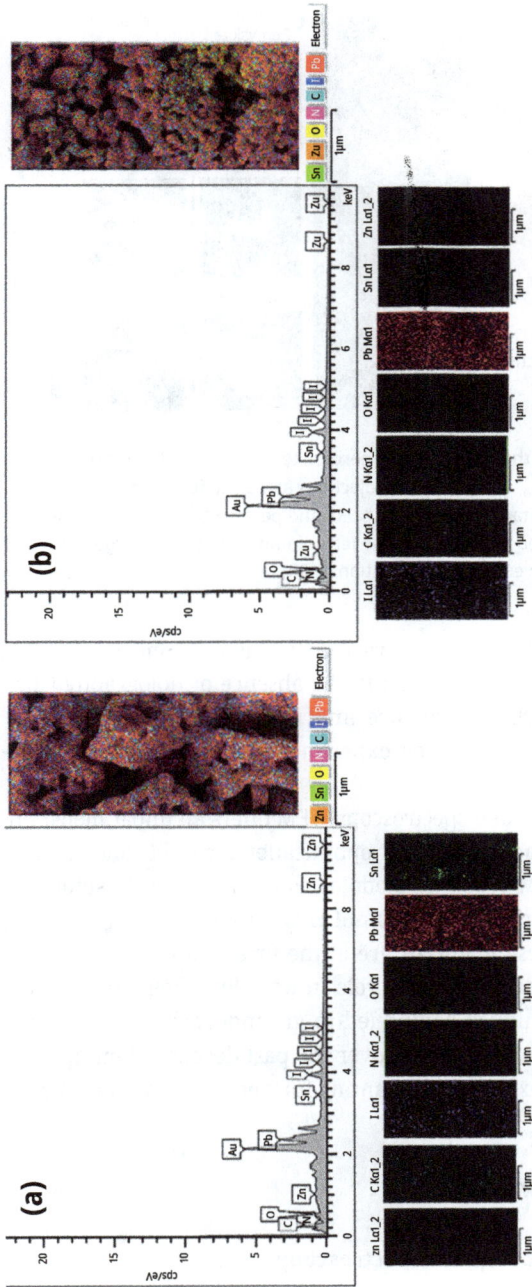

Figure 2.14: EDX spectra and elemental mapping analysis of perovskite films formed by spin coating over **(a)** SnO$_2$@ZnO nanocomposite and **(b)** SnO$_2$@ZnO.PANI (4%) nanocomposite, demonstrating the elemental distribution and composition of the corresponding interfaces [104].

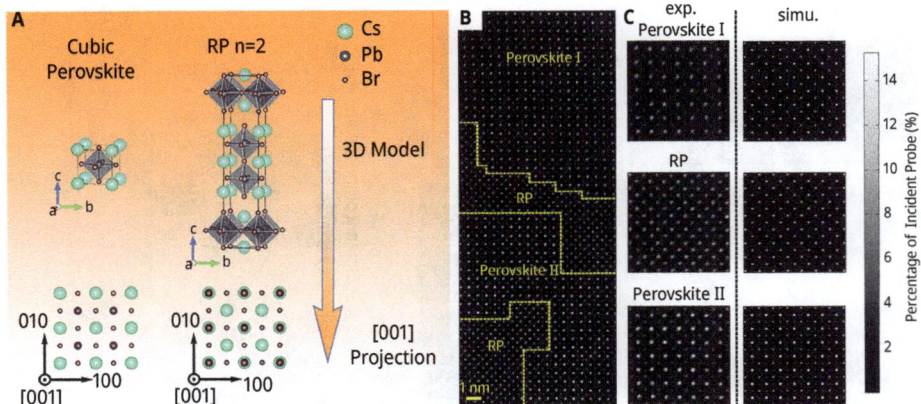

Figure 2.15: (A) Schematic representation of the cubic $CsPbBr_3$ perovskite structure and the associated Ruddlesden-Popper (RP) phase ($Cs_n + 1Pb_nBr_{3n+1}$, $n = 2$), with projections along the [001] zone axis. **(B)** HAADF-STEM picture illustrating the simultaneous presence of RP and perovskite phases in $CsPbBr_3$. **(C)** Comparison of experimental and simulated pictures, with the left column displaying enlarged portions from **(B)** and the right column illustrating the equivalent simulations [108].

a deeper comprehension of charge transport characteristics. Phase-contrast imaging can also be achieved through focus variation, even in the absence of holography [107]. Figure 2.15 illustrates the cubic $CsPbBr_3$ structure and its RP phase, supported by HAADF-STEM images and comparisons between experimental and simulated projections along the [001] axis.

By integrating imaging, diffraction, and spectroscopy, TEM offers an unparalleled understanding of perovskite materials, facilitating their optimization for stable and efficient optoelectronic applications [109]. TEM-based methodologies are particularly suited for site-specific structural and analytical characterization of HPs across atomic, nanometer, and micrometer length scales. As these materials are at the forefront of solar cell and optoelectronic device research, TEM plays a pivotal role in unveiling their fundamental properties. Despite TEM-based techniques having led to groundbreaking discoveries and significant advancements in materials science over the past decades, their application to HPs remains relatively underexplored, warranting further investigation [108].

2.3.2 Optical characterization

2.3.2.1 Ultraviolet-visible (UV-Vis) absorption spectroscopy

UV-Vis spectroscopy measures the absorption coefficients and bandgap energy of perovskites. The Tauc plot method is commonly used to determine direct and indirect bandgaps, which are crucial for tailoring absorption properties for photovoltaic applications. Metal-HPs have a direct bandgap, facilitating effective light absorption in thin films, unlike indirect bandgap semiconductors such as silicon, which need larger sheets for similar absorption efficacy [110].

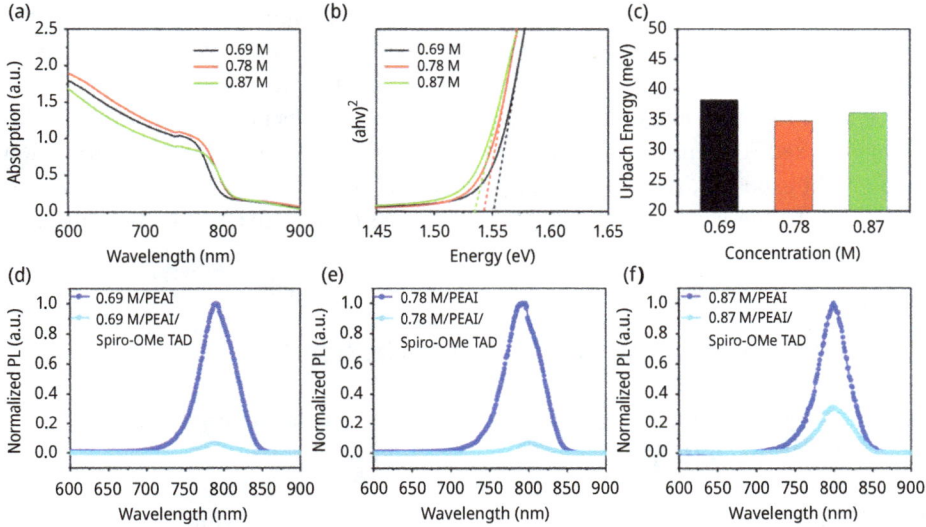

Figure 2.16: Optical and electrical characteristics of perovskite films with differing AXE concentrations. **(a)** UV–visible absorption spectra, **(b)** Tauc plots, and **(c)** Urbach energy assessment. PL spectra of samples: **(d)** 0.69 M, **(e)** 0.78 M, and **(f)** 0.87 M perovskite/PEAI, both with and without Spiro-OMeTAD. Standardized PL intensity [113].

Steady-state UV-Vis-NIR spectroscopy is often used to assess optical absorption in the ultraviolet, visible, and near-infrared spectra. Conventional UV-Vis measurements may show inaccuracies due to reflection and scattering, which are addressed using integrating sphere-based UV-Vis-NIR spectroscopy [111]. Photothermal deflection spectroscopy and Fourier transform photocurrent spectroscopy provide advanced insights into defect states and sub-bandgap absorption in perovskites [112]. Figure 2.16a -c presents the UV -visible absorption spectra, Tauc plots, and Urbach energy analysis of perovskite films with varying AXE concentrations, highlighting their optical characteristics.

2.3.2.2 Photoluminescence (PL) spectroscopy for charge recombination

PL spectroscopy evaluates charge carrier recombination, while TRPL helps determine carrier lifetimes [114]. High PL intensity indicates low defect density, while PL quenching in the presence of transport layers suggests efficient charge extraction. TRPL decay profiles reveal recombination mechanisms, with longer lifetimes correlating to better charge transport efficiency [115].

PL spectroscopy is an essential method for analyzing the emission properties of perovskite absorbers. It reveals charge recombination dynamics and identifies trap states affecting device efficiency. Exciting perovskites with high-energy photons generates emissions, with peak intensity indicative of radiative recombination. Temperature-dependent PL helps analyze intra-bandgap traps [116]. Figure 2.16d -f shows the PL spectra of perovskite/PEAI films (0.69 M, 0.78 M, and 0.87 M) with and without Spiro-OMeTAD, illustrating changes in emission intensity with composition.

Photon recycling and luminescence yield enhancements improve SC and LED efficiency. Mixed-HPs enable tunable emissions but may undergo phase segregation under light exposure, reducing stability. Potassium passivation mitigates this issue by occupying halide vacancies and improving luminescence efficiency without disrupting charge transfer [117].

2.3.3 Electronic and surface characterization

2.3.3.1 X-ray photoelectron spectroscopy (XPS) and ultraviolet photoelectron spectroscopy (UPS)

The XPS provides elemental composition and oxidation states, while UPS determines work function and valence band maximum. These techniques are critical for optimizing perovskite energy level alignment with electron and hole transport layers [118].

XPS technique is based on the photoelectric effect, where photon absorption by a material results in the emission of photoelectrons with specific kinetic energies. The energy of these emitted electrons provides critical information about the elemental composition and chemical states of the sample [119]. Each element has a unique set of core-level binding energies, making XPS a powerful tool for elemental identification. For instance, lead (Pb) exhibits core-level peaks such as Pb 4p, 4d, 4f, 5p, and 5d across the binding energy range of 0–700 eV. In practical applications, the core-level peak with the highest intensity, best energy resolution, or most significant chemical shift is selected for analysis [120].

A key advantage of XPS is its ability to detect chemical shifts arising from changes in the local chemical environment. These shifts, typically within 10 eV, can be interpreted using concepts such as electronegativity and electrostatic charge potential models. More advanced models incorporating relaxation or final-state effects enable precise predictions of chemical changes using theoretical approaches like the $(Z + 1)$ technique [121].

XPS is inherently surface-sensitive due to the inelastic scattering of emitted photoelectrons. The depth from which useful electrons originate is governed by their inelastic mean free path (IMFP), which follows an exponential decay profile. IMFP is influenced by electron–electron and electron–phonon interactions, varying with material properties and electron kinetic energy. Experimental studies have demonstrated a "universal curve" for IMFP values across diverse materials, facilitating quantitative analysis [122].

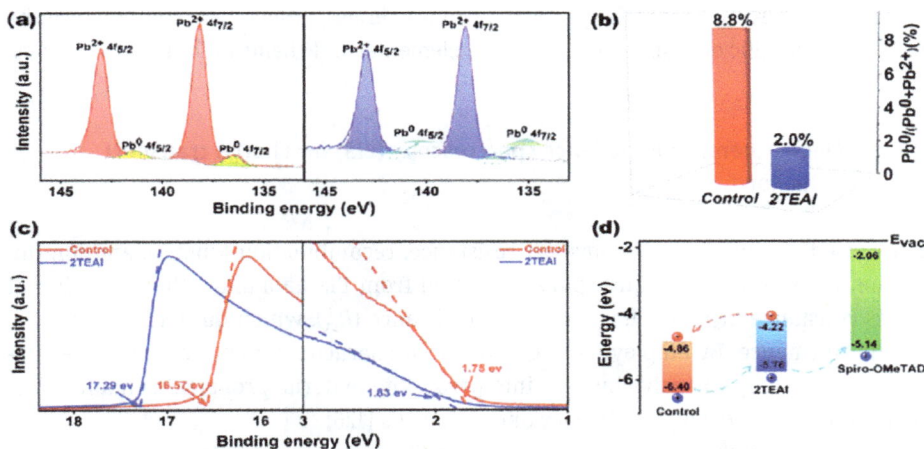

Figure 2.17: (a) High-resolution XPS spectra of Pb 4f in perovskite films with and without 2TEAI treatment. **(b)** Analysis of the fitted $Pb^0/(Pb^0 + Pb^{2+})$ ratio. **(c)** Comparison of UPS spectra between control and 2TEAI-modified films. **(d)** Energy-level diagram of control and 2TEAI-treated perovskite films [124].

The accuracy of quantitative XPS analysis is affected by variations in IMFP values, which can differ by an order of magnitude for the same element at identical energy levels. To minimize errors, researchers often calculate atomic fractions using a range of IMFP values before finalizing data interpretation. While absolute quantification remains challenging, relative quantification – comparing component ratios such as A to B – yields more reliable results, especially at high photon energies where IMFP values converge across multiple elements [123].

Beyond elemental identification, XPS enables valence band photoelectron spectroscopy (VPS), which provides insights into the electronic structure governing material properties. VPS is particularly valuable in solar cell research, as it determines the position of the valence band edge relative to the Fermi level, a crucial factor in understanding charge transfer and energy alignment in optoelectronic devices [26].

XPS is part of a broader class of photoelectron spectroscopy techniques categorized by excitation energy. Ultraviolet photoelectron spectroscopy (UPS) is commonly employed to study valence band density of states and work function. By integrating UPS and XPS data, researchers can construct energy-level diagrams essential for designing efficient electronic and solar devices [125]. Figure 2.17 presents XPS and UPS analysis of perovskite films with and without 2TEAI treatment, including Pb 4f spectra, $Pb^0/(Pb^0 + Pb^{2+})$ ratio, UPS comparison, and energy-level alignment.

Advancements in synchrotron radiation have expanded the capabilities of XPS, enabling photon energy adjustments from a few electron volts (eV) to several kilo electron volts (keV). This has led to the development of soft X-ray photoelectron spectroscopy and hard X-ray photoelectron spectroscopy, which allow for the analysis of materials with varying thicknesses. By carefully tuning photon energies, researchers can enhance the visibility of specific orbitals, enabling precise elemental and chemical state investigations.

2.3.3.2 Electrochemical impedance spectroscopy (EIS) for charge transport behavior

EIS is used to study charge transport resistance, recombination kinetics, and ion migration in perovskites. Nyquist plots generated from EIS data allow the extraction of series resistance (R_s) and charge transfer resistance (R_{ct}), which influence overall device performance. By employing these advanced characterization methods, researchers can gain comprehensive insights into perovskite material properties, enabling further optimization for high-efficiency and stable SCs [126].

EIS is a powerful technique for analyzing charge transport, interfacial properties, and defect states in materials, particularly in PSCs and photodetectors. EIS measures

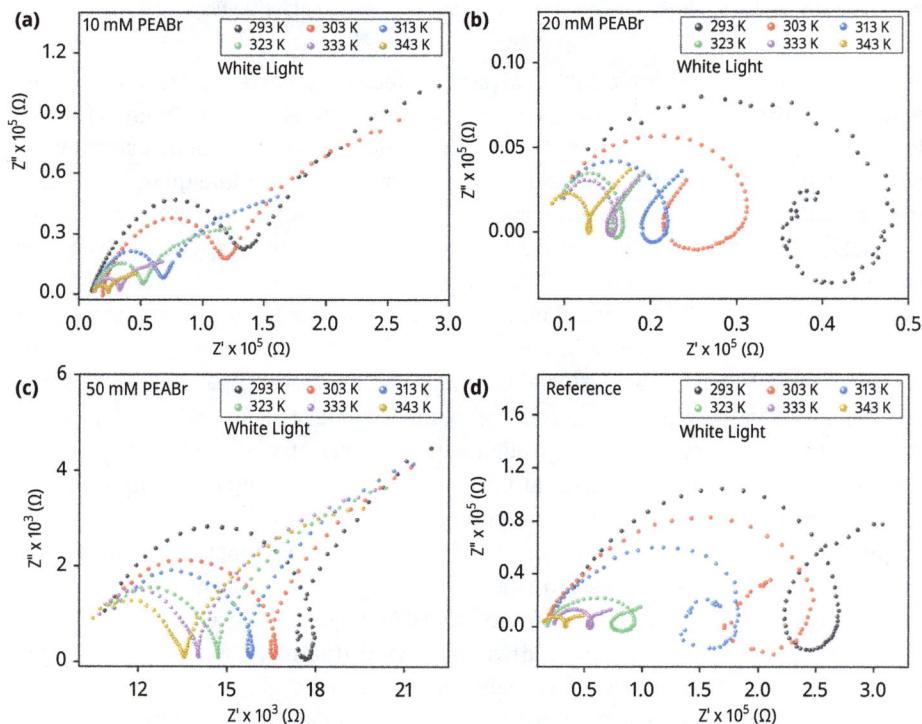

Figure 2.18: Nyquist plots of PEABr-passivated SCs measured at varying temperatures under white light illumination. Impedance response is shown for SCs treated with **(a)** 10 mM, **(b)** 20 mM, **(c)** 50 mM PEABr, and **(d)** an untreated reference SC, highlighting passivation effects on charge transport [128].

the frequency-dependent response of a material to an applied alternating current (AC) voltage, providing insights into charge transfer resistance, ion migration, and recombination kinetics [127].

EIS operates by applying a small sinusoidal voltage perturbation to a system and measuring the resulting current response. The impedance, denoted as $Z(\omega) = Z' + jZ''$, consists of a real component (Z') and an imaginary component (Z''), which are typically represented in a Nyquist diagram. A typical Nyquist plot for PSCs features a semicircle, where the high-frequency region corresponds to R_{ct}, while the low-frequency region reflects recombination resistance and ion diffusion processes. In perovskite-based devices, EIS is used to assess the effects of surface passivation in reducing trap states and enhancing charge transport. For example, (as shown in Figure 2.18) in MAPbBr$_3$ single-crystal photodetectors, passivation with PEABr at a concentration of 0.2 M significantly reduced charge recombination resistance, increasing from $R_{ct} = 4.5 \times 10^4$ Ω in the unpassivated device to $R_{ct} = 1.2 \times 10^5$ Ω. This threefold improvement indicates reduced defect-assisted recombination, which correlates with enhanced photocurrent response [128].

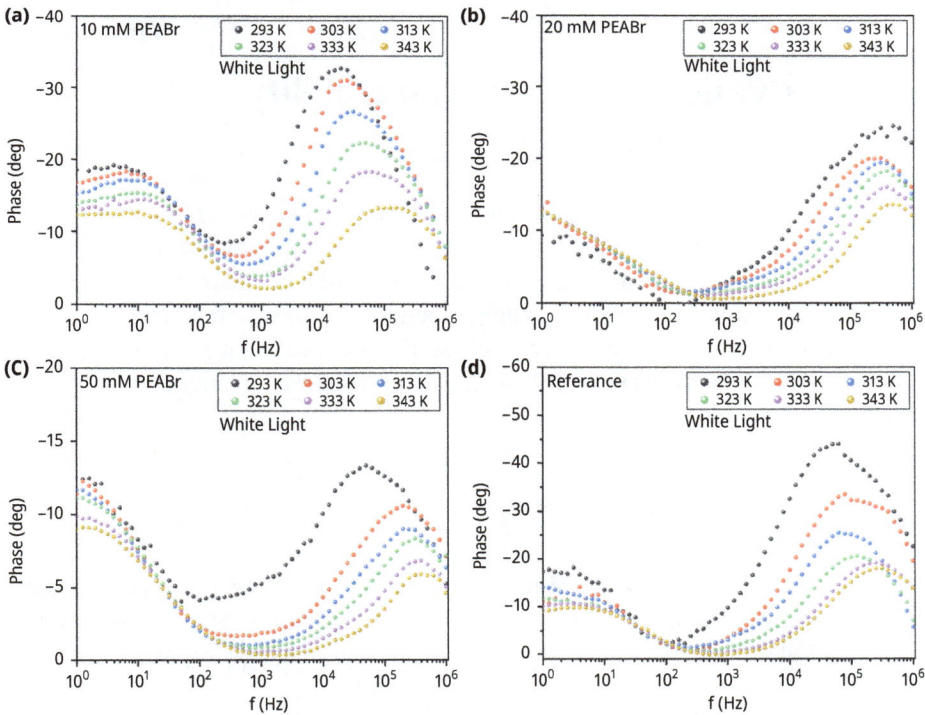

Figure 2.19: Bode graphs related to the Nyquist plots for PEABr-passivated SCs at varying temperatures under white light illumination. The frequency response is examined for SCs subjected to **(a)** 10 mM, **(b)** 20 mM, **(c)** 50 mM PEABr, and **(d)** an untreated reference SC, illustrating passivation effects [128].

The Bode plot (as shown in Figure 2.19), another representation of EIS data, illustrates phase angle (θ) and frequency-dependent impedance characteristics. A phase shift approaching $-90°$ indicates capacitive behavior, signifying effective charge accumulation at the interface. In MAPbBr$_3$ photodetectors, improved passivation increased the phase shift from $-45°$ (unpassivated) to $-80°$ (passivated), demonstrating enhanced charge separation and transport efficiency [129].

One of the key applications of EIS in PSCs is the analysis of ion migration, a crucial factor affecting device stability. The activation energy (E_a) for ion migration can be extracted from temperature-dependent EIS measurements. For unpassivated MAPbBr$_3$, E_a was determined to be 0.25 eV, while passivation increased it to 0.35 eV, indicating reduced ion movement and improved stability. By employing equivalent circuit models to fit EIS data, researchers can extract parameters such as series resistance (R_s), charge transfer resistance (R_{ct}), and interfacial capacitance (C_{dl}), providing a deeper understanding of perovskite/HTL interfaces. These insights are critical for optimizing PSCs and photodetectors, as minimizing recombination and enhancing charge extraction directly improve device performance [128].

2.4 Passivation strategies and stability improvements

Passivation strategies play a crucial role in enhancing the stability and performance of perovskite materials by reducing defect densities, suppressing ion migration, and preventing environmental degradation. Surface and grain boundary passivation techniques are essential for minimizing charge carrier recombination and improving the longevity of perovskite-based devices. This section elaborates on various passivation strategies, their mechanisms, and their impact on material stability and device efficiency.

2.4.1 Surface passivation

Surface defects, such as halide vacancies and undercoordinated metal cations, serve as non-radiative recombination centers that significantly affect PSC performance. Passivation of these defects is achieved using organic and inorganic capping agents that neutralize trap states and improve charge transport [130].

The technique addressing defects such as halide vacancies and undercoordinated lead (Pb^{2+}) ions, which otherwise act as recombination centers that degrade performance. By applying a protective chemical treatment, surface passivation significantly improves charge carrier dynamics and reduces leakage currents [131]. For example, in MAPbBr$_3$ single-crystal photodetectors (as shown in Figure 2.20), treatment with

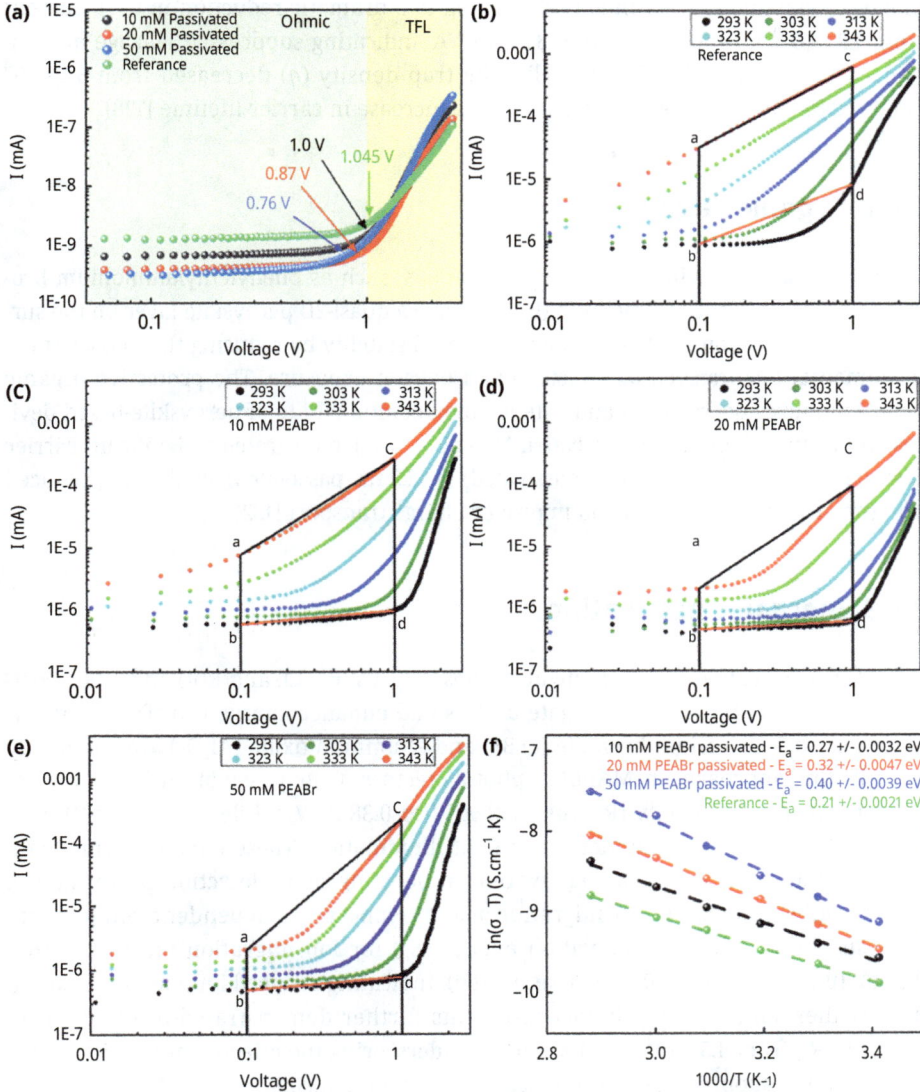

Figure 2.20: (a) Dark current-voltage (*I–V*) characteristics depicting the trap-filled limit voltage (VTFL) for both passivated and reference SCs. Distinct dark *I–V* curves are shown for **(b)** the reference SC, **(c)** the 10 mM PEABr-passivated SC, **(d)** the 20 mM PEABr-passivated SC, and **(e)** the 50 mM PEABr-passivated SC, emphasizing differences in charge transport characteristics. An Arrhenius plot illustrating the activation energy of both passivated and reference SCs, elucidating the influence of PEABr passivation on charge carrier dynamics and the diminution of trap states in SC-based optoelectronic devices [128].

phenylethylammonium bromide (PEABr) led to a dramatic reduction in dark current from 1.2×10^{-9} A (unpassivated) to 3.5×10^{-1} A, indicating suppressed leakage and enhanced charge transfer. Additionally, the trap density (η) decreased from 5.1×10^9 cm^{-3} to 2.82×10^9 cm^{-3}, resulting in a notable increase in carrier lifetime [128].

2.4.2 Organic passivation

Organic passivation utilizes large organic cations, such as phenylethylammonium bromide (PEABr) and butylammonium (BA), to form a quasi-2D perovskite layer on the surface. This passivation method enhances material stability by reducing the impact of environmental factors such as moisture and oxygen exposure. The protective organic layer minimizes degradation and extends the operational life of perovskite-based devices. TRPL studies on PEABr-passivated $MAPbBr_3$ SCs demonstrated a significant carrier lifetime increase from 4.8 ns (unpassivated) to 12.2 ns (passivated), confirming reduced non-radiative recombination and improved charge transport [132].

2.4.3 Inorganic passivation

Inorganic passivation relies on metal halides (e.g., CsBr, KCl, and RbI) and oxide coatings (e.g., Al_2O_3 and ZnO) to mitigate defects and enhance charge transfer efficiency. Materials such as cesium bromide (CsBr) and aluminum oxide (Al_2O_3) are commonly used in this approach. In $MAPbBr_3$ photodetectors, CsBr passivation led to an improvement in responsivity (R) from 0.21 A/W to 0.38 A/W, while detectivity (D^*) increased to 2.5×10^{11} Jones under green light illumination. These enhancements highlight the impact of inorganic passivation in boosting photodetection performance. Additionally, studies on ion migration using temperature-dependent conductivity analysis showed that the activation energy (E_a) for ion migration increased from 0.25 eV (unpassivated) to 0.35 eV (passivated), indicating reduced ion mobility and improved thermal stability. EIS measurements further demonstrated a threefold increase in R_{ct} from 4.5×10^4 Ω to 1.2×10^5 Ω, underscoring the effectiveness of inorganic passivation in suppressing defect-assisted recombination [128].

2.4.4 Lewis base treatment

Lewis base passivation is another effective approach that involves using molecules with electron-donating functional groups, such as thiophene, pyridine, and thiocyanate (SCN^-) and amines, to interact with undercoordinated lead (Pb^{2+}) ions and neutralize surface defects. This strategy significantly reduces trap densities and enhances carrier lifetimes, thereby improving the overall efficiency of perovskite-based devices.

The presence of Lewis base molecules also strengthens the material's resistance to environmental degradation, further extending device stability [133].

2.4.2 Grain boundary passivation

While surface passivation focuses on mitigating defects at the outermost layers of perovskite films, grain boundary passivation is essential for stabilizing the internal structure of polycrystalline perovskites. Grain boundaries in polycrystalline perovskite films act as charge trap sites, reducing efficiency and stability. Grain boundaries often serve as recombination sites and pathways for ion migration, leading to performance losses in optoelectronic applications. Effective grain boundary passivation techniques include halide additives, polymer coatings, and molecular bridging strategies [134].

2.4.2.1 Halide additives

Excess halide sources, such as methylammonium iodide (MAI) or cesium bromide (CsBr), fill halide vacancies and mitigate ion migration, thereby improving operational stability. Incorporating halide additives, such as iodide (I^-) or bromide (Br^-), into the perovskite structure stabilizes grain boundaries by compensating for halide vacancies and reducing defect densities. These additives improve charge carrier transport and suppress recombination losses, leading to enhanced photovoltaic performance. For example, bromide incorporation in mixed-HPs has been shown to increase the open-circuit voltage (V_{oc}) and improve photostability under continuous illumination [135].

2.4.2.2 Polymer coatings

Polymer coatings provide both mechanical reinforcement and chemical passivation at grain boundaries. Conductive and insulating polymers, such as poly(methyl methacrylate) (PMMA) and polyethylene oxide (PEO), serve as protective barriers that reduce exposure to environmental degradation. These coatings prevent moisture ingress, inhibit ion migration, and enhance overall structural stability. Studies have demonstrated that polymer-passivated PSCs exhibit prolonged operational stability, with power conversion efficiency (PCE) retention exceeding 90% after 1,000 h of exposure to ambient conditions, compared to only 60% retention in unpassivated devices [136].

2.4.2.3 Molecular bridging

Using small molecules such as formamidinium thiocyanate (FA-SCN) to bridge grain boundaries reduces grain-to-grain defects and enhances film crystallinity. The strategy is employed by small functional molecules to form strong interactions at grain boundaries. These molecules, often containing carboxyl, amine, or thiol groups, chemically bond with adjacent grains, reducing defect densities and facilitating charge transport. By effectively linking grain boundaries, molecular bridging enhances both mechanical stability and electrical conductivity, resulting in higher efficiency and longer-lasting perovskite devices [137].

2.4.3 Interface engineering for stability enhancement

Interfaces between the perovskite absorber and charge transport layers significantly affect device performance and long-term stability [138]. Effective passivation at these interfaces includes:

– **Self-assembled monolayers (SAMs):** Functional SAMs, such as carbazole and benzothiophene derivatives, improve energy level alignment and reduce charge recombination at the perovskite/ETL and perovskite/HTL interfaces [139].
– **Metal oxide interlayers:** Thin layers of TiO_2, SnO_2, or NiO act as buffer layers, minimizing interfacial charge recombination and enhancing device durability [140].
– **2D/3D heterostructures:** Engineering a layered 2D perovskite capping layer on top of a 3D perovskite absorber enhances moisture resistance while maintaining high charge carrier mobility [141].

2.4.4 Chemical and environmental stability improvements

PSCs suffer from environmental instability due to moisture, oxygen, UV exposure, and thermal stress. Stability improvements include:

– **Hydrophobic coatings:** Applying fluorinated materials or perovskite encapsulation with polymeric barriers prevents moisture penetration and oxidative degradation [142].
– **UV stabilization:** Introducing UV-absorbing layers or encapsulating the perovskite with wide-bandgap metal oxides (e.g., ZnO and MgO) mitigates UV-induced degradation [143].
– **Thermal stabilization:** Compositional engineering using mixed-cation perovskites (e.g., FA-Cs-Pb-I-Br) enhances thermal robustness by preventing phase segregation [144].

2.4.5 Impact of passivation on device performance

Passivation strategies have led to significant improvements in PSC efficiencies and operational stability:

- **Efficiency enhancement:** Incorporating surface and grain boundary passivation has increased PCEs from ~18% to over 25% [145].
- **Operational lifetime:** Encapsulated and passivated devices maintain over 80% of their initial efficiency for more than 1,000 h under continuous operation [146].
- **Reduced hysteresis:** Passivation minimizes charge trapping at interfaces, reducing hysteresis effects in current-voltage ($J–V$) characteristics [147].

Surface and grain boundary passivation are fundamental to improving the stability, charge transport properties, and efficiency of perovskite materials. Organic and inorganic passivation techniques effectively suppress surface defects, while Lewis base treatments further enhance carrier lifetimes. Grain boundary passivation strategies, including halide additives, polymer coatings, and molecular bridging, provide additional reinforcement to mitigate recombination losses and structural degradation.

By implementing these advanced passivation strategies, perovskite materials can achieve improved stability and efficiency, bringing them closer to commercial viability for large-scale photovoltaic applications.

2.5 Conclusion and future perspectives

Perovskite materials have demonstrated remarkable potential in photovoltaic applications due to their superior optoelectronic properties, including high absorption coefficients, tunable band gaps, and efficient charge transport mechanisms. These attributes have enabled PSCs to achieve PCEs exceeding 25%, making them competitive with traditional silicon-based photovoltaics. However, challenges such as stability, scalability, and defect passivation remain critical barriers to their widespread commercialization.

2.5.1 Key achievements in perovskite research

- **Advancements in synthesis techniques:** The development of precise synthesis methods, including hot injection for QDs, solvothermal growth for NWs, and inverse temperature crystallization for bulk single crystals, has significantly improved material quality and performance.
- **Enhanced characterization approaches**: Comprehensive characterization techniques, such as XRD for phase purity analysis, SEM/TEM for morphological stud-

ies, and EIS for charge transport analysis, have provided deeper insights into perovskite material behavior and degradation mechanisms.
- **Breakthroughs in passivation strategies:** Surface and grain boundary passivation techniques, including organic ligand engineering, polymer coatings, and 2D/3D hybrid structures, have been instrumental in mitigating defects and improving long-term stability.

2.5.2 Challenges and future directions

Despite these advancements, several key challenges need to be addressed:
1. **Stability and environmental resilience:**
   - Perovskites are highly sensitive to moisture, oxygen, UV exposure, and thermal fluctuations, leading to rapid degradation. Future research must focus on advanced encapsulation techniques, compositional engineering, and protective coatings to enhance environmental stability.
2. **Scalability and manufacturing feasibility:**
   - Large-scale production remains a major hurdle due to the complexity of deposition processes and the reliance on toxic solvents. Roll-to-roll processing, slot-die coating, and solvent-free deposition methods are potential solutions for improving scalability while maintaining high film quality.
3. **Defect and ion migration control:**
   - Ion migration leads to phase segregation and hysteresis in PSCs, affecting device reliability. Strategies such as mixed-cation engineering, interface modifications, and dopant incorporation can help reduce ion mobility and improve operational stability.
4. **Integration with emerging technologies:**
   - Perovskite materials hold promise for integration into tandem SCs, flexible electronics, and multi-junction architectures. Hybrid perovskite–silicon and perovskite–perovskite tandem cells can surpass the Shockley-Queisser limit, further enhancing efficiency.
5. **Sustainability and lead-free alternatives:**
   - The toxicity of lead-based perovskites raises environmental concerns. The development of lead-free perovskites, such as tin-based or bismuth-based materials, is essential for sustainable photovoltaic applications.

2.5.3 Conclusion

PSCs are at the forefront of next-generation photovoltaic technology, offering a viable pathway to high-efficiency, cost-effective, and scalable solar energy solutions. Ongoing research in material engineering, device architecture optimization, and stability improvements will play a crucial role in advancing perovskite technology toward commercial viability. By addressing the remaining challenges, perovskite photovoltaics can revolutionize the renewable energy sector, paving the way for sustainable and high-performance solar energy systems.

References

[1]   Limbani TA, Mahesh A. Recent advancements in materials for colored and semi-transparent perovskite solar cell applications, Emergent Mater. 2023; 6(2): 483–497.

[2]   Elangovan NK, Kannadasan R, Beenarani BB, Alsharif MH, Kim MK, Hasan Inamul Z. Recent developments in perovskite materials, fabrication techniques, band gap engineering, and the stability of perovskite solar cells, Energy Reports. 2024; 11: 1171–1190.

[3]   Kim EB, Akhtar MS, Shin HS, Ameen S, Nazeeruddin MK. A review on two-dimensional (2D) and 2D-3D multidimensional perovskite solar cells: Perovskites structures, stability, and photovoltaic performances, Journal of Photochemistry and Photobiology C: Photochemistry Reviews. 2021; 48: 100405.

[4]   Zhang Y, Abdi-Jalebi M, Larson BW, Zhang F. What matters for the charge transport of 2D perovskites?, Advanced Materials. 2024; 36(31): 2404517.

[5]   Li X, Aftab S, Hussain S, Kabir F, Henaish AMA, Al-Sehemi AG, et al. Dimensional diversity (0D{,} 1D{,} 2D{,} and 3D) in perovskite solar cells: Exploring the potential of mixed-dimensional integrations, Journal of Materials Chemistry A. 2024; 12(8): 4421–4440.

[6]   Limbani TA, Mahesh A, Bharucha SR. The influence of peabr passivation concentration, light exposure, and temperature on the electrochemical behavior and charge transfer resistance in MAPbBr3 single crystals, ChemistrySelect. 2025; 10(9): e202500273.

[7]   Li Q, Wu K, Zhu H, Yang Y, He S, Lian T. Charge transfer from quantum-confined 0D, 1D, and 2D Nanocrystals, Chemical Reviews. 2024 May 8; 124(9): 5695–5763.

[8]   Oh K, Jung K, Park D, Lee MJ. Highly luminescent CH3NH3PbBr3 quantum dots with 96.5% photoluminescence quantum yield achieved by synergistic combination of single-crystal precursor and capping ligand optimization, Journal of Alloys and Compounds. 2021; 859: 157842.

[9]   Chen Q, De Marco N, Yang Y (Michael), Song TB, Chen CC, Zhao H, et al. Under the spotlight: The organic–inorganic hybrid halide perovskite for optoelectronic applications, Nano Today. 2015; 10(3): 355–396.

[10]   Yang E, Zhang M, Wei S, Liang D, Zeb M, Zhang L, et al. Controllable synthesis and heterogeneous tailoring of 1D perovskites, emerging properties and applications, Advanced Powder Materials. 2025; 4(1): 100250.

[11]   Yu M, Zhang D, Xu Y, Lin J, Yu C, Fang Y, et al. Surface ligand engineering of CsPbBr3 perovskite nanowires for high-performance photodetectors, J Colloid Interface Sci. 2022; 608: 2367–2376.

[12]   Kore BP, Jamshidi M, Gardner JM. The impact of moisture on the stability and degradation of perovskites in solar cells, Materials Advances. 2024; 5(6): 2200–2217.

[13]   Han M, Xiao Y, Zhou C, Yang Y, Wu X, Hu Q, et al. Recent advances on two-dimensional metal halide perovskite x-ray detectors, Materials Futures. 2023; 2(1): 012104.

[14]   2022-大面积钙钛矿制备方法统计对比. 大面积钙钛矿太阳能电池制备方法的最新进展. 2022. 2 p.

[15]   Kim EB, Akhtar MS, Shin HS, Ameen S, Nazeeruddin MK. A review on two-dimensional (2D) and 2D-3D multidimensional perovskite solar cells: Perovskites structures, stability, and photovoltaic performances, Journal of Photochemistry and Photobiology C: Photochemistry Reviews. 2021; 48: 100405.

[16] Hoye RLZ, Eyre L, Wei F, Brivio F, Sadhanala A, Sun S, et al. Fundamental carrier lifetime exceeding 1 µs in Cs2AgBiBr6 double perovskite, Advanced Materials Interfaces. 2018; 5(15): 1800464.

[17] Xu Q, Datta A, Becla K, Becla P, Motakef S. Development of continuous solution growth method for growth of large and high-quality perovskite single crystals, Chemical Engineering Journal. 2023; 475: 146155.

[18] Liu D, Jiang X, Wang H, Chen H, Lu YB, Dong S, et al. Perovskite single crystals by vacuum evaporation crystallization, Advanced Science. 2024; 11(22): 2400150.

[19] MdH M, Khandaker MU, Aminul Islam M, Nur-E-Alam M, Osman H, Ullah M. Perovskite materials in X-ray detection and imaging: Recent progress{,} challenges{,} and future prospects, RSC Advances. 2024; 14(10): 6656–6698.

[20] Yang TCJ, Fiala P, Jeangros Q, Ballif C. High-bandgap perovskite materials for multijunction solar cells, Joule. 2018 Aug 15; 2(8): 1421–1436.

[21] Afre RA, Pugliese D. Perovskite solar cells: A review of the latest advances in materials, fabrication techniques, and stability enhancement strategies, Micromachines (Basel). 2024; 15(2): 192.

[22] Montoya-Escobar N, Ospina-Acero D, Velásquez-Cock JA, Gómez-Hoyos C, Serpa Guerra A, Gañan Rojo PF, et al. Use of Fourier series in X-ray Diffraction (XRD) Analysis and Fourier-Transform Infrared Spectroscopy (FTIR) for estimation of crystallinity in cellulose from different sources, Polymers (Basel). 2022; 14(23): 5199.

[23] Tombe S, Adam G, Heilbrunner H, Apaydin DH, Ulbricht C, Sariciftci NS, et al. Optical and electronic properties of mixed halide (X = I{,} Cl{,} Br) methylammonium lead perovskite solar cells, Journal of Materials Chemistry C. 2017; 5(7): 1714–1723.

[24] Cherrette VL, Chou KC, Zeitz D, Guarino-Hotz M, Khvichia M, Barnett J, et al. Ultrafast exciton dynamics of CH3NH3PbBr3 perovskite nanoclusters, J Phys Chem Lett. 2024 May 16; 15(19): 5177–5182.

[25] Toth D, Hailegnaw B, Richheimer F, Castro FA, Kienberger F, Scharber MC, et al. Nanoscale charge accumulation and its effect on carrier dynamics in Tri-cation perovskite structures, ACS Appl Mater Interfaces. 2020 Oct 21; 12(42): 48057–48066.

[26] Whitten JE. Ultraviolet photoelectron spectroscopy: Practical aspects and best practices, Applied Surface Science Advances. 2023; 13: 100384.

[27] Raju TD, Murugadoss V, Nirmal KA, Dongale TD, Kesavan AV, Kim TG. Advancements in perovskites for solar cell commercialization: A review, Advanced Powder Materials. 2025; 4(2): 100275.

[28] Wang Y, Cheng Z, Li J, Lv K, Li Z, Zhao H. Innovative approaches to large-area perovskite solar cell fabrication using slit coating, Molecules. 2024; 29(20): 4976.

[29] Jalled O, Homri A, Ordonez-Miranda J, Dhahri J. Deep insights into structural, morphological, optical and electrical properties of nanocrystalline SrLaFeCo0.5Mo0.5O6 Synthesised by the sol-gel combustion method, Ceramics International. 2025; 51(14): 18701–18712.

[30] Kumar D, Sagar Yadav R, Monika, Kumar Singh A, Bahadur Rai S. Synthesis Techniques and Applications of Perovskite Materials. Perovskite Materials, Devices and Integration. IntechOpen; 2020. http://dx.doi.org/10.5772/intechopen.86794

[31] He J, Li H, Liu C, Wang X, Zhang Q, Liu J, et al. Hot-injection synthesis of cesium lead halide perovskite nanowires with tunable optical properties, Materials. 2024; 17(10): 2173.

[32] Lv H, Zheng Y, Geng Y, Xu S, Geng C. Insight into the growth kinetics of CsPbBr3 perovskite nanocrystals using an oil-water droplet fluidic synthesis route, Chemical Engineering Journal. 2024; 480: 148315.

[33] Chen J, Zhou Y, Fu Y, Pan J, Mohammed OF, Bakr OM. Oriented halide perovskite nanostructures and thin films for optoelectronics, Chemical Reviews. 2021 Oct 27; 121(20): 12112–12180.

[34] Tang X, Zu Z, Shao H, Hu W, Zhou M, Deng M, et al. All-inorganic perovskite CsPb(Br/I)3 nanorods for optoelectronic application, Nanoscale. 2016; 8(33): 15158–15161.

[35] Wang S, Yu J, Zhang M, Chen D, Li C, Chen R, et al. Stable, strongly emitting cesium lead bromide perovskite nanorods with high optical gain enabled by an intermediate monomer reservoir synthetic strategy, Nano Letters. 2019 Sep 11; 19(9): 6315–6322.

[36] Xuan T, Xie RJ. Recent processes on light-emitting lead-free metal halide perovskites, Chemical Engineering Journal. 2020; 393: 124757.

[37] Duan D, Ge C, Rahaman MZ, Lin CH, Shi Y, Lin H, et al. Recent progress with one-dimensional metal halide perovskites: From rational synthesis to optoelectronic applications, NPG Asia Materials. 2023; 15(1): 8.

[38] Pan Z, Liu B, Wang B, Liu Y, Si T, Yi W, et al. Lead-free Cs2SnX6 (X = Cl, Br, I) nanocrystals in mesoporous SiO$_2$ with more stable emission from VIS to NIR light, Chemical Physics Letters. 2021; 782: 139023.

[39] Tang CY, Yang Z. Chapter 8 – Transmission Electron Microscopy (TEM), In: Hilal N, Ismail AF, Matsuura T, Oatley-Radcliffe D (eds.) Membrane Characterization. 2017; Elsevier, 145–159.

[40] Fu S, Sun N, Xian Y, Chen L, Li Y, Li C, et al. Suppressed deprotonation enables a durable buried interface in tin-lead perovskite for all-perovskite tandem solar cells, Joule. 2024 Aug 21; 8(8): 2220–2237.

[41] Lu Y, Qu K, Zhang T, He Q, Pan J. Metal halide perovskite nanowires: Controllable synthesis, mechanism, and application in optoelectronic devices, Nanomaterials. 2023; 13(3): 419.

[42] Sanchez SL, Tang Y, Hu B, Yang J, Ahmadi M. Understanding the ligand-assisted repreciplitation of CsPbBr$_3$ nanocrystals via high-throughput robotic synthesis approach, Materials. 2023 Sep 6; 6(9): 2900–2918.

[43] Mo X, Wang M, Song H, Li J, Wu Z, Lin Z. Upcycling of Pb from lead smelting residues to lead halide perovskite for piezocatalytic hydrogen production, Chemical Engineering Journal. 2024; 498: 155105.

[44] Fu P, Shan Q, Shang Y, Song J, Zeng H, Ning Z, et al. Perovskite nanocrystals: Synthesis, properties and applications, Science Bulletin (Beijing). 2017; 62(5): 369–380.

[45] Liu L, Pan K, Xu K, Zhang JZ. Impact of molecular ligands in the synthesis and transformation between Metal Halide perovskite quantum dots and magic sized clusters, ACS Physical Chemistry Au. 2022 May 25; 2(3): 156–170.

[46] Ma HH, Imran M, Dang Z, Hu Z. Growth of metal halide perovskite, from nanocrystal to micron-scale crystal: A Review, Crystals (Basel). 2018; 8(5): 182.

[47] Yang GL, Zhong HZ. Organometal halide perovskite quantum dots: Synthesis, optical properties, and display applications, Chinese Chemical Letters. 2016; 27(8): 1124–1130.

[48] Rong SS, Faheem MB, Li YB. Perovskite single crystals: Synthesis, properties, and applications, Journal of Electronic Science and Technology. 2021 Jun; 19(2): 100081.

[49] Park J, Kim J, Yun HS, Paik MJ, Noh E, Mun HJ, et al. Controlled growth of perovskite layers with volatile alkylammonium chlorides, Nature. 2023 Apr 27; 616(7958): 724–730.

[50] Thanh NTK, Maclean N, Mahiddine S. Mechanisms of nucleation and growth of nanoparticles in solution, Chemical Review and Letters. 2014 Aug 13; 114(15): 7610–7630.

[51] Ma F, Huang Z, Ziółek M, Yue S, Han X, Rong D, et al. Template-assisted synthesis of a large-area ordered perovskite nanowire array for a high-performance photodetector, ACS Appl Mater Interfaces. 2023 Mar 8; 15(9): 12024–12031.

[52] Jung M, Ji SG, Kim G, Seok SI. Perovskite precursor solution chemistry: From fundamentals to photovoltaic applications, Chemical Society Reviews. 2019; 48(7): 2011–2038.

[53] Rosales BA, Hanrahan MP, Boote BW, Rossini AJ, Smith EA, Vela J. Lead halide perovskites: challenges and opportunities in advanced synthesis and spectroscopy, ACS Energy Letters. 2017 Apr 14; 2(4): 906–914.

[54] Hu S, Hou P, Duan C, Dou Y, Deng X, Xiong W, et al. Vapor–solid reaction techniques for the growth of organic–inorganic hybrid perovskite thin films, Small. 2025; 21(6): 2410865.

[55] Bandara RMI, Silva SM, Underwood CCL, Jayawardena KDGI, Sporea RA, Silva SRP. Progress of Pb-Sn mixed perovskites for photovoltaics: A review, Energy \& Environmental Materials. 2022; 5(2): 370–400.

[56] Mu Y, He Z, Wang K, Pi X, Zhou S. Recent progress and future prospects on halide perovskite nanocrystals for optoelectronics and beyond, iScience. 2022 Nov 18; 25(11): 105371.

[57] Abzieher T, Moore DT, Roß M, Albrecht S, Silvia J, Tan H, et al. Vapor phase deposition of perovskite photovoltaics: Short track to commercialization?, Energy & Environmental Science. 2024; 17(5): 1645–1663.

[58] Ávila J, Momblona C, Boix PP, Sessolo M, Bolink HJ. Vapor-deposited perovskites: The route to high-performance solar cell production?, Joule. 2017 Nov 15; 1(3): 431–442.

[59] Tan L, Zhou J, Zhao X, Wang S, Li M, Jiang C, et al. Combined vacuum evaporation and solution process for high-efficiency large-area perovskite solar cells with exceptional reproducibility, Advanced Materials. 2023; 35(13): 2205027.

[60] Liu M, Johnston MB, Snaith HJ. Efficient planar heterojunction perovskite solar cells by vapour deposition, Nature. 2013; 501(7467): 395–398.

[61] Dastgeer G, Nisar S, Zulfiqar MW, Eom J, Imran M, Akbar K. A review on recent progress and challenges in high-efficiency perovskite solar cells, Nano Energy. 2024; 132: 110401.

[62] Limbani TA, Mahesh A, Gajera DC. Brush coating and spin coating analysis of stannic oxide thin film for electron transport layer in ambient atmosphere for perovskite solar cell, Mater Today Proc. 2023; 89: 44–48.

[63] Schulze PSC, Wienands K, Bett AJ, Rafizadeh S, Mundt LE, Cojocaru L, et al. Perovskite hybrid evaporation/ spin coating method: From band gap tuning to thin film deposition on textures, Thin Solid Films. 2020; 704: 137970.

[64] Limbani T, Mahesh A. Study of brush coating and spin coating on stannic oxide thin film for the electron transport layer in a perovskite solar cell, Nano-Structures & Nano-Objects. 2023; 35: 101010.

[65] Jiao J, Yang C, Wang Z, Yan C, Fang C. Solvent engineering for the formation of high-quality perovskite films: a review, Results in Engineering. 2023; 18: 101158.

[66] Lee UG, Kim WB, Han DH, Chung HS. A modified equation for thickness of the film fabricated by spin coating, Symmetry (Basel). 2019; 11(9): 1–20.

[67] Tsai CH, Lin CM, Kuei CH. Investigation of the effects of various organic solvents on the PCBM electron transport layer of perovskite solar cells, Coatings. 2020; 10(3): 237.

[68] Patil GC. Doctor blade: A promising technique for thin film coating, In: Sankapal BR, Ennaoui A, Gupta RB, Lokhande CD (eds.) Simple Chemical Methods for Thin Film Deposition: Synthesis and Applications. 2023; Singapore: Springer Nature Singapore, 509–530.

[69] Hsu HC, Wu SH, Tung YL, Shih CF. Long-term stable perovskite solar cells prepared by doctor blade coating technology using bilayer structure and non-toxic solvent, Org Electron. 2022; 101: 106400.

[70] Huang KW, Li MH, Chen YT, Wen ZX, Lin CF, Chen P. Fast fabrication of μm-thick perovskite films by using a one-step doctor-blade coating method for direct X-ray detectors, Journal of Materials Chemistry C. 2024; 12(4): 1533–1542.

[71] Li J, Munir R, Fan Y, Niu T, Liu Y, Zhong Y, et al. Phase transition control for high-performance blade-coated perovskite solar cells, Joule. 2018 Jul 18; 2(7): 1313–1330.

[72] Alanazi TI. Current spray-coating approaches to manufacture perovskite solar cells, Results in Physics. 2023; 44: 106144.

[73] Jung M, Ji SG, Kim G, Seok SI. Perovskite precursor solution chemistry: From fundamentals to photovoltaic applications, Chemical Society Reviews. 2019; 48(7): 2011–2038.

[74] Bishop JE, Smith JA, Lidzey DG. Development of spray-coated perovskite solar cells, ACS Appl Mater Interfaces. 2020 Oct 28; 12(43): 48237–48245.

[75] Lu H, Liu Y, Ahlawat P, Mishra A, Tress WR, Eickemeyer FT, et al. Vapor-assisted deposition of highly efficient, stable black-phase FAPbI$_3$ perovskite solar cells, Science (1979). 2020; 370(6512): eabb8985.

[76] Alanazi TI. Current spray-coating approaches to manufacture perovskite solar cells, Results in Physics. 2023; 44: 106144.

[77] Sanders S, Stümmler D, Pfeiffer P, Ackermann N, Simkus G, Heuken M, et al. Chemical vapor deposition of organic-inorganic bismuth-based perovskite films for solar cell application, Scientific Reports. 2019; 9(1): 9774.

[78] Laalioui S, Alaoui KB, Dads HA, Assali KE, Ikken B, Outzourhit A. No Title, Reviews on Advanced Materials Science. 2020; 59(1): 10–25.

[79] Khorasani A, Mohamadkhani F, Marandi M, Luo H, Abdi-Jalebi M. Opportunities, challenges, and strategies for scalable deposition of metal halide perovskite solar cells and modules, Advanced Energy and Sustainability Research. 2024 Jul 1; 5(7): 2300275.

[80] Maculan G, Sheikh AD, Abdelhady AL, Saidaminov MI, Haque MA, Murali B, et al. CH3NH3PbCl3 single crystals: Inverse temperature crystallization and visible-blind UV-photodetector, Journal of Physical Chemistry Letters. 2015 Oct 1; 6(19): 3781–3786.

[81] Rong SS, Faheem MB, Li YB. Perovskite single crystals: Synthesis, properties, and applications, Journal of Electronic Science and Technology. 2021; 19(2): 100081.

[82] Saidaminov MI, Abdelhady AL, Murali B, Alarousu E, Burlakov VM, Peng W, et al. High-quality bulk hybrid perovskite single crystals within minutes by inverse temperature crystallization, Nature Communications. 2015; 6(1): 7586.

[83] Fateev SA, Petrov AA, Khrustalev VN, Dorovatovskii PV, Zubavichus YV, Goodilin EA, et al. Solution processing of methylammonium lead iodide perovskite from γ-Butyrolactone: Crystallization mediated by solvation equilibrium, Chemistry of Materials. 2018 Aug 14; 30(15): 5237–5244.

[84] Lou Y, Zhang S, Gu Z, Wang N, Wang S, Zhang Y, et al. Perovskite single crystals: Dimensional control, optoelectronic properties, and applications, Materials Today. 2023; 62: 225–250.

[85] Ramukutty S, Ramachandran E. Crystal growth by solvent evaporation and characterization of metronidazole, Journal of Crystal Growth. 2012; 351(1): 47–50.

[86] Cho Y, Jung HR, Jo W. Halide perovskite single crystals: Growth{,} characterization{,} and stability for optoelectronic applications, Nanoscale. 2022; 14(26): 9248–9277.

[87] Liu D, Jiang X, Wang H, Chen H, Lu YB, Dong S, et al. Perovskite single crystals by vacuum evaporation crystallization, Advanced Science. 2024; 11(22): 2400150.

[88] Soultati A, Tountas M, Armadorou KK, Yusoff Abdr bin M, Vasilopoulou M, Nazeeruddin MK. Synthetic approaches for perovskite thin films and single-crystals, Energy Advances. 2023; 2(8): 1075–1115.

[89] Panwar AS, Singh A, Sehgal S. Material characterization techniques in engineering applications: A review, Materials Today: Proceedings. 2020; 28: 1932–1937.

[90] Bharucha SR, Dave MS, Giri RK, Chaki SH, Limbani TA. Synthesis of NbSe$_2$ nanoparticles: An insight into their structural, morphological and optical characteristics †, Engineering Proceedings. 2023; 56(1): 1–7.

[91] Ullah A, Iftikhar Khan M, Ihtisham-ul-haq ABS, AlResheedi DB N, Choi JR. Bandgap engineering and enhancing optoelectronic performance of a lead-free double perovskite Cs$_2$AgBiBr$_6$ solar cell via Al doping, ACS Omega. 2024 Apr 23; 9(16): 18202–18211.

[92] Laberty-Robert C, Fontaine ML, Mounis T, Mierzwa B, Lisovytskiy D, Pielaszek J. X-ray diffraction studies of perovskite or derived perovskite phase formation, Solid State Ion. 2005; 176(13): 1213–1223.

[93] Bharucha SR, Dave MS, Giri RK, Chaki SH, Limbani TA. Synthesis and mechanistic approach to investigate crystallite size of NbSe2 nanoparticles, Advances in Natural Sciences: Nanoscience and Nanotechnology. 2024; 15(1): 015002.

[94] Bo F, Wang K, Liang J, Zhao T, Wang J, He Y, et al. Recent advances in the application of in situ X-ray diffraction techniques to characterize phase transitions in Fischer–Tropsch synthesis catalysts, Green Carbon. 2025; 3 (1): 22–35.

[95] Chang SL. Thin-film characterization by grazing incidence X-ray diffraction and multiple beam interference, Journal of Physics and Chemistry of Solids. 2001; 62(9): 1765–1775.

[96] Rondahl SH, Pointurier F, Ahlinder L, Ramebäck H, Marie O, Ravat B, et al. Comparing results of X-ray diffraction, μ-Raman spectroscopy and neutron diffraction when identifying chemical phases in seized nuclear material, during a comparative nuclear forensics exercise, Journal of Radioanalytical and Nuclear Chemistry. 2018; 315(2): 395–408.

[97] Marin-Villa P, Gaboardi M, Joseph B, Alabarse F, Armstrong J, Drużbicki K, et al. Methylammonium Lead Iodide across physical space: Phase boundaries and structural collapse, Journal of Physical Chemistry Letters. 2025 Jan 9; 16(1): 184–190.

[98] Imran M, Khan NA. Perovskite phase formation informamidinium–methylammonium lead iodide bromide(FAPbI3)1-x(MAPbBr3)xmaterials and their morphological, optical and photovoltaic properties, Applied Physics A. 2019; 125(8): 575.

[99] Whitfield PS, Herron N, Guise WE, Page K, Cheng YQ, Milas I, et al. Structures, phase transitions and tricritical behavior of the hybrid perovskite Methyl Ammonium Lead Iodide, Scientific Reports. 2016; 6(1): 35685.

[100] Min HS. Scanning electron microscopy analysis of thin films: A review, Research Aspects in Chemical and Materials Sciences. 2022; 5: 16–28.

[101] Demchyshyn S, Verdi M, Basiricò L, Ciavatti A, Hailegnaw B, Cavalcoli D, et al. Designing ultraflexible perovskite X-Ray detectors through interface engineering, Advanced Science. 2020; 7(24): 2002586.

[102] Dang Z, Shamsi J, Palazon F, Imran M, Akkerman QA, Park S, et al. In situ transmission electron microscopy study of electron beam-induced transformations in colloidal Cesium Lead Halide perovskite nanocrystals, ACS Nano. 2017 Feb 28; 11(2): 2124–2132.

[103] Akhtar K, SA K, SB K, Asiri AM. Scanning electron microscopy: Principle and applications in nanomaterials characterization, In: Sharma SK (ed.) Handbook of Materials Characterization. 2018; Cham: Springer International Publishing, 113–145.

[104] Arjmand F, Golshani Z, Maghsoudi S, Naeimi A, Fatemi SJ. SnO$_2$@ZnO nanocomposites doped polyaniline polymer for high performance of HTM-free perovskite solar cells and carbon-based, Scientific Reports. 2022; 12(1): 21188.

[105] Singh R. Introduction to Transmission Electron Microscopy (TEM), In: Transmission Electron Microscopy Sample Preparation: From Specimen to Micrograph, 2024; Cham: Springer Nature Switzerland, 1–20.

[106] Rosenauer A, Krause FF, Müller K, Schowalter M, Mehrtens T. Conventional transmission electron microscopy imaging beyond the diffraction and information limits, Phys Rev Lett. 2014 Aug; 113(9): 96101.

[107] Yao L, Tian L, Zhang S, Tian Y, Xue J, Peng S, et al. Low-dose transmission electron microscopy study on halide perovskites: Application and challenges, EnergyChem. 2023; 5(5): 100105.

[108] Zhou Y, Sternlicht H, Padture NP. Transmission electron microscopy of halide perovskite materials and devices, Joule. 2019 Mar 20; 3(3): 641–661.

[109] Verbeeck J, Van Dyck D, Van Tendeloo G. Energy-filtered transmission electron microscopy: An overview, Spectrochim Acta Part B At Spectrosc. 2004; 59(10): 1529–1534.

[110] Chouhan L, Ghimire S, Subrahmanyam C, Miyasaka T, Biju V. Synthesis{,} optoelectronic properties and applications of halide perovskites, Chemical Society Reviews. 2020; 49(10): 2869–2885.

[111] Gao F, Li H, Jiao B, Tan L, Deng C, Wang X, et al. Perovskite facet heterojunction solar cells, Joule. 2025 Feb 19; 9(2): 101787.

[112] Van Gorkom Bt, Fun SHW, van der Pol TPA, Remmerswaal WHM, Aalbers GJW, Wienk MM, et al. Identifying the nature and location of defects in n–i–p Perovskite cells with highly sensitive sub-bandgap photocurrent spectroscopy, Solar RRL. 2024; 8(16): 2400316.

[113] Luan F, Li H, Gong S, Chen X, Shou C, Wu Z, et al. Precursor engineering for efficient and stable perovskite solar cells, Nanotechnology. 2022 Nov; 34(5): 55402.

[114] Abdi-Jalebi M, Ibrahim Dar M, Sadhanala A, Johansson EMJ, Pazoki M. Chapter 3 – Optical absorption and photoluminescence spectroscopy, In: Pazoki M, Hagfeldt A, Edvinsson T (eds.) Characterization Techniques for Perovskite Solar Cell Materials, 2020; Elsevier, 49–79. Micro and Nano Technologies.

[115] Van der Pol Tpa, Datta K, Wienk MM, Janssen RAJ. The intrinsic photoluminescence spectrum of perovskite films, Advanced Optical Materials. 2022; 10(8): 2102557.

[116] Srivastava S, Ranjan S, Yadav L, Sharma T, Choudhary S, Agarwal D, et al. Advanced spectroscopic techniques for characterizing defects in perovskite solar cells, Communications Materials. 2023; 4(1): 52.

[117] Yeom KM, Cho C, Jung EH, Kim G, Moon CS, Park SY, et al. Quantum barriers engineering toward radiative and stable perovskite photovoltaic devices, Nature Communications. 2024; 15(1): 4547.

[118] Krishna DNG, Philip J. Review on surface-characterization applications of X-ray photoelectron spectroscopy (XPS): Recent developments and challenges, Applied Surface Science Advances. 2022; 12: 100332.

[119] Hellmann T, Das C, Abzieher T, Schwenzer JA, Wussler M, Dachauer R, et al. The electronic structure of MAPI-based perovskite solar cells: Detailed band diagram determination by photoemission spectroscopy comparing classical and inverted device stacks, Advanced Energy Materials. 2020; 10(42): 2002129.

[120] Cacovich S, Dally P, Vidon G, Legrand M, Gbegnon S, Rousset J, et al. In-depth chemical and optoelectronic analysis of triple-cation perovskite thin films by combining XPS profiling and PL imaging, ACS Applied Materials & Interfaces. 2022 Aug 3; 14(30): 34228–34237.

[121] Jablonski A, Powell CJ. Relationships between electron inelastic mean free paths, effective attenuation lengths, and mean escape depths, Journal of Electron Spectroscopy and Related Phenomena. 1999; 100(1): 137–160.

[122] Shard AG. Practical guides for x-ray photoelectron spectroscopy: Quantitative XPS, Journal of Vacuum Science & Technology A. 2020; 38(4): 041201.

[123] Chambers SA. Probing perovskite interfaces and superlattices with X-ray photoemission spectroscopy, In: Woicik J (ed.) Hard X-ray Photoelectron Spectroscopy (HAXPES). 2016; Cham: Springer International Publishing, 341–380.

[124] Xu C, Zuo L, Hang P, Guo X, Pan Y, Zhou G, et al. Synergistic effects of bithiophene ammonium salt for high-performance perovskite solar cells, Journal of Materials Chemistry A. 2022; 10(18): 9971–9980.

[125] Zhang Y, Li C, Zhao H, Yu Z, Tang X, Zhang J, et al. Synchronized crystallization in tin-lead perovskite solar cells, Nature Communications. 2024; 15(1): 6887.

[126] Chen X, Shirai Y, Yanagida M, Miyano K. Effect of light and voltage on electrochemical impedance spectroscopy of perovskite solar cells: An empirical approach based on modified randles circuit, The Journal of Physical Chemistry C. 2019 Feb 21; 123(7): 3968–3978.

[127] Lazanas ACh, Prodromidis MI. Electrochemical impedance spectroscopy—A tutorial, ACS Measurement Science Au. 2023 Jun 21; 3(3): 162–193.

[128] Limbani TA, Mahesh A, Bharucha SR. The influence of PEABr passivation concentration, light exposure, and temperature on the electrochemical behavior and charge transfer resistance in MAPbBr3 single crystals, ChemistrySelect. 2025; 10(9): e202500273.

[129] Guerrero A, Bisquert J, Garcia-Belmonte G. Impedance spectroscopy of metal halide perovskite solar cells from the perspective of equivalent circuits, Chemical Reviews. 2021 Dec 8; 121(23): 14430–14484.

[130] Jiang Q, Zhao Y, Zhang X, Yang X, Chen Y, Chu Z, et al. Surface passivation of perovskite film for efficient solar cells, Nature Photonics. 2019; 13(7): 460–466.

[131] Pan Y, Wang J, Sun Z, Zhang J, Zhou Z, Shi C, et al. Surface chemical polishing and passivation minimize non-radiative recombination for all-perovskite tandem solar cells, Nature Communications. 2024; 15(1): 7335.

[132] Péan EV, Dimitrov S, De Castro CS, Davies ML. Interpreting time-resolved photoluminescence of perovskite materials, Physical Chemistry Chemical Physics. 2020; 22(48): 28345–28358.

[133] Cheng W, Zhou R, Peng S, Wang C, Chen L. Research on passivation of perovskite layer in perovskite solar cells, Materials Today Communications. 2024; 38, 107879.

[134] Gui Y, Shen J, Zhou Y, Wang R, Yang Y, Xue J. Inorganic surface passivation strategies of metal halide perovskites, Information \& Functional Materials. 2024; 1(2): 207–219.

[135] Shen X, Kang K, Yu Z, Jeong WH, Choi H, Park SH, et al. Passivation strategies for mitigating defect challenges in halide perovskite light-emitting diodes, Joule. 2023 Feb; 7(2): 272–308.

[136] Melentiev R, Yudhanto A, Tao R, Vuchkov T, Lubineau G. Metallization of polymers and composites: State-of-the-art approaches, Materials & Design. 2022 Sep; 221: 110958.

[137] Yang S, Liu W, Zuo L, Zhang X, Ye T, Chen J, et al. Thiocyanate assisted performance enhancement of formamidinium based planar perovskite solar cells through a single one-step solution process, Journal of Materials Chemistry A. 2016; 4(24): 9430–9436.

[138] Regalado-Pérez E, Díaz-Cruz EB, Villanueva-Cab J. Impact of the hole transport layer on the space charge distribution and hysteresis in perovskite solar cells analysed by capacitance–voltage profiling, Sustain Energy Fuels. 2025; 9(5): 1225–1235.

[139] Shi ZE, Cheng TH, Lung CY, Lin CW, Wang CL, Jiang BH, et al. Achieving over 42 % indoor efficiency in wide-bandgap perovskite solar cells through optimized interfacial passivation and carrier transport, Chemical Engineering Journal. 2024 Oct; 498: 155512.

[140] Dedova T, Krautmann R, Rusu M, Katerski A, Krunks M, Unold T, et al. Sb2S3 solar cells with TiO2 electron transporting layers synthesized by ALD and USP methods, Solar Energy Materials and Solar Cells. 2025 Jan; 280: 113279.

[141] Wang S, Cao F, Wu Y, Zhang X, Zou J, Lan Z, et al. Multifunctional 2D perovskite capping layer using cyclohexylmethylammonium bromide for highly efficient and stable perovskite solar cells, Materials Today Physics. 2021 Nov; 21: 100543.

[142] MdH M, MdB R, Nur-E-Alam M, Islam MA, Shahinuzzaman M, MdR R, et al. Key degradation mechanisms of perovskite solar cells and strategies for enhanced stability: Issues and prospects, RSC Advances. 2025; 15(1): 628–654.

[143] Chen T, Xie J, Gao P. Ultraviolet photocatalytic degradation of perovskite solar cells: Progress, challenges, and strategies, Advanced Energy and Sustainability Research. 2022 Jun 26; 3(6): 2100218.

[144] Schwenzer JA, Hellmann T, Nejand BA, Hu H, Abzieher T, Schackmar F, et al. Thermal stability and cation composition of hybrid organic–inorganic perovskites, ACS Appl Mater Interfaces. 2021 Apr 7; 13(13): 15292–15304.

[145] Wang H, Ye F, Liang J, Liu Y, Hu X, Zhou S, et al. Pre-annealing treatment for high-efficiency perovskite solar cells via sequential deposition, Joule. 2022 Dec; 6(12): 2869–2884.

[146] Ma S, Bai Y, Wang H, Zai H, Wu J, Li L, et al. 1000 h operational lifetime perovskite solar cells by ambient melting encapsulation, Advances Energy Mater. 2020 Mar 30; 10(9): 1902472.

[147] Wang M, Lei Y, Xu Y, Han L, Ci Z, Jin Z. The *J* – *V* hysteresis behavior and solutions in perovskite solar cells, Solar RRL. 2020 Dec 9; 4(12): 2000586.

R. K. Shukla, Anchal Srivastava*, and Nidhi Singh

Chapter 3
Interface engineering and charge transport layers

Abstract: Perovskite solar cells (PSCs) have garnered immense attention due to their exceptional power conversion efficiencies and low fabrication costs. However, their long-term operational stability remains a significant hurdle for commercial viability. This chapter emphasizes the critical role of interface engineering and charge transport layers (CTLs) in improving both efficiency and durability of PSCs. It provides a comprehensive overview of strategies employed to modify interfacial properties, reduce non-radiative recombination, enhance charge extraction, and suppress ion migration. Special focus is placed on energy level alignment, trap passivation, and the development of effective electron and hole transport materials such as TiO, SnO_2, Spiro-OMeTAD, and NiO. Additionally, the chapter reviews advanced interfacial designs using quantum dots, molecular dipoles, and hybrid 2D perovskites, offering promising pathways to overcome stability challenges. Through detailed discussion of material selection, device architecture, and charge carrier dynamics, the chapter presents a roadmap for designing high-performance PSCs with enhanced operational lifetimes.

Keywords: Interface engineering, charge transport layers (CTLs), energy-level alignment, trap passivation, ion migration, photovoltaic efficiency, device stability

3.1 Introduction

Photovoltaics have shown a great deal of interest in perovskite solar cells because of their unique optoelectronic properties and possibilities for economical engineering and manufacturing designs [1]. The existing two main configurations of PSCs are inverted configuration (also referred as *p-i-n*) and the conventional configuration, which is referred as *n-i-p*). The main components of both types of PSCs are the perovskite layer, cathode, anode, electron transport layer (ETL), and hole transport layer (HTL). Normal PSCs typically use the following: cathodes (such as FTO and ITO), anodes (such as Au, Ag, carbon material, etc.), ETLs (similar to TiO_2, SnO_2, ZnO, C_{60}, etc.),

*Corresponding author: **Anchal Srivastava**, Department of Physics, University of Lucknow, Lucknow 226007, Uttar Pradesh, India, e-mail: asrivastava.lu@gmail.com
R. K. Shukla, Nidhi Singh, Department of Physics, University of Lucknow, Lucknow 226007

https://doi.org/10.1515/9783111726847-003

HTLs (such as Spiro-MeOTAD, PTAA, P3HT, CuSCN, CuPc, etc.), and perovskite layers (such as $MAPbI_3$, $FAPbI_3$, mixed cation/mixed anion perovskites, etc.). In the *n-i-p* configuration, the structure initiates front transparent electrode such as the ITO and FTO from where light enters the ETL. Then comes the perovskite layer, finally the HTL, and then back contact. The *p-i-n* configuration is the converse of *n-i-p* structure. These configurations can have either mesoporous or planar design; the mesoporous version usually has a scaffold supporting the perovskite layer, which improves stability and makes charge transport easier. Researchers usually reported high PCEs using *n-i-p* type devices [2].

A significant factor in influencing the overall stability and PSCs performance, is the interfaces between the different layer such as perovskite layer and its neighboring carrier transport layers (ETL and HTL) [3]. Charge extraction takes place at the interfaces of PVSCs, which are especially subjected to recombination because of any potential interfacial defects and corresponding particular charge distributions [4]. Interface engineering continues to have a significant, unrealized part to play in improving PVSC stability and efficiency, which seems inevitable. To increase the durability and performance of PSCs, it is crucial to comprehend the characteristics and behavior of these interfaces.

Furthermore, interface engineering methods enhance the interfacial photoelectrical characteristics. These methods include altering the electrode's work function (WF) and utilizing more effective electron collection and blocking layers [5, 6]. The objectives of these interface engineering techniques are to decrease interface charge carrier recombination, increase charge separation and transportation, and boost overall device stability [7].

After reviewing recent developments in PVSCs, such as their material design, construction, and operation, we will concentrate on interface engineering and charge transport layers (CTLs). This review explores the vital role of interfaces, playing in determining the stability and efficiency of PSCs in this context. We also analyze how specific interface features in perovskite structures affect the performance and durability cells.

3.2 Interface engineering

Interactions between the both charge extraction layers in relation to the two electrodes and perovskite layer can have a remarkable impact on both device stability and efficiency. Interface nature is essential for the performance and stability of a device since it affects both the device ability to convert light into electricity and its stability while operating. The carrier dynamics, namely the injection and extraction behaviors, are inherently connected to the quality of these interfaces at these interfacial areas. Energy-level mismatch, non-radiative recombination (NRR) and optical losses at the interfaces may be the main causes of the efficiency gap, in addition to bulk recombination and optical losses within the perovskite layer. To mitigate interfacial losses, interface engineering is a useful method for tuning the characteristics of the required interfaces and interlayers. More importantly, cutting-edge practical interfacial design may shield perovskites from deteriorating which might significantly improve device stability [8].

The efficiency of charge transfer is based on the alignment of energy levels at each interface, which impacts critical parameters like fill factor (FF) and open-circuit voltage (V_{OC}) of solar cells [9]. Defect density, especially ionic vacancies at the interfaces, can be equally crucial, though this is linked to trap-assisted charge carrier recombination (also known as Shockley-Read-Hall recombination), which can hasten device degradation and reduce efficiency to limit the durability of device [10]. In PSCs, defects and barrier to carrier transport at interfaces are unavoidable because of the different lattice constants of the substrate and overlayers. This leads to recombination and accumulation of charge.

Furthermore, the defects may help in speeding up the deterioration of perovskite films providing a pathway for the interactions between absorber layers and oxygen and water molecules in the air [11]. Thus, improving these interfaces is essential to the development of PSC technology. Until now, numerous researches have tried to attain smooth contact between layers, eliminate defect states, and precise energy band alignment in order to guarantee that PSCs not only accomplish but maintain high performance levels over time [12]. Resolving these interrelated issues is essential to advancing PSCs toward commercial feasibility and guaranteeing that they are able to fulfill their commitment to provide reliable, economical, and efficient solar energy [13].

Consequently, early research focused on improving perovskite film technology, specifically on passivation of defect of grain boundary, using different additive engineering techniques, material compositions, and processing methods [14, 15]. Recent years have seen a rise in sophisticated research on interfacial engineering, which has greatly improved PSC stability and efficiency. Interfacial engineering is an important modification technique that reduces NRR at the interface, prevents the undesired movement of heterogeneous charge carriers, and ensures the cohesiveness of neighboring layers. Optimized energy band alignment can be attained by careful interfacial management, which will lessen the density of surface defects, enhance charge carrier movement, and strengthen the cell against environmental factors like temperature changes, oxygen, and moisture. Therefore, these treatments not only increase device durability by blocking degradation pathways but also increase PSC efficiency by minimizing interfacial energy losses.

To combat the challenges of environmental instability including moisture, UV radiation, and oxygen, researchers are focusing on developing protective encapsulating approaches in order to safeguard the perovskite substances from potential degradation. A number of intriguing methods include encapsulation using inorganic, organic, or hybrid organic–inorganic materials [16]. In addition to providing a physical barrier, these materials provide refractive index matching and high dielectric breakdown properties that prevent degradation of the device. The processes of passivation and encapsulation are crucial for boosting the durability and effectiveness of PSCs, which are known for their exceptional efficiency but are susceptible to degradation brought on by heat, moisture, and light [17]. To improve durability and efficacy of the device, passivation involves treating the surface of the perovskite layer to reduce defects and entrapment sites. The movement and recombination of charges in a perovskite device can be greatly influenced by interfaces between different layers. By modifying these interfaces via interface layers, chemical treatments, or surface passivation methods, it is possible to reduce recombination and improve charge extraction [18]. Interfaces in perovskite films are modified and engineered using a variety of techniques. This includes depositing interfacial layers (ILs), applying chemical treatments, and surface functionalization to strengthen the bond between CTLs and the perovskite layer [19]. Interfaces have the potential to induce flaws and trap states that affect electrical characteristics of perovskite films.. By altering the interface, it is possible to passivate these defects, which reduces non-radiative recombination, increases carrier lifetime, and enhances device performance [20]. Interface chemistry affects how light interacts with perovskite, based on photoluminescence, emission, and absorption. Thus, interface modification can increase light–matter interactions and improve device efficiency. Researchers have employed a number of cutting-edge interface engineering techniques to improve stability and efficiency in their quest to advance PSC technology. Min et al. made a major advancement when they created PSCs with coherent atomic interlayers on SnO_2 electrodes, which help in optimizing charge extraction and transport mechanisms while reducing interfacial imperfections. The outcome was a notable rise in PCE to 25.8% [21].

In another work, Zhang et al. integrated bifunctional alkyl chain obstacles at the critical intersection of perovskite and the HTL to introduce another layer of passivation. This significantly improved efficiency and stability by successfully preventing recombination of electrons and shielding the device from moisture. In addition to improving electrical performance of PSCs, alkyl chain barriers help solve environmental durability [22]. Dong et al. developed PSCs, which achieved 22.2% efficiency and notable improvements in mechanical robustness and operational stability by the addition of interface layers. Its addition emphasizes the significance of seamless connections between layers, guaranteeing effective charge transport [23]. Another study by Chen et al. emphasizes in situ production of two-dimensional perovskite layer at the interface of CuSCN and mixed perovskites, which improved moisture and photostability while increasing PCE from 13.72% to 16.75% [24]. The importance of adding $CsPbI_3$ quantum dots (QDs) as an interface layer was illustrated by Liu et al., who showed that doing so improved the stability and increased the PCE from 15.17% to 18.56% of PSCs. This method demonstrates how QD technologies can be used for fine-tuning of the electrical and optical properties of PSCs, providing a means of achieving high efficiency and stability simultaneously [25]. Kim et al. [26] succeeded in achieving a PCE of 25.7% by utilizing QD–SnO$_2$ layers as ETL. This method not only enhanced charge extraction but also demonstrated how combining QDs with conventional ETLs can increase PSC efficiency. Li et al. [27] improved PSC performance by altering interfaces with a multifunctional fullerene derivative for passivating the surface of TiO$_2$. This technique significantly enhanced charge extraction, improving PCE by 20.7% and highlighting the significance of surface passivation in producing high-efficiency PSCs. By using thiazolium iodide for interface engineering, Salado et al. were able to decrease thermal diffusion and greatly enhance PCE. This approach offers a viable path for further development since it shows how surface functionalization can boost the stability and efficiency of PSCs [28]. By passivating surface defects via interface ion exchange techniques, Li et al. achieved an efficiency of 20.32% with exceptionally high open-circuit voltage of 1.19 V. This method not just fixes surface defects but it also creates opportunities to enhance PSC photovoltaic performance via ion exchange mechanisms [29].

To improve efficiency and stability, the study included bismuth telluride (Bi$_2$Te$_3$) nanoplates as an interlayer in PSCs. This interlayer considerably decreased trap states and charge recombination when it was placed between the CsPbBrI$_2$ layer and HTL. An efficient method of enhancing PSC performance was demonstrated by the optimized usage of the Bi$_2$Te$_3$ interlayer, which increased PCE to 11.96% from 7.46% and after 50 days without further encapsulation, it retained almost 70% of its original PCE [22]. In another work, the efficiency and stability of planar PSCs was considerably increased adding a BiI$_3$ passivation layer between the perovskite layer and compact TiO$_2$ ETL. With a peak efficiency of 17.79%, this interface engineering strategy increased PCE from 13.85% to 16.15%. The use of the BiI$_3$ layer significantly enhances the performance optimization of PSCs by facilitating reducing hysteresis and electron extraction [21].

In addition to achieving notable stability and efficiency, researchers have also produced an outline for resolving few enduring issues in the area of photovoltaics by optimizing interfaces between distinct layers within PSCs.

The following perspectives will be used to discuss the significance of interface engineering:-

3.3 Carrier dynamics

The carrier dynamics are known to be the core of performance of photovoltaic, and needed to be carefully studied and altered to minimize NRR losses and maximize device performance in order to approach their theoretical PCE, which necessitates for regulation of the carrier dynamics throughout the whole device [30]. Device efficiency is influenced by carrier dynamics in terms of charge creation, charge diffusion, charge separation, charge collection, and charge recombination in the perovskite layer and at the interfaces between the perovskite layer and the two carrier transport layers. The perovskite in the devices absorbs photons and produces excitons when irradiated by sunlight. The excitons break down into electrons and holes at room temperature because of their low exciton binding energy. The typical timescale for the charge carrier lifetime in a perovskite light harvester is 10–100 ns for multicrystalline films while for single crystals it is >1,000 ns for [31]. In the presence of carrier concentration gradients and electric fields, charge carrier i.e., electrons and holes will transport or diffuse toward the perovskite/HTL interface and the ETL/perovskite interface, respectively. Additionally, the process of photocurrent hysteresis in PSCs is caused by carriers being separated at the interlayers over a timescale of 1 s [32]. The area of interface exhibits an uninterrupted potential drop due to offsets between the valence and conduction bands, which assures that charge carriers are extracted to ETL or HTL prior to their recombination, usually at internal quantum efficiency (IQE) approaching 100%. The typical IQE of PSCs is about 90%, because of carrier recombination that takes place along grain boundaries of perovskite at the interfaces or in the bulk [33]. The main force behind the separation and movement of charges is V_{OC}, which is directly linked to the energy difference between EVB or the highest occupied molecular orbital (HOMO) level of the ETL and ECB or the lowest unoccupied molecular orbital (LUMO) level of the HTL. J_{SC} for the performance-enhanced device can be improved and energy barriers for charge transportation can be decreased by optimization or proper interface design [34]. In PSCs, the three primary carrier recombination pathways are: radiative, defect-aided, and Auger recombination, the last two being non-radiative recombination. The defects on the surface of perovskite film or at interfaces between the different layers must be passivated in order to lower carrier recombination. Grain boundaries, crystal impurities, and a few other amorphous/crystalline which can result from unreacted precursors, may generally be the cause of hybrid perovskite defects. The important factor for effective, hysteresis-free, and stable PSCs is the reduction of interfacial NRR losses by interface engineering, particularly for interfaces linked with perovskite, that is, ETL/perovskite and perovskite/HTL interfaces. The primary components of carrier dynamics in PSCs are charge transport, charge dissociation, charge extraction, charge recombination (i.e., recombination caused by interface trap states and back charge transfer at the interfaces), charge accumulation (ionic and electronic carriers), and charge collection [30]. Interfacial carrier dynamics is a prominent component of kinetic analysis in PSCs, as evidenced by the direct correlation between interfaces and the majority of charge dynamics components, including charge extraction, charge recombination, charge accumulation, and charge collection. High carrier mobility (e.g., 24–105 for $cm^2\,V^{-1}\,s^{-1}$ $MAPbI_3$ single crystal, 1–10 for $cm^2\,V^{-1}\,s^{-1}$ $MAPbI_3$ polycrystalline film, etc.) and low exciton binding energy are responsible for the transport and dissociation of charge carriers [35, 36]. Therefore, the transport and dissociation of charge carriers are governed by the features of perovskite such as composition, crystal type and film quality. These factors can be modified through additive engineering, composition engineering, and the sensible control of perovskite nucleation and crystallization (deposition techniques, annealing, and anti-solvent and precursor solvent engineering). Through careful selection of additional materials and regulation of the perovskite layer formation, we can maintain excellent extraction of carriers at the electrodes, inhibit carrier recombination in the perovskite layer, and ease carrier injection into the CTLs [37].

The process of extracting the charge is too fast as it takes hundreds of picoseconds to remove photo-generated electrons and holes to the ETM and HTM. Nevertheless, several studies showed that the process of charge extraction can be slowed via residual PbI_2 at the contact [38]. As compared to the charge extraction, charge transfer and its recombination at the interface have far greater effect on device parameters, such as J_{sc} and V_{oc}. Typically, interfacial structural and electronic mismatches serve as the energy barriers for the transport and recombination of charge. To get rid of this effect, interface engineering is required.

By adding a $HOCO\text{-}R\text{-}NH_3^+ I^-$ group between the ETL and perovskite contact, Ogomi et al. significantly reduced the timescale of charge recombination to tens of microseconds from picoseconds [39]. Additionally, Chen's research revealed that the creation of hybrid interfaces greatly inhibited the recombination process at the *p-i-n* PVSCs [40]. In one of studies, Zhu Z. L. et al. altered the interface between ETM and perovskite using carbon QDs. This led to an expedited charge extraction process that raised the extraction rate from approximately 300 ps to about 100 ps [41].

3.4 Energy-level alignment

In PSCs, appropriate band alignment between the perovskite and ILs is intimately related to carrier dynamics of the device, FF, and V_{OC}. It has been found that altering the energy levels of the ILs by compositional alloying or chemical doping effectively improves device performance [42]. For the solar cell device to be optimized, energy-level alignment at the interfaces is essential; J_{sc} and FF can be raised by boosting V_{oc} and facilitating charge transfer and extraction by appropriate energy-level tailoring. Perovskite with HTL and ETL forms the two key interfaces for energy-level alignment. The performance of the device can be boosted by well-matched energy level alignment at the perovskite/CTLs. Energy loss within devices can be reduced and carrier extraction and transport enhanced by adjusting the energy level between the perovskite and contact layers, particularly the two crucial transport layers, that is, ETL and HTL [43]. The HOMO or valence band maximum of perovskite needs to be lower than that of HTL, and the LUMO or conduction band minimum (CBM) needs to be greater than that of ETM in order to promote the carrier transfer.

Interface engineering is anticipated to be effective in regulating the energy level and reducing energy gap to control the energy level mismatch and further reduce interfacial recombination; for instance, adenine, a tiny organic molecule, which was used as an interface modifier on the NiO$_x$ film's surface. A thin layer of adenine clearly boosted V_{oc} while reducing the energy imbalance by 0.1 eV between perovskite and HTL. In the meantime, it enhanced the device's resistance to light and moisture [44]. Additionally, the 3-aminopropyltriethoxysilane (APTES) layer utilized at the interface as a passivation layer between SnO$_2$ and perovskites [45]. The APTES terminal functional groups adjusted the SnO$_2$ WF by forming dipoles on the ETL surface. Therefore, in order to get highly effective and stable PSCs, energy band offsets must be modulated through interface engineering. Empirically, an effective charge extraction at the perovskite/ETM contacts requires a band offset of about 0.2 eV [46]. Additionally, Murata and Minemoto's theoretical research also suggested that in order to create an effective perovskite solar cell device, the HTM and perovskite band offsets must be within 0.2 eV [47]. In planar n-i-p PVSCs, interface engineering can be used to effectively tailor the work functions of adjoining layers, which should be comparable with one another.

The monotonic band edge shift of a semiconductor close to junction resulting from the energy offset with respect to its junction partner is known as band bending. This kind of band bending action was also seen at the perovskite-transporting layer interface. The band bending at the interface between 2,2′,7,7′-tetrakis[N,N-di(4-methoxyphenyl)amino]-9,9′-spirobifluorene (Spiro-OMeTad) and perovskite was measured by Kahn and team. They discovered that Spiro-OMeTAD develops a nonoptimal energy-level alignment with perovskite due to its lower ionization energy. This might reduce the V_{oc} of solar cells devices by causing upward band bending toward the interface with perovskite. For same reason, they proposed that hole extraction should be made easier by creating an ideal energy-level alignment. In order to further minimize V_{oc} losses at the junction, interface engineering is anticipated to be helpful in adjusting the interfacial energy level as well as for reducing difference in energy between the carrying material and perovskite [48].

3.5 Trap passivation

It has been noted that the hysteresis phenomena can be eliminated by passivating trap states at the perovskite surface and different interfaces, which might cause charge recombination and its accumulation losses within the device [49]. The existence of hole traps over the surface of the perovskite film was directly demonstrated by Zhu and coworkers. The under-coordinated halide anions caused these hole traps, which will cause a large charge accumulation at the perovskite/HTM contact [50]. Iodopenta-fluorobenezene was added by Snaith and colleagues to passivate the hole trap states of perovskite surface, thereby significantly preventing charge recombination [51]. Lewis base was also used to passivate the hole trap states in a similar manner. Interface engineering has been used to remove the trap states at the interfaces in order to lower the charge transfer barriers and charge recombination.

3.6 Ion migration

Since the constituent ions (such as MA^+, I^-, and Pb^{2+}) in hybrid perovskites play crucial part in the PSCs deterioration, interface, and hysteresis, their migration has drawn a lot of attention [52]. Several studies have demonstrated that the ions in perovskite materials exhibit modest ionic diffusion coefficients and low activation barriers to migrate within PSCs, particularly under the exposure of light irradiation or external bias [53]. The anomalous hysteresis in the PVSCs was the first indication of the ion migration, which was later discovered to be more widespread in explaining phenomena including halide redistribution, the light-soaking effect. Moreover, the stability of PVSCs is negatively impacted by ionic migration since the moving iodide ions may react with metal electrodes, which could lead to PVSC degradation [54]. Controlling ion migration and enhancing the stability and efficiency of PVSCs require interface engineering. Interfacial chemical reactions can also be hindered by suppression of diffusion-induced interfacial reactions and ion migration. It is widely known that the physical isolation of two intimate interface materials or the physical/chemical inhibition of molecule/ion diffusion or migration are two practical and efficient ways to prevent interfacial chemical reactions through the incorporation of a suitable interface layer. Diffusion or ion migration can also be successfully inhibited by passivating interface and bulk defects to eliminate its migration paths, since vacancy point defects may aid in ion migration [55]. To achieve effective, stable, and hysteresis-free PSCs, it is essential to minimize interfacial NRR losses by interface engineering.

It has been shown that ion migration can cause phase segregation, interfacial band bending, interfacial defects, and interfacial reactions. These phenomena have an important influence on PCE, stability, and hysteresis.

3.7 Permeation barrier

Long-term stability continues to be largest barrier to PSC commercialization. The stability problem originates from the deteriorated interfaces as well as the properties of the HTL, ETL, and perovskite materials themselves. Iodine was recently found from the perovskite layer at the Ag electrode by Kato et al. following a prolonged period of aging in the dark under N_2 environment [56]. They used X-ray diffraction (XRD) and X-ray photoelectron spectroscopy (XPS) to investigate the corrosion of Ag electrode at the spiro-OMeTAD/Ag and $CH_3NH_3PbI_3$/spiro-OMeTAD interfaces in order to propose a mechanism for the creation of AgI [56]. Additionally, the following points briefly describe the Ag electrode corrosion process: (1) An environmental H_2O molecule penetrates the permeable spiro-OMeTAD layer into the perovskite layer; (2) $CH_3NH_3PbI_3$ breaks down into CH_3NH_3I and PbI_2, and CH_3NH_3I breaks down further into CH_3NH_2 and HI; (3) the iodine-containing compound moves toward the Ag electrode; (4) the volatile iodine-containing compound diffuses on the surface; and (5) AgI is created [56]. Furthermore under the conditions of humidity, thermal stressing, and light soaking, the corrosion kinetics of PSCs were shown to be significantly severe to a greater extent. Therefore, it was recommended that creating permeation barriers is a practical means of enhancing operational stability of PSCs. The performance of the PSCs is greatly determined by the four primary types of interfaces seen in normal or inverted PSCs which are discussed below for normal PSCs.

3.8 TCO–ETL interface

High-performance devices and high-efficiency ETL still require improved physicochemical interaction at the transparent conducting oxide (TCO) interface. Snaith et al., for instance, used graphene as an interlayer between the TiO_2 ETL and FTO substrate, which had a positive impact on transport of electron because of appropriate band location of graphene for electron cascade [57]. Changing the WF of TCO is another way to design the TCO/ETL interface. Because of the little electron transfer to the TCO surface from the amine-containing molecules, Zhou et al. showed that surface modification using an amine-containing polyethylenimine ethoxylated (PEIE) may result in the formation of an interfacial dipole with a preferred orientation. This helped to reduce the WF of TCO and, consequently, the improved electron transport [38, 58].

When the Schottky barrier at the interface between the TCO and TiO_2 gets too big, it can appear to reduce the maximum power output [59]. Electrons will accumulate at the interface more easily if the electrode is modified to produce a WF that is close to or less than the Fermi energy of TiO_2. An ultrathin layer was physi-sorbed on the ITO surface using a PEIE solution in methoxyethanol (~0.1% by weight) in a study by Zhou, H. et al. The WF of ITO was successfully decreased from 4.6 to 4.0 eV by the introduced molecular dipole interactions. This change most likely contributed to better electron transport between the TiO_2 and ITO layers. The repression of a barrier for carrier injected into the electrode is consistent with the rise in fill factor seen in the PEIE-modified device. Through PEIE modification, a PSC with PCE greater than 15% is made possible, with FF = 73.28%. In contrast, the device without PEIE modification showed FF = 65.25%. Furthermore, we noticed that the use of PEIE on the ITO resulted in small decrease in the surface roughness of the perovskite films with pertinent morphological changes [38].

3.9 ETL-perovskite interface

In PSCs, the ETLs, which move electrons and inhibit holes, are essential. The metal oxides such as SnO_2, ZnO and TiO_2 are currently often used ETLs in *n-i-p* PSCs, whereas PCBM and C_{60} are widely utilized ETLs in *p-i-n* structures. The ETLs and the associated contact heterojunction interface have numerous drawbacks, despite the fact that the strong electron mobility of these material efficiently aids in light raw electronic-hole separation: (1) During the fabrication of metal oxide ETLs, oxygen vacancy defects are unavoidable and damage the interfacial contact between the perovskite heterojunction and ETLs [60]. Moreover, photo-generated carriers recombine due to vacancy defects, which accumulate an imbalanced charge at the interface. (2) Under ultraviolet light, the photocatalytic activity of O_2 vacancies on the surface of metal oxides increases, speeding up the perovskite interface deterioration that impacts the stability of PSCs under operation. (3) The built-up charge alters the internal electric field balance of interface, deteriorates the inherent stability of the device, and causes the hysteresis phenomena [61]. Additionally, the trap states of the perovskite layer at contact interface and the interfacial energy level matching are critical for overall performance parameters in PSCs. For PSCs to be effective, stable, and with minimal hysteresis, these ETL/perovskite interface challenges must be resolved. In accordance with the previously mentioned fundamental interface turning mechanism, the interface modification improved the performance of PSCs by customizing the interface energy level of ETL, WF, and trap states as well as the passivation of defects and energy-level matching on the surface of perovskite.

In order to optimize the efficiency of perovskite solar cells, it has been found that CTLs are essential for regulating extraction, transportation, and recombination of carriers [62]. Both the ETL and the HTL need to be employed to separate and transfer charges. Metal oxide materials, whether pure or doped with other metals, are affordable, stable, and useful.

Electrons from the perovskite layer are efficiently extracted and moved into the external circuit using ETLs [62]. By tuning this interface, the overall efficiency of device can be increased, charge extraction can be enhanced, and recombination losses can be decreased [63]. Effectively improving ETL/perovskite junction contributes to better electron transport, lowers losses caused by defects or traps, and improves the overall performance of device. Therefore, the stability and long-term dependability of perovskite-based optoelectronic devices are greatly impacted by the management of the ETL/perovskite interface [64]. The duration of operation of solar cells and related optoelectronic devices can be extended by specially engineered interfaces that inhibit degradation.

Mechanisms of connection between perovskite materials and ETLs are thoroughly understood, encompassing several aspects of carrier transport, separation, and combination [65]. In order to reduce energy barriers for electron injection, this method requires a proper alignment of energy levels between the perovskite layer and ETL.

Recombination losses, in which electrons and holes recombination takes place prior to contributing to output of device, must be minimized by carrier separation at the ETL/perovskite interface. In order to enhance carrier separation efficiency and, eventually, device performance, methods like surface passivation, interlayers, or interface engineering can reduce defects or trap states at the interface, improve charge extraction and hence high current densities, and increased device efficiency [66].

Carrier dynamics are all impacted by the interface between the ETL and the perovskite layer, and these factors are crucial to the functionality of the devices. Therefore, it is crucial to focus on interface engineering techniques to alter the chemical, electrical, and structural characteristics at the ETL/perovskite junction. Since the bandgaps of ETL and perovskite differ, for instance, the conduction band at the ETL/perovskite contact must be discontinuous, specifically an energy cliff (ΔE_c) present at the interface. The electrons of ETL might readily overcome the barrier to reestablish the interface in the event of a cliff structure. Holes and electrons would annihilate in the forward bias condition because of the many defects serving as charge carrier recombination centers at the ETL/perovskite contact interface [67].

3.10 Perovskite/HTL interface

At the perovskite/HTM interface, defects are always accumulated. These defects readily capture water and oxygen molecules, causing perovskite degradation and lowering the long-term stability of PSCs. In addition, interface defects typically altered the matching of the interface energy level and increased interface carrier recombination, which affects the hysteresis and performance of PSCs. Furthermore, Li-TFSI and t-BP are frequently employed as p-type dopants in the HTMs. Nonetheless, the Li-TFSI is extremely hydrophilic, whereas t-BP readily coordinates with the PbI_2 of the perovskite at the interface, which speeds up the collapse of the perovskite crystal and reduces the performance of PSCs [68, 69]. For the production of stable, effective, and low hysteresis PSCs, it is crucial to adjust the hydrophobicity, interfaced effects, and energy-level alignment of the perovskite/HTM interface.

By blocking secondary electrons, interface engineering between the perovskite and HTL is a popular method of defect mitigation that improves hole extraction. The interface layer, which is located between the absorber and HTL, is a useful method for preventing the rapid deterioration of the perovskite layer caused by external factors. Additionally, it enhances the overall efficiency of the device by decreasing recombination at interfaces [70]. The addition of additional ILs between the perovskite layer and HTL as part of interface engineering greatly increases solar cell efficiency. Pyridine and thiophene chemicals were added as the IL between the two layers in a recent study, which led to a notable increase in cell efficiency, rising from 13.3% to 15.3% [71]. The use of passivation techniques has also resulted in notable decrease in

NRR pathways. In other work, the F4TCNQ was used as the IL, and higher efficiency of 18% was achieved, as compared to 15% without the IL. This improvement was ascribed to the stronger electric field, which reduced carrier losses. Utilizing ILs efficiently prevents photo-generated electrons from recombining along the perovskite/HTL interface by establishing an energy barrier [72].

Achieving optimal device performance in PSCs requires strengthening the link between the perovskite film and HTL. Various approaches have been proposed to customize this connection with the goal of improving charge transfer efficiency, reducing recombination, and improving energy alignment [73].

3.11 HTL–electrode interface

In PSCs, the HTL/electrode interface plays a crucial role in defining the performance and stability of devices. This interface must effectively prevent the passage of electrons while facilitating hole extraction from the HTL and into the electrode. To control charge transfer and stop recombination, optimization of this contact requires careful material selection, interface engineering, and maybe the addition of ILs.

Even in enclosed and desiccated circumstances, PSCs are clearly susceptible to deterioration despite protective efforts. In particular, the degradation of Ag electrode may be accelerated through interaction between I and Ag ions. Additionally, prolonged light and positive bias might hasten the migration of I ions and diffusion of atoms of Au or Ag, hence reducing the PCE of PSCs [74]. To strengthen the stability of the device, researchers have created different interfacial barriers between the metal electrode and the HTL. Arora et al. made a significant breakthrough by employing copper thiocyanate (CuSCN) as a hole extraction layer in an effective PSC. They developed a reduced graphene oxide (rGO) barrier layer that is positioned between the CuSCN and Au in order to mitigate the instability resulting from potential-induced deterioration at the CuSCN/Au contact. With this change, the PSC became extremely stable, retaining more than 95% of original PCE over 1,000 h of aging at 60 °C while using maximum power point tracking (MPPT) [75]. Additionally, metal oxides have been used as the intermediate layers between the electrode and HTL. For instance, Yang and coworkers [76] used MoO_3-encapsulated Au nanonets with significant MoO_3 surface tension. The wettability of gold was evidently improved, which promoted the Frank-van der Merwe growth mode to produce ultrathin gold nanolayers. In addition to having outstanding electrical conductivity, these layers also retain adequate optical transparency. Furthermore, the top MoO_3 layer reduces reflections from the Au layer, increasing light transmittance and, in turn, the effectiveness of the device. Despite the potential of these approaches, there is still a dearth of study on the HTL/electrode interface, which necessitates more investigation and improvement in order to maximize PSC stability and efficiency simultaneously. A similar efficiency to that of precious metals was attained when MoO_x was used as a buffer layer to remove the charge transport barrier at the HTM/Al interface for the use of inexpensive metal back electrodes [77]. Similarly, an IL has been added to the PCBM/electrode interface to facilitate charge transmission [78].

3.12 Charge transport layers

As the quality of the perovskite layer increased, attention turned to optimizing the transport layers (TLs). In classical PSCs, the perovskite layer is placed between two TLs and electrodes in a simple *n-i-p* or *p-i-n* configuration. An effective TL needs to meet a number of requirements, such as (a) a favorable energy level alignment, which allows one type of charge carrier to transfer efficiently while successfully blocking the other [79]; (b) appropriate chemical and physical characteristics, to prevent harmful reactions with environment and surrounding, and also appropriate surface characteristics [80]; (c) high transparency of the TLs to optimize absorption of the perovskite layer [81]; and (d) good transport characteristics to guarantee quick transport of the charge carriers in the direction of electrode [82].

The CTLs have two main purposes: to collect electrons from the active layer and stop them from recombining with holes. For the best possible device performance, it is essential to minimize interface defects and guarantee the safety of the perovskite layer in addition to charge carrier transport and extraction. When compared to those from other layers, the NRR loss from metal electrodes (such as Au, Ag, and Cu) and TCO electrodes (e.g., FTO and ITO) is insignificant and can be disregarded. Various attempts have been made to develop and optimize the CTLs in order to reduce defects and improve carrier transport, because bulk charge NRR losses in the CTLs may arise from the defects and low electrical properties such as carrier concentration, conductivity, and carrier mobility of the ETL and HTL layers [83]. Numerous parameters like thickness of the CTL affect the PSC performance. An appropriate thicker CTL can boost charge carrier extraction and lead to notable performance improvements by better reflecting charge carriers at the smooth interface of HTL [84].

3.13 Electron transport layers

ETLs facilitate electron migration away from a contact by blocking holes. Although a vacancy is not considered a physical particle in itself, it can pass between atoms in a semiconductor system to prevent the creation of carriers. The ETLs have sufficient HOMO energies to obstruct holes and triplet energies high enough to prevent quenching excitons [85]. Due to their capacity to carry electrons, these materials were used as the ETL/HTL. Semiconductor metal oxides of the *n*-type are often employed as ETLs in conventional configurations. Both organic and inorganic materials are vital in electronic applications. ZnOS, TiO_2, ZnO, and WO_3 are examples of inorganic compounds that have a variety of uses in electronic devices, such as conducting, insulating, and semi-conductivity [86]. On the other hand, organic materials like C_{60}, $PC_{61}BM$, and C_{70}, used as ETL and are necessary for the movement of electrons in devices [87]. They improve the functionality and efficiency of devices by facilitating electron transfer processes. As an electron acceptor, PCBM is unique as it efficiently extracts electrons from the light-absorbing layer. The exceptional electron mobility of fullerene (C_{60}) facilitates their quick movement within the cell. Charge carrier transport and transparency to light are balanced by titanium dioxide (TiO_2), which serves as a flexible ETL. SnO_2 works as a transparent conductive oxide, allowing light to flow through while quickly removing accumulated electrons. PCBM is frequently utilized in PSCs owing to its advantageous electron transport characteristics. Another substance with superior electron transport qualities that is used in PSC devices is fullerene (C_{60}). Titanium Dioxide (TiO_2) is frequently employed as ETLs in certain PSC designs as compact layers or nanoparticles [88]. Notably, TiO_2 has a high electron affinity, strong UV-light stability, and outstanding electron mobility, all of which reduce recombination losses and enable effective extraction and transport of electrons [89]. Beyond its remarkable performance, TiO_2 is a cost-effective and environmentally friendly option for solar applications due to its non-toxicity, chemical stability, and abundance. Owing to excellent electron transport properties of tin oxide (SnO_2), it can also be utilized as an ETL and is sometimes employed as an alternative to TiO_2. Jiang et al. emphasized the more effective band alignment and high electron mobility of SnO_2 as reasons for its superiority as an ETL [90]. Examples of materials that are frequently used as ETL include titanium dioxide (TiO_2), zirconium dioxide (ZrO_2), and tin oxide (SnO_2) because of their superior electron mobility and appropriate energy levels that enable efficient charge transport [91].

Benefits of ZnO over the commonly used TiO_2 include low processing temperatures and significantly better electron mobility with bulk mobility of 205–300 $cm^2 V^{-1} s^{-1}$) [92]. ZnO nanorods and nanoparticles used as ETM in hybrid PSCs demonstrated PCEs of 11.13% and 15.7%, respectively [93]. However, because of the chemical interactions that take place at the perovskite/ZnO interface, HPSCs that use ZnO ETLs have low stability. SnO_2 is a more attractive ETL in this regard because of its larger bandgap, excellent transparency, strong electron mobility (bulk mobility: 240 $cm^2 V^{-1} s^{-1}$), and superior chemical stability.

3.14 Hole transport layers

The hole transporting layer, or HTL, is essential for the functioning of ordinary PSCs because it effectively extracts holes from the perovskite film and transfers them to the metal electrode. Additionally, HTLs act as barriers to obstruct direct contact between the metal electrode and the perovskite layer, reducing the possibility of electron and hole recombination in conventional PSCs [94]. Specifically, in inverted PSCs, the HTL has a substantial impact on the perovskite layer's grain boundaries and grain sizes, which in turn significantly affects the long-term stability and PCEs of the PSCs. Inevitably, the commonly employed organic HTL is hygroscopic, has defect diffusion, and deteriorates due to bias, light, and heat [95]. To overcome these issues in conventional HTLs, inorganic materials with improved stability under challenging conditions, such as high temperatures and extended illumination, have been developed. In addition to being cheap, these materials exhibit excellent carrier mobility. These include NiO, PEDOT.PSS, Cu_2O, CuSCN, etc. Spiro-OMeTAD is commonly used as a hole transport material in PSCs, especially when paired with the right dopants and additives [96]. In addition to having a high hole mobility, which is necessary for good charge transport, its ionization potential is in good alignment with the perovskite to enable efficient charge extraction [97]. Comparing Spiro-OMeTAD with other hole transport materials, it is also more economical and ecologically stable due to its high yield of synthesis and purity [98]. Crucially, it is possible to alter its chemical structure to improve heat stability and prolong its working lifespan [97].

In addition to serving as HTL, PEDOT: PSS can also be used as an IL to improve stability and adhesion [99]. Zinc oxide (ZnO) or aluminum oxide (Al_2O_3) can be used as interlayers or modifications to enhance device stability or charge extraction [100]. These materials are essential for improving PSC stability and performance because they reduce recombination losses at their interfaces and facilitate effective charge transfer. According to research, by resolving interfacial defects and reducing surface recombination, these materials can enhance charge extraction and improve device stability. Using inorganic materials as hole transport layers, such as metal oxides like MoO_3 and NiO, is an alternative method of interface modification [101]. Additionally, methods such as adding interlayers like polyethyleneimine or doping the hole transport layer with organic amines are promising in modifying the WF at the electrode and improving carrier extraction in the ETL [102]. The process of pulsed laser deposition is widely used to produce perovskite and HTL thin films [103].

3.15 Interface engineering challenges and promising approaches for extremely stable and efficient PSCs

Despite significant advancements in interface engineering at different PSC interfaces, issues with interface material selection, assessment, deposition, and recycling still require immediate attention:

1. The guidelines for developing and synthesizing interface materials are inadequate. The majority of current research is based on the trial-and-error approach to find appropriate interface materials because of the intricate connections between perovskite and interface. This approach is time-taking for the practical application of PSCs. Therefore, machine learning needs to be developed to speed up the search for new and efficient interlayers [104]. Additionally, certain fundamental ideas for ion immobilization and defect passivation using function groups or Lewis acids and bases can be considered [105, 106]. Molecular dyes have garnered a lot of attention lately and have been used in a variety of solar cell types as well as at different PSC interfaces. This is because of their easy synthesis, extreme absorption and emission in the near-infrared and visible spectrum, and stable photochemical characteristics under heat and humidity [107, 108]. In particular, squaraine dyes with a reasonably stable zwitterionic structure are attractive passivator candidates for use in PSCs because they can passivate both shallow and deep defects on the perovskite surface [109].

2. The assessment of interface materials ought to be more rigorous and practical, in compliance with the International Summit on Organic PV Stability (ISOS) protocols and the International Electrotechnical Commission's (IEC) standards (IEC 61 215 and IEC 61 646) for commercial terrestrial and thin-film PV modules [110]. Interface engineering studies mostly focus on improving initial efficiency and shelf stability (in air or inert gas), ignoring the photothermal stability of interface materials for potential commercialization. Hence, in order to guide the study on interface materials, criteria that align with IEC and ISOS should be developed.

3. The creation of cost-effective and multipurpose interface materials is essential to the competitiveness of PSC technology. Interface engineering is an extra step in the fabrication process of PSCs that unavoidably raises the overall cost. To ensure a low levelized cost of electricity (LCOE) value comparable with other energy resources, novel interface materials with multiple functions, such as increasing the mobility of CTLs, passivating interface defects, cross-linking perovskite and CTLs, releasing residual stress inside perovskite, mitigating ion migration and air ingress, etc., should be synthesized and studied.

4. The current study focusses on interface engineering of lab-scale small-area PSCs (usually active area less than 1 cm^2), which are primarily produced by spin coating. For large-area PSCs (active area ranging from 1 cm^2 to 100 cm^2) in mass manufacturing, this approach is incompatible. In order to solve this problem, techniques for deposition such as atomic layer deposition, screen printing, thermal evaporation, and doctor blading need to be developed in order to deposit a compact and uniform interface layer for large-area PSCs and high efficiency.

5. The widespread use of PSCs will raise concerns about interface material contamination or pollution during the deposition process. Strict and organized waste recycling, particularly of solvent and interface materials, is necessary for environmentally friendly and sustainable development.

6. It is difficult to produce stable and high-efficiency PSCs with interface engineering at a single interface because of the complexity of PSCs. To further increase PSC stability and efficiency, multi-interface engineering – that is, concurrently altering several interfaces in PSCs with interlayers – should be developed [111]. In this regard, the most advantageous combination is multi-interface engineering and perovskite additive engineering.

3.16 Conclusion

The evolution of perovskite solar cell technology has reached a stage where further enhancements in device performance hinge critically on effective interface engineering and the optimization of charge transport layers. The interfaces between perovskite and adjacent layers (ETL, HTL, and electrodes) not only dictate the charge extraction and transport processes but also govern the mechanisms of degradation and stability. By tailoring these interfaces through energy level alignment, defect passivation, and barrier layer insertion, significant improvements in both efficiency and longevity have been realized. Innovations such as the incorporation of QDs, multifunctional fullerene derivatives, and inorganic interlayers have demonstrated substantial gains in power conversion efficiency and environmental robustness. Moreover, controlling ion migration and interfacial recombination through targeted engineering strategies is pivotal for achieving hysteresis-free, thermally stable devices. While substantial progress has been made at the laboratory scale, future work must address the challenges of scalable fabrication, cost-effective materials, and environmentally sustainable processing. Multi-interface engineering, combined with advanced material design and characterization protocols, is anticipated to be the key to unlocking the full potential of PSCs for commercial applications.

References

[1]   Han C, Xiao X, Zhang W, Gao Q, Qi J, Liu J, Xiang J, Cheng Y, Du J, Qiu C, Mei A. Impact and role of epitaxial growth in metal halide perovskite solar cells, ACS Materials Letters. 2023 Aug 10; 5(9): 2445–2463.

[2]   Zhang F, Zheng D, Yu D, Wu S, Wang K, Peng L, Liu SF, Yang D. Perovskite photovoltaic interface: From optimization towards exemption, Nano Energy. 2024 Mar; 16: 109503.

[3]   Li J, Ye Z, Mo P, Pang Y, Gao E, Zhang C, Du G, Sun R, Zeng X. Compliance-tunable thermal interface materials based on vertically oriented carbon fiber arrays for high-performance thermal management, Composites Science and Technology. 2023 Mar 22; 234: 109948.

[4]   Shi J, Xu X, Li D, Meng Q. Interfaces in perovskite solar cells, Small. 2015 Jun; 11(21): 2472–2486.

[5]   Ma Y, Gong J, Zeng P, Liu M. Recent progress in interfacial dipole engineering for perovskite solar cells, Nano-Micro Letters. 2023 Dec; 15(1): 173.

[6]   Tang H, Bai Y, Zhao H, Qin X, Hu Z, Zhou C, Huang F, Cao Y. Interface engineering for highly efficient organic solar cells, Advanced Materials. 2024 Apr; 36(16): 2212236.

[7]   Palei S, Murali G, Kim CH, In I, Lee SY, Park SJ. A review on interface engineering of MXenes for perovskite solar cells, Nano-Micro Letters. 2023 Dec; 15(1): 123.

[8]   Boyd CC, Cheacharoen R, Leijtens T, McGehee MD. Understanding degradation mechanisms and improving stability of perovskite photovoltaics, Chemical Reviews. 2018 Nov 16; 119(5): 3418–3451.

[9]   Wang S, Sakurai T, Wen W, Qi Y. Energy level alignment at interfaces in metal halide perovskite solar cells, Advanced Materials Interfaces. 2018 Nov; 5(22): 1800260.

[10]   Tress W, Yavari M, Domanski K, Yadav P, Niesen B, Baena JP, Hagfeldt A, Graetzel M. Interpretation and evolution of open-circuit voltage, recombination, ideality factor and subgap defect states during reversible light-soaking and irreversible degradation of perovskite solar cells, Energy & Environmental Science. 2018; 11(1): 151–165.

[11]   Daboczi M, Hamilton I, Xu S, Luke J, Limbu S, Lee J, McLachlan MA, Lee K, Durrant JR, Baikie ID, Kim JS. Origin of open-circuit voltage losses in perovskite solar cells investigated by surface photovoltage measurement, ACS Applied Materials & Interfaces. 2019 Nov 18; 11(50): 46808–46817.

[12]   Liu C, Yuan J, Masse R, Jia X, Bi W, Neale Z, Shen T, Xu M, Tian M, Zheng J, Tian J. Interphases, interfaces, and surfaces of active materials in rechargeable batteries and perovskite solar cells, Advanced Materials. 2021 Jun; 33(22): 1905245.

[13]   Bi D, Yi C, Luo J, Décoppet JD, Zhang F, Zakeeruddin SM, Li X, Hagfeldt A, Grätzel M. Polymer-templated nucleation and crystal growth of perovskite films for solar cells with efficiency greater than 21%, Nature Energy. 2016 Sep 19; 1(10): 1–5.

[14]   Ke W, Xiao C, Wang C, Saparov B, Duan HS, Zhao D, Xiao Z, Schulz P, Harvey SP, Liao W, Meng W. Employing lead thiocyanate additive to reduce the hysteresis and boost the fill factor of planar perovskite solar cells, Advanced Materials. 2016 May 4; 28(26): 5214–5221.

[15]   Sherkar TS, Momblona C, Gil-Escrig L, Avila J, Sessolo M, Bolink HJ, Koster LJA. Recombination in perovskite solar cells: Significance of grain boundaries, interface traps, and defect ions, ACS Energy Letters. 2017; 2(5): 1214–1222.

[16]   Kumar A, Chang DW, Baek JB. Current status and future of organic–inorganic hybrid perovskites for photoelectrocatalysis devices, Energy & Fuels. 2023 Oct 30; 37(23): 17782–17802.

[17]   Getachew G, Wibrianto A, Rasal AS, Dirersa WB, Chang JY. Metal halide perovskite nanocrystals for biomedical engineering: Recent advances, challenges, and future perspectives, Coordination Chemistry Reviews. 2023 May 1; 482: 215073.

[18]   Al-Dhahir I, Niu X, Yu M, McNab S, Lin Y, Altermatt PP, Patrick CE, Bonilla RS. Ion-charged dielectric nanolayers for enhanced surface passivation in high efficiency photovoltaic devices, Advanced Materials Interfaces. 2023 Jun; 10(16): 2300037.

[19]   Chen B, Guo R, He Z, Peng C, Su H, Sun L, Li X, Zhang Q, Wang L. Self-assembled monolayers as hole transport layers for efficient thermally evaporated blue perovskite light-emitting diodes, Chemical Engineering Journal. 2023 Nov 15; 476: 146476.

[20]   Park H, Heo J, Jeong BH, Lee S, Lee KT, Park S, Park H. Interface modification of perovskite solar cell for synergistic effect of surface defect passivation and excited state property enhancement, Journal of Alloys and Compounds. 2023 Oct 15; 960: 170606.

[21]   Min H, Lee DY, Kim J, Kim G, Lee KS, Kim J, Paik M, Kim Y, Kim KS, Kim M, Shin TJ. Perovskite solar cells with atomically coherent interlayers on SnO_2 electrodes, Nature. 2021 Oct 21; 598(7881): 444–450.

[22]   Zhang J, Hu Z, Huang L, Yue G, Liu J, Lu X, Hu Z, Shang M, Han L, Zhu Y. Bifunctional alkyl chain barriers for efficient perovskite solar cells, Chemical Communications. 2015; 51(32): 7047–7050.

[23]   Dong Q, Zhu C, Chen M, Jiang C, Guo J, Feng Y, Dai Z, Yadavalli SK, Hu M, Cao X, Li Y. Interpenetrating interfaces for efficient perovskite solar cells with high operational stability and mechanical robustness, Nature Communications. 2021 Feb 12; 12(1): 973.

[24] Chen J, Seo JY, Park NG. Simultaneous improvement of photovoltaic performance and stability by in situ formation of 2D perovskite at (FAPbI$_3$) 0.88 (CsPbBr$_3$) 0.12/CuSCN interface, Advanced Energy Materials. 2018 Apr; 8(12): 1702714.

[25] Noman M, Sherwani T, Jan ST, Ismail M. Exploring the impact of kesterite charge transport layers on the photovoltaic properties of MAPbI3 perovskite solar cells, Physica Scripta. 2023 Nov 8; 98(12): 125507.

[26] Kim M, Jeong J, Lu H, Lee TK, Eickemeyer FT, Liu Y, Choi IW, Choi S, Jo Y, Kim HB, Mo SI. Conformal quantum dot–SnO$_2$ layers as electron transporters for efficient perovskite solar cells, Science. 2022 Jan 21; 375(6578): 302–306.

[27] Li Y, Zhao Y, Chen Q, Yang Y, Liu Y, Hong Z, Liu Z, Hsieh YT, Meng L, Li Y, Yang Y. Multifunctional fullerene derivative for interface engineering in perovskite solar cells, Journal of the American Chemical Society. 2015 Dec 16; 137(49): 15540–15547.

[28] Salado M, Andresini M, Huang P, Khan MT, Ciriaco F, Kazim S, Ahmad S. Interface engineering by thiazolium iodide passivation towards reduced thermal diffusion and performance improvement in perovskite solar cells, Advanced Functional Materials. 2020 Apr; 30(14): 1910561.

[29] Li Z, Wang L, Liu R, Fan Y, Meng H, Shao Z, Cui G, Pang S. Spontaneous interface ion exchange: Passivating surface defects of perovskite solar cells with enhanced photovoltage, Advanced Energy Materials. 2019 Oct; 9(38): 1902142.

[30] Yang Z, Dou J, Wang M. Interface engineering in n-i-p metal halide perovskite solar cells, Solar RRL. 2018 Dec; 2(12): 1800177.

[31] Shi J, Li Y, Li Y, Li D, Luo Y, Wu H, Meng Q. From ultrafast to ultraslow: Charge-carrier dynamics of perovskite solar cells, Joule. 2018 May 16; 2(5): 879–901.

[32] Yang Z, Dou J, Wang M. Interface engineering in n-i-p metal halide perovskite solar cells, Solar RRL. 2018 Dec; 2(12): 1800177.

[33] Pan H, Shao H, Zhang XL, Shen Y, Wang M. Interface engineering for high-efficiency perovskite solar cells, Journal of Applied Physics. 2021 Apr 7; 129(13): 1–9.

[34] Li G, Zhu R, Yang Y. Polymer solar cells, Nature Photonics. 2012 Mar; 6(3): 153–161.

[35] Miyata A, Mitioglu A, Plochocka P, Portugall O, Wang JT, Stranks SD, Snaith HJ, Nicholas RJ. Direct measurement of the exciton binding energy and effective masses for charge carriers in organic–inorganic tri-halide perovskites, Nature Physics. 2015 Jul; 11(7): 582–587.

[36] Brenner TM, Egger DA, Kronik L, Hodes G, Cahen D. Hybrid organic—inorganic perovskites: Low-cost semiconductors with intriguing charge-transport properties, Nature Reviews Materials. 2016 Jan 11; 1(1): 1–6.

[37] Zhou H, Chen Q, Li G, Luo S, Song TB, Duan HS, Hong Z, You J, Liu Y, Yang Y. Interface engineering of highly efficient perovskite solar cells, Science. 2014 Aug 1; 345(6196): 542–546.

[38] Wang L, McCleese C, Kovalsky A, Zhao Y, Burda C. Femtosecond time-resolved transient absorption spectroscopy of CH$_3$NH$_3$PbI$_3$ perovskite films: Evidence for passivation effect of PbI2, Journal of the American Chemical Society. 2014 Sep 3; 136(35): 12205–12208.

[39] Ogomi Y, Morita A, Tsukamoto S, Saitho T, Shen Q, Toyoda T, Yoshino K, Pandey SS, Ma T, Hayase S. All-solid perovskite solar cells with HOCO-R-NH$_3$+ I–anchor-group inserted between porous titania and perovskite, The Journal of Physical Chemistry C. 2014 Jul 31; 118(30): 16651–16659.

[40] Chen W, Wu Y, Liu J, Qin C, Yang X, Islam A, Cheng YB, Han L. Hybrid interfacial layer leads to solid performance improvement of inverted perovskite solar cells, Energy & Environmental Science. 2015; 8(2): 629–640.

[41] Zhu Z, Ma J, Wang Z, Mu C, Fan Z, Du L, Bai Y, Fan L, Yan H, Phillips DL, Yang S. Efficiency enhancement of perovskite solar cells through fast electron extraction: The role of graphene quantum dots, Journal of the American Chemical Society. 2014 Mar 12; 136(10): 3760–3763.

[42] Qin P, Paek S, Dar MI, Pellet N, Ko J, Grätzel M, Nazeeruddin MK. Perovskite solar cells with 12.8% efficiency by using conjugated quinolizino acridine based hole transporting material, Journal of the American Chemical Society. 2014 Jun 18; 136(24): 8516–8519.

[43] Cao Q, Li Z, Han J, Wang S, Zhu J, Tang H, Li X, Li X. Electron transport bilayer with cascade energy alignment for efficient perovskite solar cells, Solar RRL. 2019 Dec; 3(12): 1900333.

[44] Xie L, Cao Z, Wang J, Wang A, Wang S, Cui Y, Xiang Y, Niu X, Hao F, Ding L. Improving energy level alignment by adenine for efficient and stable perovskite solar cells, Nano Energy. 2020 Aug 1; 74: 104846.

[45] Yang G, Wang C, Lei H, Zheng X, Qin P, Xiong L, Zhao X, Yan Y, Fang G. Interface engineering in planar perovskite solar cells: Energy level alignment, perovskite morphology control and high performance achievement, Journal of Materials Chemistry A. 2017; 5(4): 1658–1666.

[46] Zhou Z, Pang S, Liu Z, Xu H, Cui G. Interface engineering for high-performance perovskite hybrid solar cells, Journal of Materials Chemistry A. 2015; 3(38): 19205–19217.

[47] Minemoto T, Murata M. Theoretical analysis on effect of band offsets in perovskite solar cells, Solar Energy Materials and Solar Cells. 2015 Feb 1; 133: 8–14.

[48] Sum TC, Chen S, Xing G, Liu X, Wu B. Energetics and dynamics in organic–inorganic halide perovskite photovoltaics and light emitters, Nanotechnology. 2015 Aug 3; 26(34): 342001.

[49] Chen B, Yang M, Priya S, Zhu K. Origin of J–V hysteresis in perovskite solar cells, The Journal of Physical Chemistry Letters. 2016 Mar 3; 7(5): 905–917.

[50] Wu X, Trinh MT, Niesner D, Zhu H, Norman Z, Owen JS, Yaffe O, Kudisch BJ, Zhu XY. Trap states in lead iodide perovskites, Journal of the American Chemical Society. 2015 Feb 11; 137(5): 2089–2096.

[51] Abate A, Saliba M, Hollman DJ, Stranks SD, Wojciechowski K, Avolio R, Grancini G, Petrozza A, Snaith HJ. Supramolecular halogen bond passivation of organic–inorganic halide perovskite solar cells, Nano Letters. 2014 Jun 11; 14(6): 3247–3254.

[52] Yuan Y, Huang J. Ion migration in organometal trihalide perovskite and its impact on photovoltaic efficiency and stability, Accounts of Chemical Research. 2016 Feb 16; 49(2): 286–293.

[53] Azpiroz JM, Mosconi E, Bisquert J, De Angelis F. Defect migration in methylammonium lead iodide and its role in perovskite solar cell operation, Energy & Environmental Science. 2015; 8(7): 2118–2127.

[54] Li J, Dong Q, Li N, Wang L. Direct evidence of ion diffusion for the silver-electrode-induced thermal degradation of inverted perovskite solar cells, Advanced Energy Materials. 2017 Jul; 7(14): 1602922.

[55] Chen J, Park NG. Causes and solutions of recombination in perovskite solar cells, Advanced Materials. 2019 Nov; 31(47): 1803019.

[56] Kato Y, Ono LK, Lee MV, Wang S, Raga SR, Qi Y. Silver iodide formation in methyl ammonium lead iodide perovskite solar cells with silver top electrodes, Advanced Materials Interfaces. 2015 Sep; 2(13): 1500195.

[57] Wang JT, Ball JM, Barea EM, Abate A, Alexander-Webber JA, Huang J, Saliba M, Mora-Sero I, Bisquert J, Snaith HJ, Nicholas RJ. Low-temperature processed electron collection layers of graphene/TiO2 nanocomposites in thin film perovskite solar cells, Nano Letters. 2014 Feb 12; 14(2): 724–730.

[58] Qiu J, Yang S. Material and interface engineering for high-performance perovskite solar cells: A personal journey and perspective, The Chemical Record. 2020 Mar; 20(3): 209–229.

[59] Snaith HJ, Grätzel M. The role of a "Schottky Barrier" at an electron-collection electrode in solid-state dye-sensitized solar cells, Advanced Materials. 2006 Jul 18; 18(14): 1910–1914.

[60] Tan H, Jain A, Voznyy O, Lan X, García De Arquer F, Fan JZ, Quintero-Bermudez R, Yuan M, Zhang B, Zhao Y, Fan F. Efficient and stable solution-processed planar perovskite solar cells via contact passivation, Science. 2017 Feb 17; 355(6326): 722–726.

[61] Xia J, Luo J, Yang H, Zhao F, Wan Z, Malik HA, Shi Y, Han K, Yao X, Jia C. Vertical phase separated cesium fluoride doping organic electron transport layer: A facile and efficient "bridge" linked heterojunction for perovskite solar cells, Advanced Functional Materials. 2020 Jul; 30(27): 2001418.

[62] Choi S, Yong T, Choi J. Towards scalability: Progress in metal oxide charge transport layers for large-area perovskite solar cells, Inorganic Chemistry Frontiers. 2024; 11(1): 50–70.

[63] Mohammad A, Mahjabeen F. Promises and challenges of perovskite solar cells: A comprehensive review, BULLET: Jurnal Multidisiplin Ilmu. 2023; 2(5): 1147–1157.

[64] Nazir G, Lee SY, Lee J, Rehman A, Lee JK, Seok Sİ, Park SJ. Stabilization of perovskite solar cells: Recent developments and future perspectives, Advanced Materials. 2022 Dec; 34(50): 2204380.

[65] Subudhi P, Punetha D. Pivotal avenue for hybrid electron transport layer-based perovskite solar cells with improved efficiency, Scientific Reports. 2023 Nov 9; 13(1): 19485.

[66] Katta VS, Waheed M, Kim JH. Recent advancements in enhancing interfacial charge transport for perovskite solar cells, Solar RRL. 2024 Apr; 8(7): 2300908.

[67] Ding C, Zhang Y, Liu F, Kitabatake Y, Hayase S, Toyoda T, Wang R, Yoshino K, Minemoto T, Shen Q. Understanding charge transfer and recombination by interface engineering for improving the efficiency of PbS quantum dot solar cells, Nanoscale Horizons. 2018; 3(4): 417–429.

[68] Xia J, Zhang R, Luo J, Yang H, Shu H, Malik HA, Wan Z, Shi Y, Han K, Wang R, Yao X. Dipole evoked hole-transporting material p-doping by utilizing organic salt for perovskite solar cells, Nano Energy. 2021 Jul 1; 85: 106018.

[69] Xia J, Zhang Y, Xiao C, Brooks KG, Chen M, Luo J, Yang H, Klipfel NI, Zou J, Shi Y, Yao X. Tailoring electric dipole of hole-transporting material p-dopants for perovskite solar cells, Joule. 2022 Jul 20; 6(7): 1689–1709.

[70] Wu S, Chen R, Zhang S, Babu BH, Yue Y, Zhu H, Yang Z, Chen C, Chen W, Huang Y, Fang S. A chemically inert bismuth interlayer enhances long-term stability of inverted perovskite solar cells, Nature Communications. 2019 Mar 11; 10(1): 1161.

[71] Noel NK, Abate A, Stranks SD, Parrott ES, Burlakov VM, Goriely A, Snaith HJ. Enhanced photoluminescence and solar cell performance via Lewis base passivation of organic–inorganic lead halide perovskites, ACS Nano. 2014 Oct 28; 8(10): 9815–9821.

[72] Song D, Wei D, Cui P, Li M, Duan Z, Wang T, Ji J, Li Y, Mbengue J, Li Y, He Y. Dual function interfacial layer for highly efficient and stable lead halide perovskite solar cells, Journal of Materials Chemistry A. 2016; 4(16): 6091–6097.

[73] Song D, Ramakrishnan S, Xu Y, Yu Q. Designing effective hole transport layers in tin perovskite solar cells, ACS Energy Letters. 2023 Sep 14; 8(10): 4162–4172.

[74] Chen J, Lee D, Park NG. Stabilizing the Ag electrode and reducing J–V hysteresis through suppression of iodide migration in perovskite solar cells, ACS Applied Materials & Interfaces. 2017 Oct 18; 9(41): 36338–36349.

[75] Arora N, Dar MI, Hinderhofer A, Pellet N, Schreiber F, Zakeeruddin SM, Grätzel M. Perovskite solar cells with CuSCN hole extraction layers yield stabilized efficiencies greater than 20%, Science. 2017 Nov 10; 358(6364): 768–771.

[76] Wang Z, Zhu X, Zuo S, Chen M, Zhang C, Wang C, Ren X, Yang Z, Liu Z, Xu X, Chang Q. 27%-Efficiency four-terminal perovskite/silicon tandem solar cells by sandwiched gold nanomesh, Advanced Functional Materials. 2020 Jan; 30(4): 1908298.

[77] Zhao Y, Nardes AM, Zhu K. Effective hole extraction using MoOx-Al contact in perovskite CH3NH3PbI3 solar cells, Applied Physics Letters. 2014 May 26; 104(21).

[78] Zhang H, Azimi H, Hou Y, Ameri T, Przybilla T, Spiecker E, Kraft M, Scherf U, Brabec CJ. Improved high-efficiency perovskite planar heterojunction solar cells via incorporation of a polyelectrolyte interlayer, Chemistry of Materials. 2014 Sep 23; 26(18): 5190–5193.

[79] Stolterfoht M, Wolff CM, Márquez JA, Zhang S, Hages CJ, Rothhardt D, Albrecht S, Burn PL, Meredith P, Unold T, Neher D. Visualization and suppression of interfacial recombination for high-efficiency large-area pin perovskite solar cells, Nature Energy. 2018 Oct; 3(10): 847–854.

[80]   Zhang S, Stolterfoht M, Armin A, Lin Q, Zu F, Sobus J, Jin H, Koch N, Meredith P, Burn PL, Neher D. Interface engineering of solution-processed hybrid organohalide perovskite solar cells, ACS Applied Materials & Interfaces. 2018 Jun 1; 10(25): 21681–21687.

[81]   Marinova N, Tress W, Humphry-Baker R, Dar MI, Bojinov V, Zakeeruddin SM, Nazeeruddin MK, Grätzel M. Light harvesting and charge recombination in CH3NH3PbI3 perovskite solar cells studied by hole transport layer thickness variation, ACS Nano. 2015 Apr 28; 9(4): 4200–4209.

[82]   Stolterfoht M, Wolff CM, Amir Y, Paulke A, Perdigón-Toro L, Caprioglio P, Neher D. Approaching the fill factor Shockley–Queisser limit in stable, dopant-free triple cation perovskite solar cells, Energy & Environmental Science. 2017; 10(6): 1530–1539.

[83]   Shin SS, Lee SJ, Seok SI. Metal oxide charge transport layers for efficient and stable perovskite solar cells, Advanced Functional Materials. 2019 Nov; 29(47): 1900455.

[84]   Bhattarai S, Sharma A, Muchahary D, Gogoi M, Das TD. Carrier transport layer free perovskite solar cell for enhancing the efficiency: A simulation study, Optik. 2021 Oct 1; 243: 167492.

[85]   Tian R, Xu H. Solution-processable electron transporting materials, In InSolution-Processed Organic Light-emitting Devices. 2024 Jan 1; Woodhead Publishing, 151–174.

[86]   AlZoubi T, Mourched B, Al Gharram M, Makhadmeh G, Abu Noqta O. Improving photovoltaic performance of hybrid organic-inorganic MAGeI3 perovskite solar cells via numerical optimization of carrier transport materials (HTLs/ETLs), Nanomaterials. 2023 Jan; 13(15): 2221.

[87]   Krishna BG, Ghosh DS, Tiwari S. Hole and electron transport materials: A review on recent progress in organic charge transport materials for efficient, stable, and scalable perovskite solar cells, Chemistry of Inorganic Materials. 2023 Dec 1; 1: 100026.

[88]   ALmehmadi FG, Mkawi EM, Al-Hadeethi Y, Alajlan Y, Bekyarova E. Efficient methylammonium lead triiodide CH3NH3PbI3 perovskite solar cells with improved surface properties using PM6 (PBDB-T-2F) polymer, Results in Physics. 2024 Jan 1; 56: 107270.

[89]   Fujishima A, Honda K. Electrochemical photolysis of water at a semiconductor electrode, Nature. 1972 Jul 7; 238(5358): 37–38.

[90]   Jiang Q, Zhang X, You J. SnO2: A wonderful electron transport layer for perovskite solar cells, Small. 2018 Aug; 14(31): 1801154.

[91]   Nazeeruddin MK, Vasilopoulou M. Perovskite Solar Cells, Glob. Energy. 2023; 29(1): 21–79.

[92]   Shao S, Zheng K, Zidek K, Chabera P, Pullerits T, Zhang F. Optimizing ZnO nanoparticle surface for bulk heterojunction hybrid solar cells, Solar Energy Materials and Solar Cells. 2013 Nov 1; 118: 43–47.

[93]   Zhang P, Wu J, Zhang T, Wang Y, Liu D, Chen H, Ji L, Liu C, Ahmad W, Chen ZD, Li S. Perovskite solar cells with ZnO electron-transporting materials, Advanced Materials. 2018 Jan; 30(3): 1703737.

[94]   Isikgor F, Zhumagali S, T. Merino LV, De Bastiani M, McCulloch I, De Wolf S. Molecular engineering of contact interfaces for high-performance perovskite solar cells, Nature Reviews Materials. 2023 Feb; 8(2): 89–108.

[95]   Zhang L, Mei L, Wang K, Lv Y, Zhang S, Lian Y, Liu X, Ma Z, Xiao G, Liu Q, Zhai S. Advances in the application of perovskite materials, Nano-Micro Letters. 2023 Dec; 15(1): 177.

[96]   Vaghi L, Rizzo F. The future of spirobifluorene-based molecules as hole-transporting materials for solar cells, Solar RRL. 2023 Apr; 7(7): 2201108.

[97]   Shi D, Qin X, Li Y, He Y, Zhong C, Pan J, Dong H, Xu W, Li T, Hu W, Brédas JL. Spiro-OMeTAD single crystals: Remarkably enhanced charge-carrier transport via mesoscale ordering, Science Advances. 2016 Apr 15; 2(4): e1501491.

[98]   Lim I, Kim EK, Patil SA, Ahn DY, Lee W, Shrestha NK, Lee JK, Seok WK, Cho CG, Han SH. Indolocarbazole based small molecules: An efficient hole transporting material for perovskite solar cells, RSC Advances. 2015; 5(68): 55321–55327.

[99] Sharma GK, Joseph SL, James NR. Recent progress in Poly (3, 4-Ethylene Dioxythiophene): Polystyrene sulfonate based composite materials for electromagnetic interference shielding, Advanced Materials Technologies. 2024 Jan; 9(1): 2301203.

[100] Lan A, Li Y, Zhu H, Zhu J, Lu H, Do H, Lv Y, Chen Y, Chen Z, Chen F, Huang W. Improved efficiency and stability of organic solar cells by interface modification using atomic layer deposition of ultrathin aluminum oxide, Energy & Environmental Materials. 2024 May; 7(3): e12620.

[101] Fernandes JM, Joseph DP, Kovendhan M. Metal oxide charge transport layers for halide perovskite light-emitting diodes, In Metal Oxides for Next-Generation Optoelectronic, Photonic, and Photovoltaic Applications. 2024 Jan 1; Elsevier, 301–342.

[102] Krishna BG, Ghosh DS, Tiwari S. Hole and electron transport materials: A review on recent progress in organic charge transport materials for efficient, stable, and scalable perovskite solar cells, Chemistry of Inorganic Materials. 2023 Dec 1; 1: 100026.

[103] Rimal G, Comes RB. Advances in complex oxide quantum materials through new approaches to molecular beam epitaxy, Journal of Physics D: Applied Physics. 2024 Feb 15; 57(19): 193001.

[104] Li J, Pradhan B, Gaur S, Thomas J. Predictions and strategies learned from machine learning to develop high-performing perovskite solar cells, Advanced Energy Materials. 2019 Dec; 9(46): 1901891.

[105] Zuo L, Guo H, deQuilettes DW, Jariwala S, De Marco N, Dong S, DeBlock R, Ginger DS, Dunn B, Wang M, Yang Y. Polymer-modified halide perovskite films for efficient and stable planar heterojunction solar cells, Science Advances. 2017 Aug 23; 3(8): e1700106.

[106] Bi D, Li X, Milić JV, Kubicki DJ, Pellet N, Luo J, LaGrange T, Mettraux P, Emsley L, Zakeeruddin SM, Grätzel M. Multifunctional molecular modulators for perovskite solar cells with over 20% efficiency and high operational stability, Nature Communications. 2018 Oct 26; 9(1): 4482.

[107] He Q, Worku M, Liu H, Lochner E, Robb AJ, Lteif S, Vellore Winfred JR, Hanson K, Schlenoff JB, Kim BJ, Ma B. Highly efficient and stable perovskite solar cells enabled by low-cost industrial organic pigment coating, Angewandte Chemie International Edition. 2021 Feb 1; 60(5): 2485–2492.

[108] Qi Y, Ndaleh D, Meador WE, Delcamp JH, Hill G, Pradhan NR, Dai Q. Interface passivation of inverted perovskite solar cells by dye molecules, ACS Applied Energy Materials. 2021 Sep 3; 4(9): 9525–9533.

[109] Li Y, Wu F, Han M, Li Z, Zhu L, Li ZA. Merocyanine with hole-transporting ability and efficient defect passivation effect for perovskite solar cells, ACS Energy Letters. 2021 Feb 8; 6(3): 869–876.

[110] Khenkin MV, Katz EA, Abate A, Bardizza G, Berry JJ, Brabec C, Brunetti F, Bulović V, Burlingame Q, Di Carlo A, Cheacharoen R. Consensus statement for stability assessment and reporting for perovskite photovoltaics based on ISOS procedures, Nature Energy. 2020 Jan; 5(1): 35–49.

[111] Rajagopal A, Yao K, Jen AK. Toward perovskite solar cell commercialization: A perspective and research roadmap based on interfacial engineering, Advanced Materials. 2018 Aug; 30(32): 1800455.

Prerit Chauhan, Nitika Verma, Sahil Kumar, Neha Kumari,
Gun Anit Kaur, and Mamta Shandilya*

Chapter 4
Fabrication techniques for perovskite solar cells

Abstract: In this chapter, we discuss various fabrication techniques for perovskite solar cells (PSCs), focusing on their impact on device performance and stability. Perovskite materials have rapidly advanced photovoltaics due to their outstanding optical and electronic properties. Since their introduction in 2009, the power conversion efficiency of PSCs has improved significantly. This chapter reviews key fabrication methods, categorized into solution-based (e.g., spin coating and inkjet printing) and vacuum-based (e.g., thermal evaporation, chemical vapor deposition, and atomic layer deposition) techniques, highlighting their influence on device efficiency, scalability, and long-term stability. Solution-based methods are praised for their cost-effectiveness and suitability for flexible, large-area devices, while vacuum-based methods offer superior film uniformity and thickness control. By comparing these approaches, the chapter emphasizes the pivotal role of fabrication in enhancing the commercial viability and durability of PSCs and related optoelectronic devices.

Keywords: Solar cells, perovskite, CVD, ALD, spin coating, inkjet printing

4.1 Introduction

Energy demand is quickly increasing around the world and traditional fossil fuels can no longer meet it due to their negative environmental impact and limited resources. This makes it more critical than ever for a move to clean, renewable energy sources, such as solar [1], wind, biogas [2], etc. Among these, solar energy is the most widely available and consistent source throughout the year, which makes it an excellent alternative for sustainable energy. Solar power can be produced through thermal systems or photovoltaic (PV) converters, with PV being the more popular choice because it transforms sunlight directly into energy in a clean and efficient manner [3]. Bec-

*Corresponding author: Mamta Shandilya, School of Physics and Materials Science, Shoolini University, Bajhol, Solan 173229, Himachal Pradesh, India, e-mails: mamtashandilya@shooliniuniversity.com, mamta2882@gmail.com
Prerit Chauhan, Nitika Verma, Sahil Kumar, Neha Kumari, Gun Anit Kaur, School of Physics and Materials Science, Shoolini University, Bajhol, Solan 173229, Himachal Pradesh

https://doi.org/10.1515/9783111726847-004

querel discovered the photovoltaic effect, which Bell Labs eventually improved into the modern solar cells. It allows sunlight to be converted into power by generating and separating electron–hole pairs. The performance of solar cells is determined by factors such as material type, layer thickness, flaws, and design [4]. Efficiency, affordability, durability and environmental effect are all major barriers to commercialization, although many of these can be overcome with appropriate production techniques [5]. Perovskite is currently the most useful material for overcoming multiple challenges in solar cells. Perovskite, discovered in 1839 by Gustav Rose, is a unique crystal structure, with metal cations coordinated by oxygen atoms, which are used in solar cells, optoelectronics, electronic devices, and superconducting materials. Perovskites, a class of materials with the formula ABX_3 where "A" and "B" are cations and "X" is an anion, are being increasingly used in solar cells due to their exceptional optoelectronic properties. PSCs are also being integrated into various applications, such as tandem solar cells, Building-Integrated Photovoltaics (BIPV), flexible and wearable electronics, and space applications. These diverse applications highlight the versatility and potential of PSCs in various sectors [6]. Perovskite solar cells (PSCs) are transforming the landscape of photovoltaic technology due to their exceptional power conversion efficiencies (PCEs) and promise of cost-effective manufacturing. PSCs are a next-generation challenger in solar energy, combining innovation and practicality. This article examines the fundamentals of perovskite materials, their rapidly expanding range of applications and techniques driving their development highlighting key advantages, cutting-edge fabrication strategies, and real-world uses that are speeding up their transition from lab to market [7]. Research studies suggest that these cells could reach efficiencies of around 20%. The best possible efficiency for PSCs is estimated to be about 30%, which means there is still room for improvement and development in this exciting field of renewable energy. Further research must be performed to understand the exact degradation mechanisms, allowing appropriate stabilization and encapsulating techniques to be set up [8]. Figure 4.1 describes several ways that can be taken to increase the device stability along with overall device performance.

PSCs can be fabricated utilizing a variety of approaches, which are roughly divided into solution-based and vacuum-based methods. One of the most common solution-based approaches is the **spin coating process**, which involves depositing a precursor solution on a substrate and spinning it at high speed to build a uniform thin film. This method is commonly employed in lab-scale research due to its simplicity and cost-effectiveness [9]. Researchers used spin coating to create all-inorganic PSCs, creating perovskite layers and evaluating the impact of zinc oxide nanoparticle size and residual water on device performance. Lowering ZnO nanoparticle size increased short-circuit current density and open-circuit voltage, resulting in higher PCE. However, residual water had a harmful effect, indicating the need for moisture content management during fabrication. The spin coating technique is effective for creating perovskite layers in PSCs but precise control of nanoparticle size and moisture levels can significantly impact device performance. By improving these parameters, spin-coated PSCs can attain higher efficiencies, highlighting the

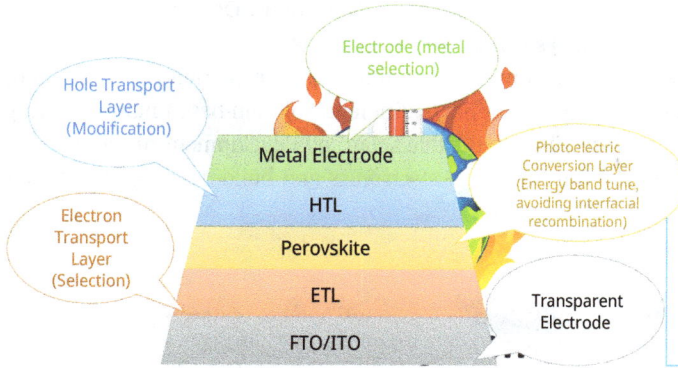

Figure 4.1: Structure of a conventional perovskite solar cell.

technique's potential for scalable and efficient solar cell manufacture [10]. Other techniques like include **doctor blading**; Chang et al. developed a two-step doctor blading procedure for producing high-quality perovskite films in atmospheric conditions. They utilized methylammonium chloride (MACl) as a component to control the crystallization and orientation of the perovskite layer for producing films with larger grain sizes with increased crystallinity and an inclined face-up orientation. The PSCs have PCE of 23.14% for small-area devices, 21.20% for 1.03 cm^2 devices, and 17.54% for 10.93 cm^2 mini-modules. After 70 days of storage at room temperature and 30% relative humidity, the unencapsulated devices retained 80% of their original PCE. The research highlights doctor blading as a cost-effective and scalable production approach for commercializing perovskite solar technology [10], **Slot-die coating**, using a two-step method, improved the crystallinity and morphology of PSCs. A porous PbI$_2$:CsI sheet is coated with an organic cation-containing solution, followed by another layer of the same solution. The method was optimized with nitrogen blowing and substrate temperature control, leading to a PCE of 18.13% for fully slot-die-coated PSCs and 13.00% for roll-to-roll (R2R) printed flexible PSCs. This process improves the morphological and crystalline features of perovskite layers to ensure scalability and stability, making it a promising approach to commercializing perovskite solar technology [11]. In the case of **inkjet printing technique**, Chen et al. identified a method for producing perovskite thin films using inkjet printing. They achieved homogenous perovskite thin films with increased crystallinity and morphology by modifying several factors such as printing voltage, distance, ink droplet size, substrate temperature and annealing temperature, which ultimately improved solar cell performance. This study shows that inkjet printing is a viable process for generating high-quality perovskite thin films for solar cell applications, showing its potential as a scalable and cost-effective option for mass production [12], and further, solution-based processes are better suited for large-area devices and scalable production. Solution-based methods have emerged as a game changer in solar cell manufacturing, especially for perovskite and other next-generation photovoltaic technologies. These technologies have multiple significant benefits, making them ideal for research and industrial applications. One of the key advantages of solution-based methodologies is their cost-effectiveness. Solution-based processes such as spin coating, doctor blading, and inkjet printing can be carried out in ambient environment conditions with relatively simple equipment. Solution-based manufacturing facilitates low-temperature processing for flexible solar cell development on plastic substrates, allowing for integration into wearable electronics and building-integrated photovoltaics applications.

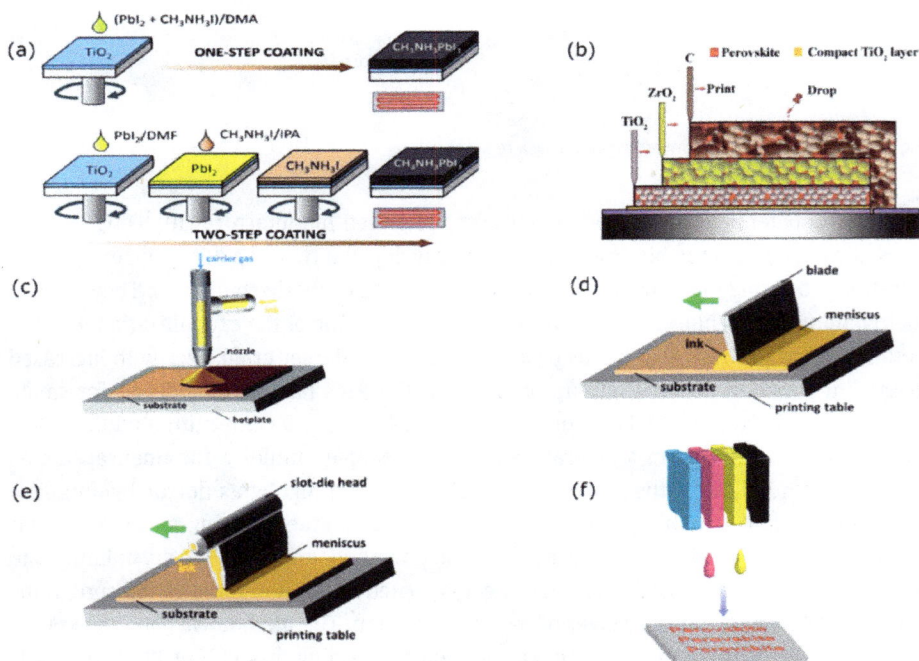

Figure 4.2: Different types of solution-based techniques like (a) spin coating, (b) drop casting, (c) spray coating, (d) doctor blade, (e) slot-die coating, and (f) inkjet printing [1].

Vacuum-based methods, such as thermal evaporation and dual-source evaporation [13], provide more control over film thickness and composition, making them ideal for producing high-quality, pinhole-free perovskite layers. In addition, chemical vapor deposition (CVD) [14], physical vapor deposition (PVD) [14], and atomic layer deposition [15] have been examined to manufacture high-purity films with exceptional uniformity and reproducibility. The fabrication technique provided significant effects on the shape, crystallinity, and overall efficiency of PSCs. Vacuum deposition processes such as PVD and CVD allow for the production of thin films with high uniformity, purity, and controlled thickness. This precision contributes to better devices performance. For example vacuum-deposited perovskite layers have higher crystallinity and fewer flaws than those produced through solution methods, resulting in higher PCEs and operational stability [16]. Sputtering and thermal evaporation techniques are commonly used in industrial settings because they are scalable and precise. Vacuum-based deposition processes have helped CIGS (copper indium gallium selenide) solar cells achieve high efficiency, making them acceptable for large-scale manufacturing [17]. Figure 4.3 showcases various techniques for fabricating PSCs.

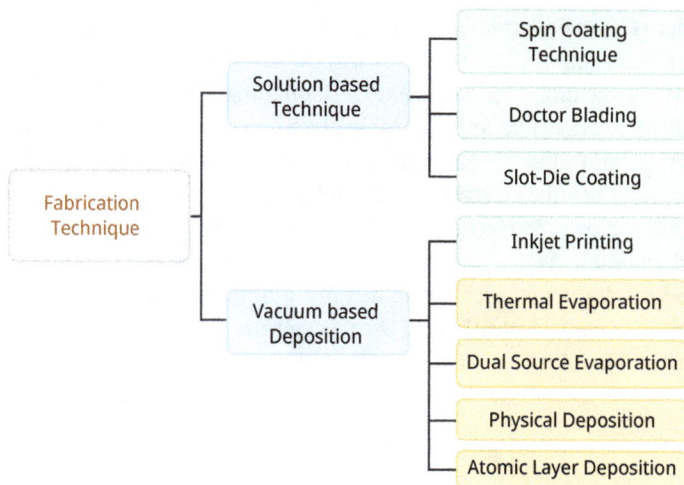

Figure 4.3: Various fabrication methods of perovskite solar cells (PSCs).

To fill the gap between solution and vacuum techniques, hybrid methods such as vapor-assisted solution processing (VASP) have been developed. Hybrid and emerging fabrication methods are revolutionizing PSC production by combining solution-based and vacuum-based techniques. One notable hybrid method involves a two step process, first one is a solution-based deposit and the second is vapor-assisted annealing, producing uniform and high-quality films, with PCEs up to 18.7%. This method uses green solvents, and operates under ambient conditions, addressing environmental and scalability concerns [18]. Another is R2R fabrication process, which combines solution processing and thermal evaporation; this has successfully produced flexible perovskite solar modules, with PCEs reaching 15.5% for small-area cells and 11.0% for larger modules [19]. Different types of manufacturing processes for PSCs, as discussed in this article, have a significant impact on their appearance, functionality, and performance. From simple processes [20] to more complex vacuum/ vapor procedures, each approach offers unique benefits in terms of film quality, scalability, and efficiency. As research advances, these techniques hold key to making PSCs not only more powerful but also ready for scaling production in a cost-effective and reliable manner [21].

4.2 Overview of perovskite solar cells

This article provides an in-depth overview of PSCs, highlighting their working principles, material composition, and structural design. It underlines the rapid improvements in efficiency and the benefits over outdated solar systems. The article explains why PSCs are leading the next-generation photovoltaic research. As we know, PSCs provide a wide range of applications, including tandem solar cells, building-integrated photovoltaics (BIPV), space applications, portable and wearable electronics, energy storage integration and photodetectors, and LEDs [6, 22]. They are combined with silicon or CIGS for increased efficiency; they are lightweight and flexible, making them excellent for apertures, exteriors, and urban infrastructure. PSCs are also being tested to power spacecraft, portable and wearable devices, and self-charging energy systems [22]. All applications are influenced by the structure; thus, to acquire a fundamental understanding of the structure and material compositions of a solar cell, see the next section that describes all of its features in brief.

4.2.1 Structure

PSCs are a type of photovoltaic device that uses perovskite-structured materials as its light-absorbing layer. Their unique properties and relatively inexpensive production procedures have caused considerable interest in the field of renewable energy. As we can see in Figure 4.4, the PSCs are made of multiple layer.

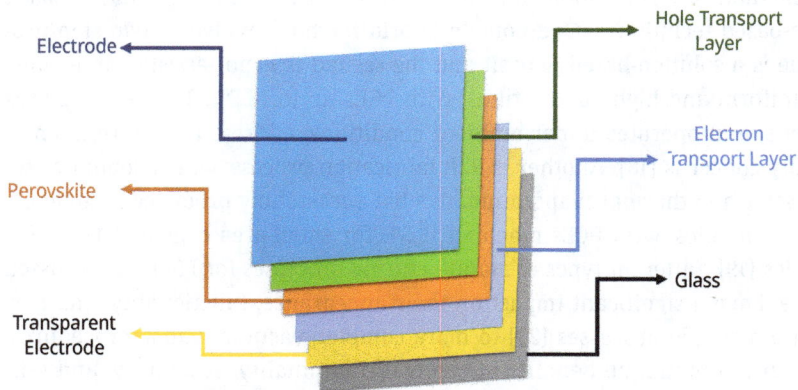

Figure 4.4: Perovskite solar cells are composed of several layers.

PSC consists of a transparent conductive substrate (TDS), electron transport layer (ETL), perovskite absorber layer (PAB), hole transport layer (HTL), and metal electrode. The TDS such as FTO and ITO serves as the front electrode, allowing light to enter the cell [23]. The ETL includes titanium dioxide (TiO_2)/tin oxide (SnO_2), which facilitates the extraction and transport of electrons from the perovskite layer to the electrode. The PSC's layer absorbs absorb sunlight and generates charge carriers. Typical perovskite materials have the formula ABX_3, where "A" is a cation like methylammonium (MA^+) or formamidinium (FA^+), "B" is a metal cation like lead (Pb^{2+}), and "X" is a halide anion such as iodide (I^-). The HTL transports holes from the perovskite layer to the back electrode, with materials like Spiro-OMeTAD or poly(triarylamine) (PTAA) [24]. The metal electrode, typically made of gold or silver, collects the holes and completes the electrical circuit. PSC architectures are often divided into two configurations, depending on the arrangement of transport materials. In the n-i-p structure, the ETL is deposited first, followed by the perovskite layer and the HTL [25]. This is the usual structure. In the p-i-n structure, the HTL is deposited first, then the perovskite layer and finally the ETL. This inverted form has advantages in several applications.

4.2.2 Working principle of the perovskite solar cell

PSCs depend on the photovoltaic effect, which converts light energy into electrical energy through a sequence of photo-induced reactions. Light absorption process occurs when sunlight enters the solar cell through the transparent electrode. The perovskite layer is usually a substance, with the formula ABX_3, such as methylammonium lead iodide absorbs photons. This produces electron–hole pairs, sometimes known as excitons, within the absorber layer. Excitons in perovskite materials can easily separate into electrons and holes due to their advantageous dielectric properties, requiring minimal energy. This is known as exciton dissociation process. Next process is charging transportation, when electrons migrate toward the ETL (e.g., TiO_2 or SnO_2) and are collected at the front electrode, which are typically transparent conductive oxides like FTO or ITO. Holes proceed in the opposite direction, toward the hole transport layer (e.g., Spiro-OMeTAD or PTAA), and gather at the back electrode, often made of Au or Ag. The final two processes are charge collection and power generation. The divided and transported charges are gathered by the various electrodes, while an external circuit allows electrons to pass, resulting in an electric current. Finally, the flow of electrons across the external circuit powers electronic devices completes the energy conversion process [26, 27].

4.2.3 Material composition

Table 4.1: Different types of materials and their properties that are commonly used in the solar cells.

Component	Materials used	Properties	References
Light absorbers	– Methylammonium lead halide – Formamidinium lead halide – Tin-based perovskites – Double perovskites	Absorb sunlight and generate excitons. Determine bandgap and light absorption range.	[28]
Exciton dissociation	– TiO_2 – SnO_2 – ZnO – Phosphorene–TiO_2 heterostructures – Single-material organic systems	Separate excitons into free charge carriers. Facilitate charge transfer.	[29, 30]
Charge transportation/ charge collection	– Spiro-OMeTAD – Poly(3-hexylthiophene) – Nickel oxide – Barium stannate – Tin oxide – Copper(I) thiocyanate – [6]-Phenyl-C_{61}-butyric acid methyl ester – Al_2O_3 – Poly(3,4-ethylenedioxythiophene): poly(styrenesulfonate)	Transport electrons and holes to the respective electrodes. Reduce recombination losses. Collect and transport charge carriers to external circuit. Maintain transparency and conductivity.	[31]

From 2015 to 2025, great progress was achieved in improving the material components of PSCs, as shown in Table 4.1, resulting in significant increases in PCE, stability, and scalability. Lead-based perovskites, such as methylammonium and formamidinium halides, have remained the most common light absorbers because of their superior optoelectronic features, although alternative materials such as tin-based and double perovskites have been explored to address toxicity issues. Metal oxides like TiO_2, ZnO, and SnO_2 are effective for exciton dissociation, while developments like phosphorene-TiO_2 heterostructures improve charge separation efficiency even more. Transmission of charges has benefited from advances in both classic metal oxide layers and novel carbon-based materials such as graphene, which provide greater mobility and better interface contact. PSCs' power generating capability has been steadily improved by modifying absorber compositions and optimizing energy band alignment across device layers. Finally, charge collection has been improved through the use of transparent conducting oxides and conductive materials such as Ag, Au, and graphene-based electrodes, which allowing for effective current extraction, while maintaining optical transparency [32]. These advancements indicate a complete materials-driven approach that is pushing PSCs closer to commercial viability.

4.3 Fabrication techniques

4.3.1 Solution-based techniques

Perovskite material development, specifically for optoelectronic and solar applications, is based on solution-based manufacturing techniques. In these techniques, perovskite precursors, including metal halides and organic/inorganic salts, are dissolved in polar solvents such as isopropanol, dimethylformamide (DMF), dimethyl sulfoxide (DMSO), or γ-butyrolactone (GBL) [33]. The resulting solutions are subsequently applied to substrates by means of electrodeposition, spin coating, blade coating, slot-die coating, or inkjet printing [34]. Low processing temperatures, ease of formulation, scalability for R2R operations, and compatibility with flexible substrates are the main features that make solution-based techniques popular [35]. These benefits have made it possible to produce solar cells at the laboratory scale, with PCEs higher than 25%, and to explore commercialization avenues using scalable printing and coating methods [36, 37]. Techniques for processing solutions can be broadly divided into two categories: single-step and multistep [38]. In the single-step process, all of the perovskite precursors are combined in a solvent system, and then direct deposition is carried out, frequently with the help of anti-solvent drips to promote crystallization [39]. Multistep techniques, on the other hand, provide for more control over morphology, stoichiometry, and film homogeneity by depositing distinct precursors in a sequential manner (e.g., PbI_2 followed by MAI or FAI) [37]. Some of the solution-based fabrication techniques are discussed as follows.

4.3.1.1 One-step spin coating

Spin coating is one of the most recognized solution-based technique for creating higher quality perovskite thin films, particularly in research laboratories [40]. This technique involves depositing a precursor solution onto a substrate, which is then quickly rotated to distribute the solution uniformly using centrifugal force. By varying the spin speed, time, solution viscosity, and concentration, the film thickness and homogeneity may be precisely controlled. Spin coating's simplicity, affordability, and capacity to create highly homogeneous films on small-area substrates are some of its key advantages. In order to create compact, pinhole-free perovskite layers that are essential for producing high-performance solar cells, this method is especially helpful [41]. Additionally, it makes anti-solvent engineering easier, which involves dropping a non-solvent onto a spinning substrate to speed up crystallization and enhance film morphology. With its superior control over film characteristics and device optimization, it remains a leading approach for rapid prototyping and scholarly research. There are two popular spin-coating protocols: one-step and two-step. All of the precursor salts are dissolved in a solvent and spin-coated straight away in the one-step procedure. By first spin-coating a metal halide, such PbI_2, and then spin-coating an organic halide, like MAI or FAI, the two-step method enables controlled crystallization and phase formation [42]. In order to improve film quality, reproducibility, and environmental friendliness, recent research have shown developments in spin coating, vacuum-assisted spin coating, and the incorporation of green solvents and additives in ambient conditions [43]. In 2018, Takeo Oku et al. enhanced $CH_3NH_3PbI_3$ PSCs by adding polysilanes. XRD (Figure 4.5a) showed improved crystallinity for the +DPPS film with a strong peak at 14.1° (110). Optical microscopy (Figure 4.5b) revealed uniform, dense grains for DPPS-modified films. J–V curves (Figure 4.5c) showed the +DPPS device had the highest current density (17.8 mA/cm^2) and voltage (~1.0 V), outperforming standard (12.4 mA/cm^2), PMPS (9.2 mA/cm^2), and PPS (7.5 mA/cm^2). DPPS also reduced series resistance and improved overall efficiency to 10.46%. Thus, DPPS was the most effective additive for performance enhancement. The improved morphology and charge transport due to DPPS led to higher shunt resistance and reduced recombination. These results confirm DPPS as a promising additive for high-performance PSCs [44].

4.3.1.2 Two-step sequential deposition

The sequential or bilayer method, which is another name for two-step sequential deposition, is a popular solution-based approach for creating high-quality perovskite films. Through the sequential deposition of precursor materials, this technique first spin coats or doctor blades a metal halide (usually PbI_2) onto the substrate [45]. Next, an organic halide (like methylammonium iodide [MAI] or formamidinium iodide [FAI]) is deposited, which reacts with the first layer to form the perovskite structure. Compared to single-step techniques, this method offers more control over crystallization, grain size, stoichiometry, and film shape [46]. Better infiltration and uniform conversion are made possible by the intermediate phase created during the reaction, resulting in dense, pinhole-free perovskite films with enhanced optoelectronic properties using ion-based and assisted techniques [47]. This method has multiple versions; for instance, solution dripping, spin-coating, dipping, or even vapor-phase exposure can be used to carry out the second stage (organic halide deposition) [48]. Variables like temperature, environmental conditions, and solvent polarity can all be used to adjust the reaction kinetics. Usually, post-annealing is employed to encourage uniformity in the film and complete crystallization. With inorganic perovskites such as $CsPbI_3$, where compositional stability is essential, the two-step process works very well [49]. Additionally, it makes it possible to add additives or grain boundary passi-

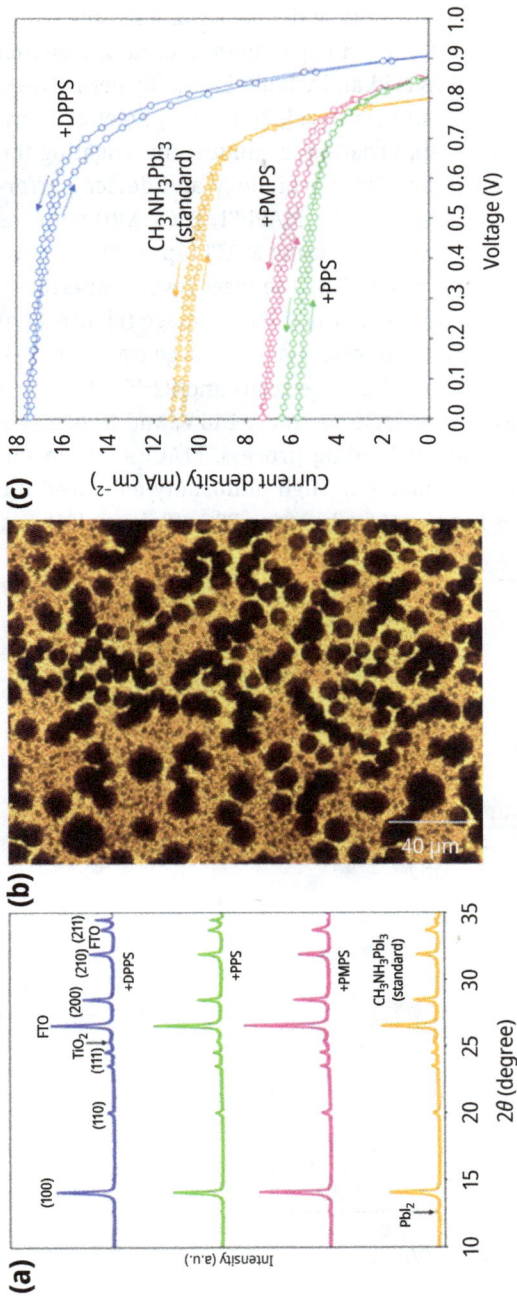

Figure 4.5: (a)–(c) XRD, ultrasonic microscopy, and *I–V* curves of CH$_3$NH$_3$PbI$_3$ perovskite-based material via spin-coating method for PSC application [44].

vators at each stage, which enhances the functionality of the device and its stability in the environment. The two-step process is still the principal method used in research and pre-commercialization studies for both hybrid and wholly inorganic perovskites, since it has been successful in reaching PCEs of over 24% [50]. In 2015, Ko et al. fabricated $CH_3NH_3PbI_3$ (MAPbI$_3$) films using the spin-coating technique, investigating the effect of substrate preheating temperature on film morphology and device performance. The figure shows the effect of MACl additive on MAPbCl$_3$ PSCs. XRD patterns (Figure 4.6a) confirm enhanced crystallinity with increasing MACl (up to 30%), showing a strong α-phase peak at ~14.1° and suppressed PbI$_2$. The inset reveals intermediate phases at low 2θ values. Scanning electron microscopy (SEM) image (Figure 4.6b) exhibits dense, uniform grains with minimal voids, essential for charge transport. $J-V$ curves (Figure 4.6c) show a high PCE of 23.14% (reverse scan) and 22.45% (forward scan). The corresponding J_{sc}, V_{oc}, and FF were 23.35 mA cm^{-2}, 1.18 V, and 84.06%, respectively. The device used a two-step doctor blading process. MACl addition improved film morphology and reduced recombination. High uniformity enhanced the device's charge transport. Thus, MACl is beneficial for boosting PSC efficiency [51].

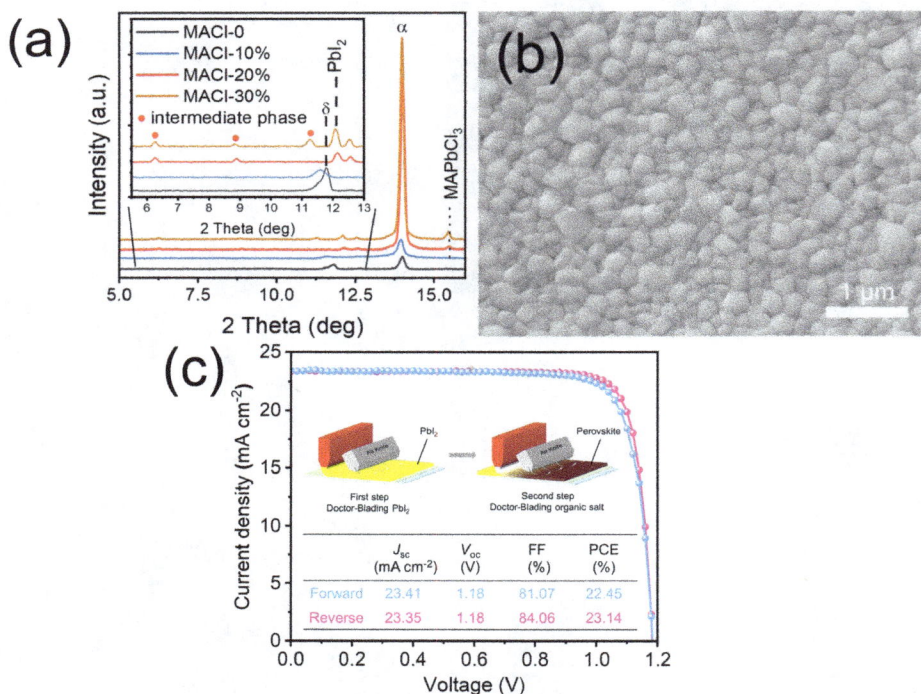

Figure 4.6: (a)–(c) XRD, SEM, and $I-V$ curve of MACl perovskite-based material via two-step sequential deposition method for PSC application [51].

4.3.1.3 Doctor blading, slot-die, and inkjet printing

Doctor blading, also known as blade coating, is a mechanical deposition method in which a perovskite precursor solution is applied to a substrate by means of a blade. The ink viscosity, coating speed, and blade height are all adjusted to control the wet film's thickness [52]. With the help of this scalable technique, high-quality films can be deposited on large surfaces without the need for complicated equipment. Green solvent-based inks, ambient condition processing, and additive-assisted formulations are examples of recent developments that have improved film crystallinity, coverage, and photovoltaic performance [53]. PCEs have been demonstrated by films made by doctor blading, and their long-term stability under heat and light stress is encouraging [54]. In 2023, Chang et al. fabricated MACl perovskite-based film using doctor blading technique for PSCs applications. PbI$_2$ and organic salt solutions were used in a two-step sequential doctor blading technique to create perovskite films. After being applied to ITO/SnO$_2$ substrates, the PbI$_2$ layer was treated with 4-tert-butylpyridine (TBP) to create a porous structure that would allow organic salts to diffuse more easily. After annealing at 150 °C, SEM pictures demonstrated that increasing the MACl concentration (0–30%) affected the final perovskite film shape and grain size, while also confirming consistent pore distribution in the PbI$_2$ layer. The perovskite films' crystallinity was confirmed by XRD analysis, carried out with a Rigaku Smart Lab SE equipment, and additional structural information was obtained using GIWAXS measurements. V_{OC} = 1.18 V, J_{SC} = 23.35 mA cm^{-2}, and FF = 84.08% were all reached by the optimized device, which was produced with 20% MACl and had a champion PCE of 23.14% [55].

Slot-die coating is a continuous, pre-metered deposition technique, in which the precursor ink is sprayed onto a moving substrate via a slot die, a tiny slit. One of the most R2R-compatible methods for processing perovskites, it offers outstanding control over film thickness and homogeneity [56]. Slot-die coating is ideal for tandem device configurations and flexible substrates, and it drastically lowers material waste, in contrast to spin coating [57]. The drying dynamics, substrate temperature, and ink texture can all be optimized to avoid defects like striations and incomplete conversion [58]. In 2023, Li et al. fabricated FAPbI$_3$ perovskite-based film using slot die technique for PSCs applications. When 46 vol% acetonitrile (ACN) was added to the precursor ink, PSCs made by slot-die coating FAPbI$_3$ thin films showed remarkable morphological and optoelectronic quality. In contrast to the porosity and rough morphologies seen with unoptimized inks, SEM demonstrated that this formulation produced uniform, pinhole-free films with big grains. Good crystallinity and consistent phase distribution across films were confirmed by X-ray diffraction (XRD) and GIWAXS, whereas a smaller peak width suggested an enhanced microstructure. The scalability and performance of the slot-die technique were validated by the optimized devices' certified PCE of 22.3% and the 17.1% PCE of mini-modules covering 12.7 cm^2. A digital, mask-free technique for depositing perovskite layers, with excellent material efficiency and spatial resolution, is inkjet printing. Because it enables the direct patterning of complex geometries, it is ideal for wearable photovoltaics, mini-modules, and multi-junction devices [59]. Film thickness and coverage can be precisely controlled with drop-on demand printing by carefully regulating droplet volume and spacing [60]. Recent fully inkjet-printed PSC demonstrations have opened the door for low-cost, decentralized manufacturing due to their excellent mechanical flexibility, operational stability, and good PCEs [61]. In 2024, Tara et al. fabricated $FASn_{1-x}Pb_xI_3$ ($x = 0.25, 0.5, 0.75$) perovskite films using inkjet printing for PSC applications. XRD patterns (Figure 4.7a) confirmed the perovskite phase, with prominent peaks at 14.5° and 28.7°, and a slight blue shift near 28° indicated lattice expansion with increasing Pb content. SEM analysis (Figure 4.7b) showed that the $FASn_{0.5}Pb_{0.5}I_3$ film had the most uniform morphology and strong crystallinity. J–V curves (Figure 4.7c) revealed that the $FASn_{0.5}Pb_{0.5}I_3$-based device achieved the best performance, with a PCE of 10.26%, V_{OC} of 0.74 V, J_{SC} of 25.66 mA/cm^2, and FF of 54.06% under AM 1.5 G illumination [55].

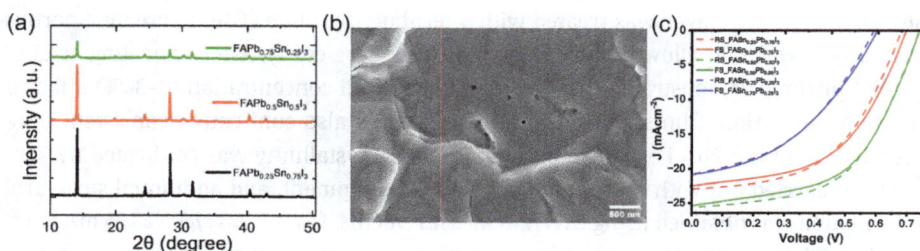

Figure 4.7: (a)–(c) XRD, SEM, and J–V curves of FAPbSnI perovskite-based material via inject printing method for PSC application [55].

4.3.2 Vapor-phase deposition techniques

For the production of perovskite thin films, vapor-phase deposition techniques have become strong substitutes for solution-based techniques [62]. The performance and long-term stability of PSCs are largely dependent on the film's uniformity, thickness, composition, and crystallinity, all of which can be better controlled with these methods [63]. In contrast to solution procedures, which use liquid precursors and frequently have problems like solvent residue, uneven wetting, or restricted scalability, vapor-phase techniques use sublimed or gaseous precursors that react or condense on the substrate surface under carefully monitored conditions. Better reproducibility and precise thickness control are also made possible by these methods, which are crucial for commercial dependability and device integration [64]. Furthermore, a lot of vapor-based techniques work in vacuum or inert environments, which greatly reduces the deterioration caused by oxygen and moisture during film development. Because of this, vapor-deposited perovskite layers frequently show better thermal and environmental stability than those made using solution-based techniques [45]. The three most common vapor-phase methods for perovskite film formation for PSCs are thermal co-evaporation, VASP, and CVD, which will be discussed in the subsequent sections. Each method has special benefits with regard to material quality, process scalability, and device compatibility.

4.3.2.1 Thermal co-evaporation

A PVD process called thermal co-evaporation sublimates many precursor materials at once in a high-vacuum environment [65]. The vapors of these precursors, which are usually inorganic metal halides (like lead iodide or cesium iodide) and organic halides (like methylammonium iodide or formamidinium iodide), co-deposit onto a substrate where they react in situ to form a perovskite layer [66]. This technique is especially useful for creating multi-cation, mixed-halide, and bandgap-tuned perovskites because it provides remarkable control over film thickness, composition, and homogeneity [67]. Thermal co-evaporation, in contrast to solution-based techniques, removes problems such as solvent residues and uneven film coverage, making it possible to fabricate superior, pinhole-free films over wide regions [68]. Scalability and reproducibility are two of thermal co-evaporation's main advantages. It is especially suitable for industrial-scale module production and tandem solar cells that need exact thickness and composition control across several layers because of its vacuum operation and highly regulated deposition rates [69]. Additionally, the method works well with low-temperature substrates, which makes it appealing for flexible solar cells. Recent research has shown that perovskite films made using thermal co-evaporation can achieve PCEs of over 20%, in addition to significant enhancements in film crystallinity, stability in the presence of light and humidity, and decreased defect density [70]. Low-temperature evaporation, in situ monitoring, and hybrid co-evaporation with blade or slot-die coating for increased throughput and tunability are further advancements in the technique. In 2023, Piot et al. fabricated MAPbI$_3$ (MAPI) perovskite films using the thermal co-evaporation technique for PSC applications. XRD patterns (Figure 4.8a) revealed that films deposited at lower rates (MAPI ×1) showed strong diffraction peaks at $2\theta = 14.2°$ and 28.5°, corresponding to the (001) and (002) planes of cubic MAPI. At higher deposition rates (MAPI x4), these peaks diminished, and a preferred orientation shift toward the (111) plane at $2\theta = 24.5°$ was observed. Cross-sectional SEM imaging (Figure 4.8b) confirmed a dense and compact film morphology. The best-performing device demonstrated efficient photovoltaic characteristics, as shown in the J–V curve (Figure 4.8c), with a PCE of 19.4%, V_{OC} of 1.12 V, J_{SC} of 22.1 mA/cm^2, and FF of 78.5% [71].

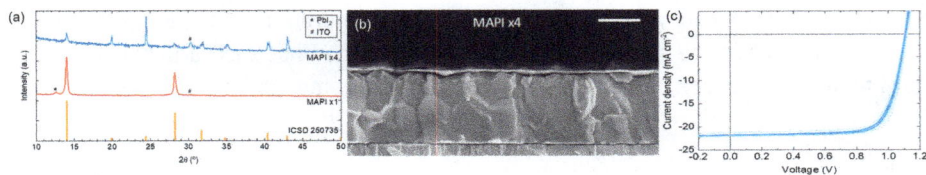

Figure 4.8: (a)–(c) XRD, SEM, and *I–V* curves of MAPbI₃ perovskite-based material via thermal co-precipitation method for PSC application [71].

4.3.2.2 Vapor-assisted solution processing (VASP)

VASP is a hybrid deposition technique that enhances perovskite films by combining the benefits of vapor-phase and solution-based method [45]. The process begins with the deposition of an initial inorganic precursor film using a solution-based technique, such as spin coating or blade coating. Under carefully controlled conditions, this precursor film is then exposed to organic halide vapor, such as formamidinium iodide (FAI) or methylammonium iodide (MAI), initiating a solid-state reaction that forms the perovskite layer [72]. Compared to conventional solution techniques, VASP has a number of advantages, such as better control over film shape, grain size, and stoichiometry, which are made possible by the separation of precursor deposition and crystallization. These properties result in increased consistency, less pinholes, and better coverage for all essential components of scalable and effective PSCs. This technique reduces solvent-related problems, including residue and fast solvent evaporation, which frequently result in uneven nucleation in films that have undergone solution preparation [73]. VASP's adaptability to ambient processing conditions is a key characteristic that makes it appropriate for both industrial and laboratory environments. VASP does not require vacuum systems, in contrast to thermal co-evaporation, and basic heating settings can be used to control the organic vapor pressure [3]. VASP is appealing for processing multilayered device designs, such as tandem and flexible perovskite modules for PSCs, because of its versatility. VASP is an effective method for future large-area, low-cost perovskite photovoltaic production, as evidenced by recent findings showing devices fabricated using this process to have improved environmental and thermal stability, and PCEs reaching 21% [74]. In 2016, Yokoyama fabricated MASnI₃ perovskite films using the vapor-assisted solution process (VASP) for PSC applications. XRD analysis (Figure 4.9a) confirmed the formation of a pseudocubic MASnI₃ phase across all samples, with strong diffraction peaks at $2\theta = 14.3°$ and $28.0°$, corresponding to the (100) and (200) planes, respectively. The LT-VASP films exhibited enhanced preferred orientation along the [001] direction, indicating improved structural alignment. SEM imaging (Figure 4.9b) revealed that one-step processed films left large areas of the TiO₂ substrate exposed, while VASP and LT-VASP films showed more continuous coverage, though still affected by sporadic large crystal formation.

The device incorporating VASP-processed films achieved stable performance without short-circuiting, and a maximum PCE of 1.86%. The IPCE spectrum (Figure 4.9c) showed a broad photo-response across the visible range, peaking near 600 nm [75].

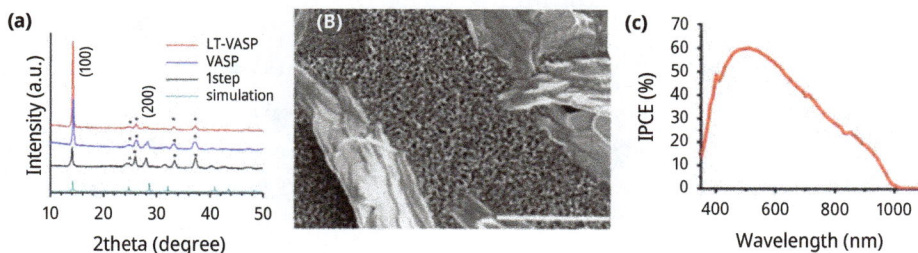

Figure 4.9: (a)–(c) XRD, SEM, and PCE curves of perovskite-based material via VASP method for PSC application [75].

4.3.2.3 Chemical vapor deposition (CVD)

CVD is a gas-phase thin-film deposition technique that creates a dense, homogeneous, and crystalline layer of perovskite by reacting or breaking down volatile precursor molecules on a heated substrate surface [76]. Organic and inorganic halide precursors are usually delivered sequentially or simultaneously during CVD in the context of perovskite production. These precursors react on the substrate surface to generate a perovskite phase [77]. The main benefits of CVD include compatibility with multi-junction and tandem device architectures, particularly on textured or structured substrates, excellent thickness control and homogeneity of film across wide areas, and elimination of residual solvents, which improves film purity and stability [78]. Additionally, precise interface engineering is made possible by CVD, which is essential for improving charge extraction and reducing recombination losses [79]. For perovskite/silicon tandem systems, where interface integrity and defect control are crucial, this makes it particularly suitable. Recent research has effectively shown that CVD-grown perovskite films have exceptional mechanical and environmental stability, in addition to PCEs of 18.9%. With the aid of innovative methods like low-temperature APCVD, multistep vapor chemistry, and in situ monitoring, CVD is becoming an attractive option for the mass production of high-performance perovskite devices [80]. In 2019, Sanders et al. fabricated perovskite films based on BiI_3 and MAI/BiI_3 using a CVD technique for PSC applications. XRD patterns (Figure 4.10a) showed peaks at 12.6°, 25.5°, and 38.8 °for pure BiI_3, consistent with its known structure, while MBI (methylammonium bismuth iodide) films matched calculated patterns, confirming successful formation. SEM analysis (Figure 4.10b) revealed a layered structure of Spiro-MeOTAD, MBI, and mesoporous TiO_2 (mp-TiO_2), with grain sizes of ~ 100 nm at lower MAI/BiI_3 ratios and larger, rougher grains (100–200 nm) at higher ratios due to partial vertical growth. Photovoltaic performance (Figure 4.10c) of the BiI$_3$-only device was poor, with a PCE of 0.02%, J_{SC} of 0.13 mA/cm^2, and V_{OC} of 0.39 V. In contrast, MBI-based devices achieved a PCE of 0.17%, V_{OC} over 0.7 V, and J_{SC} above 0.5 mA/cm^2, demonstrating the potential of MBI as a lead-free absorber for PSCs [32].

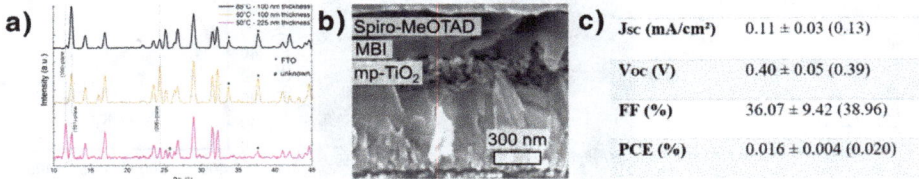

Figure 4.10: (a)–(c) XRD, SEM, and PCE values of mp-TiO$_2$ perovskite-based material via CVD method for PSC application [32].

4.4 Challenges and outlook

The future of PSC fabrication lies in developing methods that enable high efficiency, long-term stability, and commercial scalability. As PSCs move toward real-world deployment, there is a strong need for fabrication techniques that support large-area, uniform film deposition with minimal defects, particularly through scalable approaches like R2R printing, hybrid vapor-solution processes, and ambient-condition manufacturing. One of the major challenges highlighted in this chapter is the sensitivity of solution-based techniques, such as spin coating and doctor blading, to environmental factors like humidity and solvent residues, which affect film quality, crystallinity, and device performance. In the future, this issue can be mitigated by optimizing precursor formulations, introducing green solvents, and applying advanced environmental control systems or encapsulation strategies during fabrication. Additionally, lead-free and eco-friendly perovskite compositions, such as tin-based or double perovskites, should be prioritized to address toxicity concerns. Fabrication methods must also adapt to support emerging applications like tandem solar cells, flexible electronics, and BIPV. Integration of real-time monitoring and AI-driven optimization will further enhance reproducibility, defect control, and scalability, paving the way for cost-effective, sustainable, and industrially viable PSC production.

4.5 Conclusion

This chapter presents a detailed overview of the diverse fabrication techniques employed for the development of high-efficiency PSCs, focusing on their impact on performance, scalability, and structural integrity. Both solution-based methods such as spin coating, doctor blading, slot-die coating, and inkjet printing, and vacuum-based techniques, such as thermal evaporation, VASP, and CVD have been discussed with emphasis on their advantages and limitations. As we highlighted, the fabrication technique performed provides significant effects on the shape, crystallinity, and overall efficiency of PSCs. Solution-based processes are lauded for their low cost and flexibility, while vacuum-based techniques offer superior film quality and thickness control. Emerging hybrid methods aim to bridge the advantages of both. Continuous innovation in fabrication strategies is essential to address current challenges such as stability, reproducibility, and environmental safety, thereby accelerating the commercialization and broader application of PSCs in the global renewable energy landscape.

References

[1] Nema P, Nema R, Rangnekar S. A current and future state of art development of hybrid energy system using wind and PV-solar: A review, Renewable and Sustainable Energy Reviews. 2009; 13(8): 2096–2103.

[2] Panwar NL, Kaushik SC, Kothari S. Role of renewable energy sources in environmental protection: A review, Renewable and Sustainable Energy Reviews. 2011; 15(3): 1513–1524.

[3] Kajal P, Ghosh K, Powar S. Manufacturing techniques of perovskite solar cells, Applications of Solar Energy. 2018; 341–364.

[4] Furkan D, Emin MM. Critical factors that affecting efficiency of solar cells. Smart Grid And Renewable Energy. 2010; 2010; 1–4.

[5] Hasan K, et al. Effects of different environmental and operational factors on the PV performance: A comprehensive review, Energy Science & Engineering. 2022; 10(2): 656–675.

[6] Bati AS, et al. Next-generation applications for integrated perovskite solar cells, Communications Materials. 2023; 4(1): 2.

[7] Zhang C, Park N-G. Materials and methods for cost-effective fabrication of perovskite photovoltaic devices, Communications Materials. 2024; 5(1): 194.

[8] Yadav C, Kumar S. Review on perovskite solar cells via vacuum and non-vacuum solution based methods, Results in Surfaces and Interfaces. 2024; (14): 100210.

[9] Shibayama N, et al. All-inorganic inverse perovskite solar cells using zinc oxide nanocolloids on spin coated perovskite layer, Nano Convergence. 2017; 4: 1–5.

[10] Ko H-S, Lee J-W, Park N-G. 15.76% efficiency perovskite solar cells prepared under high relative humidity: Importance of PbI2 morphology in two-step deposition of CH3NH3PbI3, Journal of Materials Chemistry A. 2015; 3(16): 8808–8815.

[11] Wang L, et al. Study on process parameters of ink-jet printing perovskite solar cell film, In: Interdisciplinary Research for Printing and Packaging. 2022; Springer, 200–206.

[12] Chen L-C, et al. Fabrication and properties of high-efficiency perovskite/PCBM organic solar cells, Nanoscale Research Letters. 2015; 10: 1–5.

[13] Ávila J, et al. Vapor-deposited perovskites: The route to high-performance solar cell production?, Joule. 2017; 1(3): 431–442.

[14] Zardetto V, et al. Atomic layer deposition for perovskite solar cells: Research status, opportunities and challenges, Sustainable Energy & Fuels. 2017; 1(1): 30–55.

[15] Liu J, et al. Evolutionary manufacturing approaches for advancing flexible perovskite solar cells, Joule. 2024; 8(4): 944–969.

[16] Sessolo M, et al. Photovoltaic devices employing vacuum-deposited perovskite layers, Mrs Bulletin. 2015; 40(8): 660–666.

[17] Machkih K, Oubaki R, Makha M. A review of CIGS thin film semiconductor deposition via sputtering and thermal evaporation for solar cell applications, Coatings. 2024; 14(9): 1088.

[18] Siegrist S, et al. Triple-cation perovskite solar cells fabricated by a hybrid PVD/blade coating process using green solvents, Journal of Materials Chemistry A. 2021; 9(47): 26680–26687.

[19] Weerasinghe HC, et al. The first demonstration of entirely roll-to-roll fabricated perovskite solar cell modules under ambient room conditions, Nature Communications. 2024; 15(1): 1656.

[20] Faheem MB, et al. Insights from scalable fabrication to operational stability and industrial Opportunities for perovskite solar cells and modules, Cell Reports Physical Science. 2022; (3): 3(4).

[21] Bruening K, et al. Scalable fabrication of perovskite solar cells to meet climate targets, Joule. 2018; 2(11): 2464–2476.

[22] Zhang L, et al. Advances in the application of perovskite materials, Nano-Micro Letters. 2023; 15(1): 177.

[23] Grace T, et al. Use of carbon nanotubes in third-generation solar cells, In: Industrial Applications of Carbon Nanotubes. 2017; Elsevier, 201–249.

[24] Lekesi LP, et al. Developments on perovskite solar cells (PSCs): A critical review, Applied Sciences. 2022; 12(2): 672.

[25] Roy P, et al. Perovskite solar cells: A review of the recent advances, Coatings. 2022; 12(8): 1089.

[26] Chuang C-HM, et al. Improved performance and stability in quantum dot solar cells through band alignment engineering, Nature Materials. 2014; 13(8): 796–801.

[27] Hussain I, et al. Functional materials, device architecture, and flexibility of perovskite solar cell, Emergent Materials. 2018; 1: 133–154.

[28] Ahmadian-Yazdi M-R, Eslamian M. Fabrication of semiconducting methylammonium lead halide perovskite particles by spray technology, Nanoscale Research Letters. 2018; 13(1): 6.

[29] Cho YJ, et al. Charge transporting materials grown by atomic layer deposition in perovskite solar cells, Energies. 2021; 14(4): 1156.

[30] Wang Y, et al. Metal oxide charge transport layers in perovskite solar cells—Optimising low temperature processing and improving the interfaces towards low temperature processed, efficient and stable devices, Journal of Physics: Energy. 2020; 3(1): 012004.

[31] Ji T, et al. Charge transporting materials for perovskite solar cells, Rare Metals. 2021; 40(10): 2690–2711.

[32] Sanders S, et al. Chemical vapor deposition of organic-inorganic bismuth-based perovskite films for solar cell application, Scientific Reports. 2019; 9(1): 9774.

[33] Manjunatha S, et al. The characteristics of perovskite solar cells fabricated using DMF and DMSO/GBL solvents, Journal of Electronic Materials. 2020; 49(11): 6823–6828.

[34] Zhang Y, et al. Solution-processed transparent electrodes for emerging thin-film solar cells, Chemical Reviews. 2020; 120(4): 2049–2122.

[35] Subudhi P, Punetha D. Progress, challenges, and perspectives on polymer substrates for emerging flexible solar cells: A holistic panoramic review, Progress in Photovoltaics, Research and Applications. 2023; 31(8): 753–789.

[36] Afre RA, Pugliese D. Perovskite solar cells: A review of the latest advances in materials, fabrication techniques, and stability enhancement strategies, Micromachines. 2024; 15(2): 192.

[37] Tailor NK, et al. Recent progress in morphology optimization in perovskite solar cell, Journal of Materials Chemistry A. 2020; 8(41): 21356–21386.

[38] Valadez-Villalobos K, Davies ML. Remanufacturing of perovskite solar cells, RSC Sustainability. 2024; 2(8): 2057–2068.

[39] Kumar J, Srivastava P, Bag M. Advanced strategies to tailor the nucleation and crystal growth in hybrid halide perovskite thin films, Frontiers in Chemistry. 2022; 10: 842924.

[40] Butt MA. *Thin-film coating methods: A successful marriage of high-quality and cost-effectiveness—a brief exploration, Coatings. 2022; 12(8); **(169)**: 1115.*

[41] Wang Y, et al. Innovative approaches to large-area perovskite solar cell fabrication using slit coating, Molecules. 2024; 29(20): 4976.

[42] Kumar A, et al. Processing methods towards scalable fabrication of perovskite solar cells: A brief review, Inorganic Chemistry Communications. 2024: 113115.

[43] Hamukwaya SL, et al. A review of recent developments in preparation methods for large-area perovskite solar cells, Coatings. 2022; 12(2): 252.

[44] Oku T, et al. Fabrication and characterization of CH3NH3PbI3 perovskite solar cells added with polysilanes, International Journal of Photoenergy. 2018; 2018(1): 8654963.

[45] Yan Jiang, Sisi He, Longbin Qiu, Yixin Zhao, Yabing Qi; Perovskite solar cells by vapor deposition based and assisted methods. Appl. Phys. Rev. 1 June 2022; 9 (2): 1–52.

[46] Ma X, et al. Perovskite seeding approach of two-step sequential deposition for efficient solar cells, ACS Applied Energy Materials. 2024; 7(10): 4540–4548.

[47] Zhang L, et al. A layering technique for achieving pinhole-free organic–inorganic halide perovskite thin films through the vapor–solid reaction, Sustainable Energy & Fuels. 2024; 8(11): 2485–2493.

[48] Khorasani A, et al. Opportunities, challenges, and strategies for scalable deposition of metal halide perovskite solar cells and modules, Advanced Energy and Sustainability Research. 2024; 5(7): 2300275.

[49] Yao Z, Zhao W, Liu SF. Stability of the CsPbI 3 perovskite: From fundamentals to improvements, Journal of Materials Chemistry A. 2021; 9(18): 11124–11144.

[50] Han Y, et al. Review of two-step method for lead halide perovskite solar cells, Solar Rrl. 2022; 6(6): 2101007.

[51] Chang, J., et al., Crystallization and orientation modulation enable highly efficient doctor-bladed perovskite solar cells. Nano-micro letters, 2023. 15(1): p. 164.

[52] Patil GC. Doctor blade: A promising technique for thin film coating, In: Simple Chemical Methods for Thin Film Deposition: Synthesis and Applications. 2023; Springer, 509–530.

[53] Rana S, Akteruzzaman M. A comprehensive systematic literature review on perovskite solar cells: Advancements, efficiency optimization, and commercialization potential for next-generation photovoltaics, American Journal of Scholarly Research and Innovation. 2022; 1(01): 137–185.

[54] Kaya IC, et al. Crystal reorientation and amorphization induced by stressing efficient and stable P–I–N vacuum-processed MAPbI3 perovskite solar cells, Advanced Energy and Sustainability Research. 2021; 2(3): 2000065.

[55] Tara A, et al. Inkjet-printed FASn1–xPbxI3-based perovskite solar cells, ACS Applied Materials & Interfaces. 2024; 16(46): 63520–63527.

[56] Duarte VC, Andrade L. Recent advancements on slot-die coating of perovskite solar cells: The lab-to-fab optimisation process, Energies. 2024; 17(16): 3896.

[57] Tu Y, et al. Slot-die coating fabrication of perovskite solar cells toward commercialization, Journal of Alloys and Compounds. 2023; 942: 169104.

[58] Han X, et al. Advancements in flexible electronics fabrication: Film formation, patterning, and interface optimization for cutting-edge healthcare monitoring devices, ACS Applied Materials & Interfaces. 2024; 16(41): 54976–55010.

[59] Zhang Z, et al. Progress on inkjet printing technique for perovskite films and their optoelectronic and optical applications, Acs Photonics. 2023; 10(10): 3435–3450.

[60] Cheng Y, et al. Droplet manipulation and crystallization regulation in inkjet-printed perovskite film formation, CCS Chemistry. 2022; 4(5): 1465–1485.

[61] Angmo D. Upscaling of Indium Tin Oxide (Ito)-free Polymer Solar Cells: Performance, Scalability, Stability, and Flexibility. 2014.

[62] Abzieher T, et al. Vapor phase deposition of perovskite photovoltaics: Short track to commercialization?, Energy & Environmental Science. 2024; 17(5): 1645–1663.

[63] Li N, et al. Towards commercialization: The operational stability of perovskite solar cells, Chemical Society Reviews. 2020; 49(22): 8235–8286.

[64] Sharmile N, Chowdhury RR, Desai S. A comprehensive review of quality control and reliability research in Micro–Nano technology, Technologies. 2025; 13(3): 94.

[65] Rodkey NJ. Scalable Vacuum-Based Deposition Methods for Halide Perovskites. 2024.

[66] Swartwout R, Hoerantner MT, Bulović V. Scalable deposition methods for large-area production of perovskite thin films, Energy & Environmental Materials. 2019; 2(2): 119–145.

[67] Bae S-R, Heo D, Kim S. Recent progress of perovskite devices fabricated using thermal evaporation method: Perspective and outlook, Materials Today Advances. 2022; 14: 100232.

[68] Hwang J-K, et al. A review on dry deposition techniques: Pathways to enhanced perovskite solar cells, Energies. 2023; 16(16): 5977.

[69] Leccisi E, Fthenakis V. Life-cycle environmental impacts of single-junction and tandem perovskite PVs: A critical review and future perspectives, Progress in Energy. 2020; 2(3): 032002.

[70] Li J, et al. Highly efficient thermally co-evaporated perovskite solar cells and mini-modules, Joule. 2020; 4(5): 1035–1053.

[71] Piot M, et al. Fast coevaporation of 1 μm thick perovskite solar cells, ACS Energy Letters. 2023; 8(11): 4711–4713.

[72] Sanger A. Perovskite nanocrystals-based thin film deposition for photovoltaic devices, In: Advanced Nanomaterials for Solution-Processed Flexible Optoelectronic Devices. 2025; CRC Press, 142–178.

[73] Gao Q, et al. Halide perovskite crystallization processes and methods in nanocrystals, single crystals, and thin films, Advanced Materials. 2022; 34(52): 2200720.

[74] Soultati A, et al. Synthetic approaches for perovskite thin films and single-crystals, Energy Advances. 2023; 2(8): 1075–1115.

[75] Yokoyama T, et al. Overcoming short-circuit in lead-free CH3NH3SnI3 perovskite solar cells via kinetically controlled gas–solid reaction film fabrication process, The Journal of Physical Chemistry Letters. 2016; 7(5): 776–782.

[76] Awan TI, Afsheen S, Kausar S. Role of thin film in deposition techniques, In: Thin Film Deposition Techniques: Thin Film Deposition Techniques and Its Applications in Different Fields. 2025; Springer, 219–239.

[77] Calisi N, et al. Thin films deposition of fully inorganic metal halide perovskites: A review, Materials Science in Semiconductor Processing. 2022; 147: 106721.

[78] Wang M, Carmalt CJ. Film fabrication of perovskites and their derivatives for photovoltaic applications via chemical vapor deposition, ACS Applied Energy Materials. 2021; 5(5): 5434–5448.

[79] Ding M, et al. Surface/Interface engineering for constructing advanced nanostructured photodetectors with improved performance: A brief review, Nanomaterials. 2020; 10(2): 362.

[80] Zeb S, et al. Advanced developments in nonstoichiometric tungsten oxides for electrochromic applications, Materials Advances. 2021; 2(21): 6839–6884.

Pulkit Katiyar, Yashwant Kumar Singh, D.K. Dwivedi*, Pooja Lohia, and Pravin Kumar Singh

Chapter 5
Advanced materials for high-efficiency perovskite solar cells

Abstract: While perovskite solar cells (PSCs) are highly promising, their stability over time, especially in environments with humidity or high temperatures, stands in the way of their wide commercial use. Besides, the threat of lead exposure from advanced 3D perovskites has prompted scientists to pursue lead-free substitutes and better architecture designs. Recently, a key achievement was made in developing DJ-type 2D-3D perovskite heterojunctions that offer enhanced moisture and heat tolerance and higher performance than traditional perovskites. Differently from RP phases, where organic cations make the space between layers wider, DJ phases are composed of inorganic layers that are close together. Narrower gaps in the atomic layers make the device more stable and allow electrons to move faster, boosting conductivity and improving the device's performance. Presented simulations show that these structures have significant potential. One of the devices made of lead has the structure (FTO/ZnO/CH$_3$NH$_3$PbI$_3$/PeDAMA$_5$Pb$_6$I$_{19}$/Cu$_2$O/Ni). It involves two bandgaps: 1.58 eV (3D layer) and 1.60 eV (DJ 2D layer), leading to a strong power conversion efficiency (PCE) of 27.12% at harsh conditions, as well as high values of V_{OC}, J_{SC}, and fill factor (FF). The tin-based structure study achieved a new record configuration (FTO/ZnO/CH$_3$NH$_3$SnI$_3$/PeDAMA$_5$Pb$_6$I$_{19}$/NiCo$_2$O$_4$/Pt), where ZnO served as the electron transport layer and NiCo$_2$O$_4$ served as the hole transport layer. The structure has been developed to have bandgaps of 1.3 eV for 3D tin and 1.6 eV for DJ 2D perovskites, with materials chosen to suit the required thickness, bandgap, and defects at the interface. The device gave a superior PCE of 31.81% at 300 K (V_{OC} = 1.25 V, J_{SC} = 31.33 mA/cm^2, FF = 80.92%) while our lead-based structure (FTO/ZnO/CH$_3$NH$_3$PbI$_3$/PeDAMA$_5$Pb$_6$I$_{19}$/Cu$_2$O/Ni) performed well, with a slightly lower PCE of 27.12% at 300 K (V_{OC} = 1.43 V, J_{SC} = 22.83 mA/cm^2, FF =

*Corresponding author: **D.K. Dwivedi,** Photonics and Photovoltaic Research Lab (PPRL), Department of Physics and Material Science, Madan Mohan Malaviya University of Technology, Gorakhpur 273010, Uttar Pradesh, India, email: todkdwivedi@gmail.com
Pulkit Katiyar, Yashwant Kumar Singh, Photonics and Photovoltaic Research Lab (PPRL), Department of Physics and Material Science, Madan Mohan Malaviya University of Technology, Gorakhpur 273010, India
Pooja Lohia, Department of Electronics and Communication Engineering, Madan Mohan Malaviya University of Technology, Gorakhpur 273010, India
Pravin Kumar Singh, Institute of Advanced Materials, IAAM, Gammalkilsvägen 18, 59053 Ulrika, Sweden

https://doi.org/10.1515/9783111726847-005

82.92%), showcasing its good stability at high temperatures. According to these findings, choosing the right layer is essential and by optimizing its structure and thickness and using the DJ phase, stability is greatly improved. Creating high-performance, eco-friendly and profitable PSCs relies on making 2D–3D heterojunctions that use lead-based and lead-free materials together.

Keywords: 2D-3D perovskite, RP phase, DJ phase, lead-free perovskites

5.1 Introduction

It is projected that the world's energy consumption will rise from 15 TW in 2011 to 30 TW by 2050. As people need more energy and as climate change is a pressing problem, finding sustainable approaches is essential. Given that fossil fuels currently provide 81% of the world's energy and because of their negative impact on the environment and lack of long-term supply, we must turn to sustainable renewables. As an environmentally friendly source, solar energy is set to become the main solution for future needs.

Outcompeting other photovoltaics by more than 90%, due to their security and the chance to reach 26.7% efficiency, conventional silicon (C-Si) solar cells still present some challenges. Because they are costly to make, have a specific type of bandgap, and lack good results for long wavelengths, their adoption and overall performance are limited [1, 2]. Perovskite solar cells (PSCs) were recognized as a third-generation technology in 2009 and functioned at an initial efficiency of 3.8% [3]. Since then, they have advanced swiftly. The efficiency certified for these types of solar cells has advanced to 25.5%, the same level as silicon, while providing benefits such as a lower cost, adjustable electronic properties, and simpler development (Best Research-Cell Efficiency Chart | Photovoltaic Research | NREL, n.d.). When 2D-3D heterostructures are added to organic–inorganic hybrid materials, this allows for high theoretical conversion efficiency over 30%, and improves stability to handle thermal and moisture issues. Their ability to work more efficiently, adapt to any light, and be made affordably, has the potential to strongly improve solar energy around the world. Such advancement is important to attain the Sustainable Development Goals related to cutting down CO_2 emissions, making more efficient use of energy and caring for ecosystems [4, 5].

PSCs, representing third-generation photovoltaics, utilize materials with the general structure ABX_3. With this structure, "A" is an inorganic or organic cation (MA^+, FA^+, or Cs^+), "B" is a Pb^{2+} or Sn^{2+} metal cation, and "X" is I^-, Br^-, or Cl^-. Thanks to their unique qualities, these materials show amazing features such as bandgaps that can be changed, excellent light absorbing capabilities, long distances for carriers to move, and the opportunity for inexpensive, solution-based production methods [6, 7]. Since 2009, lead-based PSCs have climbed from about 3.8% to certified levels, close to 25.7%, rapidly nearing the 33.7% theoretical ceiling for single-junction devices (7, 8). Even so, getting paint into the market requires overcoming serious difficulties: strong moisture, heat, and light rapidly degrade it and using lead makes environmentalists worry [9]. To ensure stability, researchers are designing perovskite structures in two dimensions, mainly creating stable DJ phases. Furthermore, in DJ phases, organically modified diammonium structures connect inorganic plates, remove disruptive van der Waals gaps, and ensure the structural framework remains supported by strong hydrogen bonding, while conducting charge well [10]. Tin perovskites (such as $CH_3NH_3SnX_3$ or $MASnI_3$) are attractive because they are less harmful and have a narrow bandgap, close to 1.3 eV, making absorption of energy from deeper into the infrared possible [11, 12]. Efficiency for Sn-PSC devices used to be low, but after improvements in both materials and device development, they have reached more than 14% and may eventually reach 33% [13–15]. Improving device architecture by putting heterojunction laminated layers together with simpler hole transport layer (HTL)-free (no HTL) designs (like $FTO/MASnI_3/C_{60}/Au$) [16] absorbs more light, makes production more accessible and affordable, and allows simulated efficiencies to rise above 26% [17, 18]. For all these exciting developments, there are still significant issues: making systems work consistently under normal use, strengthening safety across a product's entire life, increasing understanding of material issues and developing approaches that can be replicated, and successfully scaling up production processes to unlock the full potential of this transformative technology.

Compared to the usual PSCs, DJ phase 2D perovskites have considerably better physical and environmental properties. The monovalent diammonium organic cations in DJ phases form covalent cross-links between inorganic layers, unlike the divalent organic cations in RP phases. As a result, weak spaces between layers are minimized; so the stack is stronger, more resistant to moisture, and offers faster movement of charge carriers along its length [19, 20]. Substances like $PeDAMA_5Pb_6I_{19}$ display this design, having a higher forbidden energy band than 3D perovskites such as $CH_3NH_3PbI_3$ but taking advantage of its 2D-3D interfaces to block defects and grain boundaries effectively.

Even though DJ phases have many advantages, the movement of charges may still be restricted. As long as the layers are not thick enough and their energies do not align well, the structure exhibits energy problems, meaning accurate and uniform fabrication is necessary. The fact that stability grows with dimensionality but charge mobility falls, is very different from what is seen in isotropic 3D perovskites. Overly thick or badly configured organic layers in the stacks can subtly hinder the device's performance [19].

These structures, which include $CH_3NH_3SnI_3$ instead of lead, are another positive sign for the environment. With bandgap adjustment and improved materials ordering, RP-phase 2D perovskites have climbed to a maximum efficiency of ~13.7%. Still, they fall well below 22.7% of the maximum efficiency in 3D perovskites because organic spacers in the structure separate the charges [21, 22]. Because DJ phases are uninterrupted, this problem does not exist. This means that, despite needing extensive care in engineering and defect management, DJ-phase 2D perovskites are the most realistic way to generate high efficiency, lasting stability, and excellent commercial potential in future PSCs [19].

The use of DJ 2D-3D perovskite heterostructures leads to an improved balance of stability and efficiency in PSCs. The study presents recent methods for engineering crystals to improve how stable PSCs, based on tin, are, in comparison to lead. Researchers have optimized novel designs for solar cells using DJ 2D perovskites and 3D absorbers with SCAPS-1D simulations. A key discovery reveals that using the 3D $CH_3NH_3SnI_3$ (MASnI$_3$), $CH_3NH_3SnI_3$ compound and the DJ 2D PeDAMA$_5$Pb$_6$I$_{19}$ perovskite (also including the PeDA^{2+} organic cation) in an FTO/ZnO/MASnI$_3$/PeDAMA$_5$Pb$_6$I$_{19}$/NiCo$_2$O$_4$/Pt structure gives an impressive simulated PCE of 31.81%, while lead-based structure gives 27.12% PCE in optimized condition . This effect comes from putting great care into optimizing the sizes and properties of the layer bandgaps, their thicknesses, junction areas, contact materials, doping levels, and the series and shunt resistance. The DJ 2D layer (PeDAMA$_5$Pb$_6$I$_{19}$) shields the underlying tin perovskite from moisture and greatly reduces its degradation, as well as helping electrons move efficiently through the solar cell. The improved cell performed well in simulations, recording V_{OC} = 1.25 V, J_{SC} = 31.33 mA/cm^2, and FF = 80.92%. Additionally, the structure is thermally stable at 330 K and only shows performance issues when exposure to higher temperatures is longer. According to these investigations, incorporating DJ 2D-3D heterojunctions into PSCs helps them reach exceptional performance (>31%), with heightened environmental resistance. Mainly, the team worked on setting ideal absorber layer properties, minimizing any interfacial defects, and adopting efficient thermal management. The research stresses that mixed perovskite heterostructures have a huge capacity to improve the reliability and energy output of solar cells for mass use.

5.2 Material and simulation modeling

The main goal of the present research involves utilizing SCAPS-1D simulator to simulate and model the performance capabilities of a PSC with structures, which includes tin-based structure (FTO/ZnO/CH$_3$NH$_3$SnI$_3$/PeDAMA$_5$Pb$_6$I$_{19}$/NiCo$_2$O$_4$/Pt) and lead-based structure (FTO/ZnO/CH$_3$NH$_3$PbI$_3$/PeDAMA$_5$Pb$_6$I$_{19}$/Cu$_2$O/Ni). The lead-free device structure uses PeDAMA$_5$Pb$_6$I$_{19}$ two-dimensional (2D) perovskite material, which was synthesized by adding penta-methylene diamine (PeDA) as an extended ammonium spacer into Dion-Jacobson (DJ) perovskite crystal layers, for integration with a 3D CH$_3$NH$_3$SnI$_3$ lead-free absorption layer, while lead based structure uses CH$_3$NH$_3$PbI$_3$ as 3D layer and 2D layer in both simulations. SCAPS-1D, operated as an open-source software and launched in 2000 through the Ghent University's Department of Electronics and Information Systems (ELIS), was used to conduct analysis on the device under standard test conditions (AM1.5G spectrum, 300 K) [23, 24]. The software implements solutions to the Poisson equation, electron–hole continuity equations for calculating carrier densities, together with band alignment, external quantum efficiency (EQE), J–V characteristics, and spectral response, along with recombination profiles [25, 26]. SCAPS-1D serves as a simulation platform that allows scientists to study seven-layer planar-graded photovoltaic structures to evaluate key performance components like dopant densities, together with material thicknesses and electrical contact effects. The software allows users to study how various parameters affect PCE and charge transport by modeling their complex interactions, but does not require actual physical device manufacturing [22]. The combination of band diagrams, alongside J-V curves and EQE spectra, enables users to identify favorable material designs with operational parameters, using SCAPS-1D, which functions as a fundamental device for advancing perovskite-based optoelectronic device development.

The SCAPS-1D simulations are fundamentally based on three core governing equations, which are discussed below:

$$\text{Poisson equation:} \frac{\partial}{\partial x}\left(\epsilon(x)\frac{\partial V}{\partial x}\right) = q[p(x)n(x) + N_D^+(x) - N_A^-(x) + p_t(x) - n_t(x)] \quad (5.1)$$

$$\text{Equation of continuity for the hole:} \frac{\partial p}{\partial t} = \frac{1}{q}\frac{\partial J_p}{\partial x} + G_p - R_p \quad (5.2)$$

$$\text{Equation of continuity for the electron:} \frac{\partial n}{\partial t} = \frac{1}{q}\frac{\partial J_n}{\partial x} + G_n - R_n \quad (5.3)$$

where q is the elementary charge of an electron, v is the electrostatic potential, (x) is the concentration of free holes, J_n is the current density of electrons, J_P is the current density of holes, (x) is the concentration of free electrons, ε is the material's dielectric permittivity, $N_D^+(x)$ is the density of donor atoms, $NA^-(x)$ is the density of acceptor atoms, $Pt(x)$ is the trap concentration for holes, $nt(x)$ is the trap concentration for electrons, G_n, G_P are generation rates of electrons and holes, respectively, and R_n, R_P are recombination rates for electrons and holes, respectively [26, 27].

The three fundamental equations (Poisson's and continuity equations) governing the transport, generation–recombination of electrons – holes and performances of the device simulate the PSC being modelled. Both multilayered structures, lead-free and lead-based, are illustrated in Figure 5.1. In reference to Figure 5.1[i](a), the structural design of the investigated PSC involves the use of an FTO (fluoride-doped tin oxide) layer, a transparent conductive oxide layer which allows light entry [25, 26], followed by a 0.2 μm ZnO electron transport layer (ETL), a 0.07 μm NiCo$_2$O$_4$ HTL, a 1 μm thick 3D perovskite absorber CH$_3$NH$_3$SnI$_3$ (Methylammonium tin triiodide), and a 0.05 μm 2D Dion-Jacobson (DJ) perovskite layer (PeDAMA$_5$Pb$_6$I$_{19}$). Platinum (Pt) serves as the back contact and also a lead-based structure in which CH$_3$NH$_3$PbI$_3$ (methylammonium lead triiodide) is 3D layer, with the same 2D layer and ETL, as mentioned above, and nickel (Ni) as a back contact electrode. The incident photon is absorbed by the active layer and electrical stabilization charge localization by the DJ 2D perovskite (PeDAMA$_5$Pb$_6$I$_{19}$). Table 5.1 gives detailed information like the thicknesses and rates of various device layers, along with mobilities of carriers, doping densities, bandgaps, and defect states of various semiconductor materials. For both structure layers' respective optical absorption spectra, refer to Figure 5.1(b). Important parameters of the materials employed for simulations were entered into SCAPS-1D software (Table 5.2) for simulation of the device at 300 K, under illumination of AM1.5G. The ZnO ETL and NiCo$_2$O$_4$ HTL help charge extraction efficiently, and the exclusion of interfacial defects in the simulation isolates the effect of bulk material properties and layer structure. A depiction shows the electronic band structure of the materials used in both device constructions, as in Figure 5.1(c). This study indicates that PCE and charge transport mechanisms of high-performance PSCs can be optimally influenced by structural parameters like layer thicknesses and doping, as well as other material choices, through analysis of outputs like band alignment, J–V characteristics, and EQE.

Table 5.1: Material parameters utilized in the current simulation study [24, 25, 28–30].

Parameters	FTO	ZnO	CH$_3$NH$_3$PbI$_3$	CH$_3$NH$_3$SnI$_3$	PeDAMA$_5$Pb$_6$I$_{19}$	Cu$_2$O	NiCo$_2$O$_4$
Thickness (μm)	0.1	0.2	1	1	0.05	0.25	0.07
E_g (eV)	3.50	3.50	1.58	1.30	1.60	2.17	2.3
χ (eV)	4.0	4.0	3.9	4.17	3.98	3.2	3.48
ε (relative)	9	9	10	6.5	25	7.11	11.9
N_C (1/cm^3)	2.2×10^{18}	2.2×10^{18}	2.2×10^{18}	1×10^{18}	7.5×10^{17}	2.02×10^{17}	2.2×10^{18}
N_V (1/cm^3)	1.8×10^{19}	1.8×10^{19}	1×10^{19}	1×10^{18}	1.8×10^{18}	1.1×10^{19}	1×10^{19}
μ_h (cm^2/Vs)	1×10^1	2.5×10^1	2.2	1.6	3×10^{-1}	8×10^1	1.61
μ_e (cm^2/Vs)	2×10^1	1×10^2	2.2	1.6	1.4	2×10^2	1.05
V_e (cm/s)	1×10^7	1×10^7	1×10^7	1×10^7	1×10^7	1×10^7	1×10^7
V_h (cm/s)	1×10^7	1×10^7	1×10^7	1×10^7	1×10^7	1×10^7	1×10^7
N_A (1/cm^3)	0	0	1×10^{12}	3.2×10^{17}	0	1×10^{18}	1×10^{18}
N_D (1/cm^3)	2×10^{19}	1×10^{18}	1×10^{13}	0	0	0	0
N_t (1/cm^3)	1×10^{15}	1×10^{13}	1×10^{14}	1×10^{12}	2.5×10^{14}	1×10^{14}	1×10^{14}

5.3 Result and discussion

5.3.1 Effect of ETL and HTL layer alterations on photovoltaic device performance

The ability of PSCs to function well largely depends on how the ETL and hole transport layer function in charge separation, transport, recombination, and efficiency. It is essential for efficient charge extraction that the perovskite aligns energy-wise with

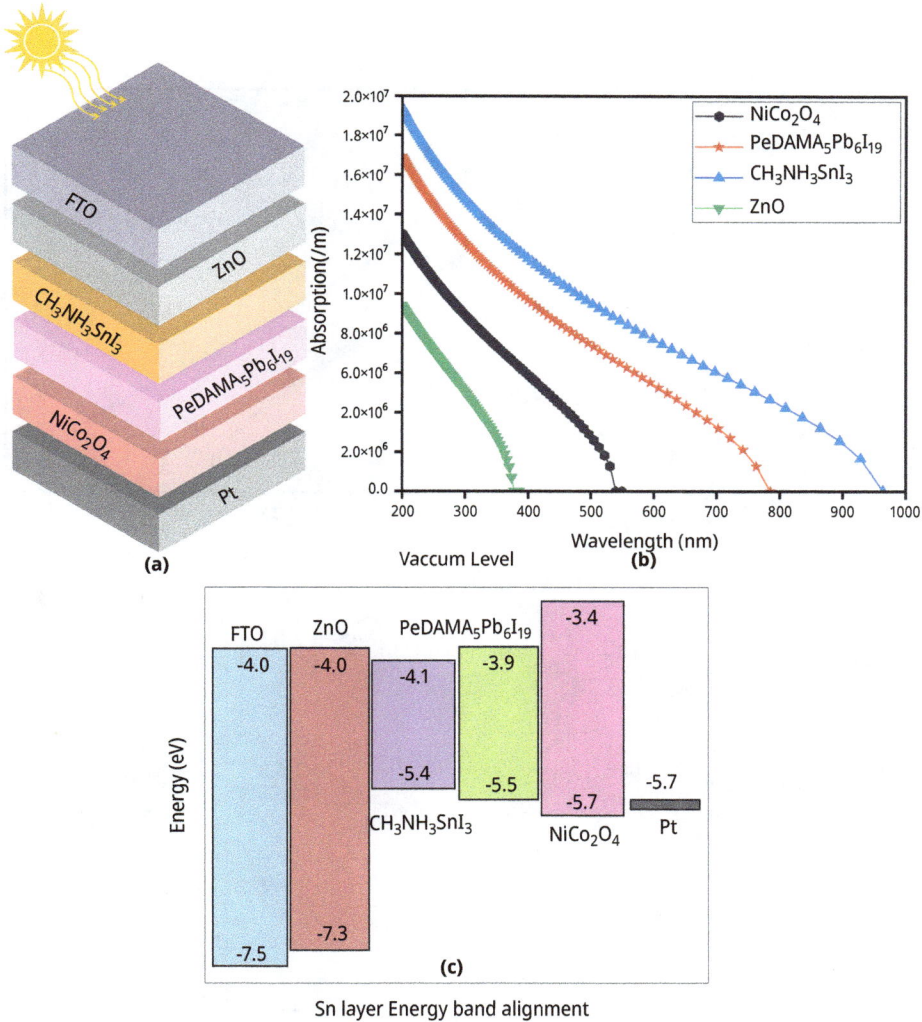

Figure 5.1: (i) (a) A proposed MASnI3-based DJ 2D-3D PSC structure working under air mass (AM) 1.5 spectrum; (b) the absorption spectra of the various materials used in simulation; and (c) energy band structure. **(ii)** (a) A proposed MAPbI3-based DJ 2D-3D PSC structure working under air mass (AM) 1.5 spectrum; (b) the absorption spectra of the various materials used in simulation; and (c) energy band structure.

the ETL/HTLs, preventing the transport of wrong polarity charges through them [19]. Efficiency, fill factor (FF), open-circuit voltage, and short-circuit current density all improve when the structures and components of these thinner layers are well optimized. By reducing the opposition to electron movement, supporting less energy loss, preventing carrier recombination, and acting as an electron-only contact, an effective

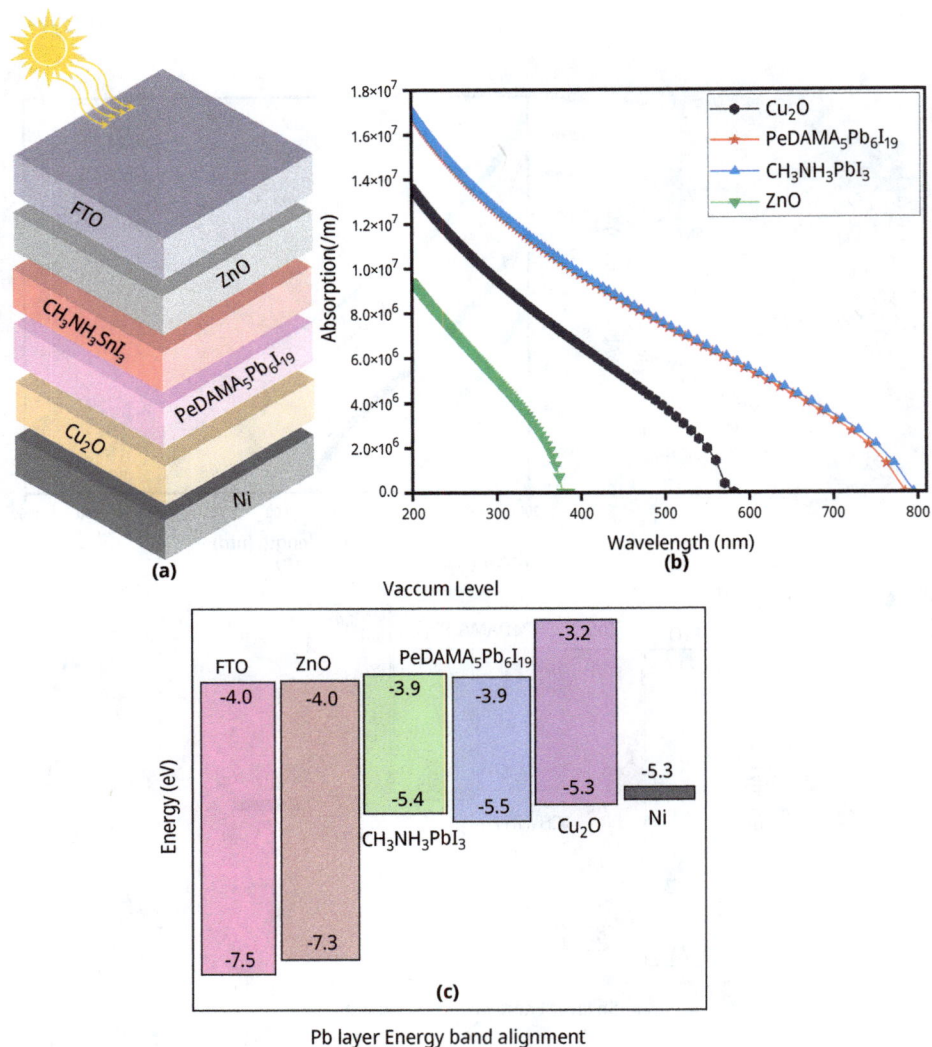

Figure 5.1 (continued)

ETL increases the power conversion efficiency of a device. When the HTL is of high quality, hole transfer to the anode is smooth, helping reduce resistance and allowing good hole mobility; ineffective hole transport, on the other hand, affects efficiency and causes charge to be stuck in the cells. In simulations, variations in ETL/HTL composition strongly impact the properties of the device.

Results from Sn-based PVK structure (Figure 5.2(i)a–d) showed that ZnO and $NiCo_2O_4$ were the best combination for a DJ 2D-3D structure, leading to a PCE of 31.81% (V_{OC} of

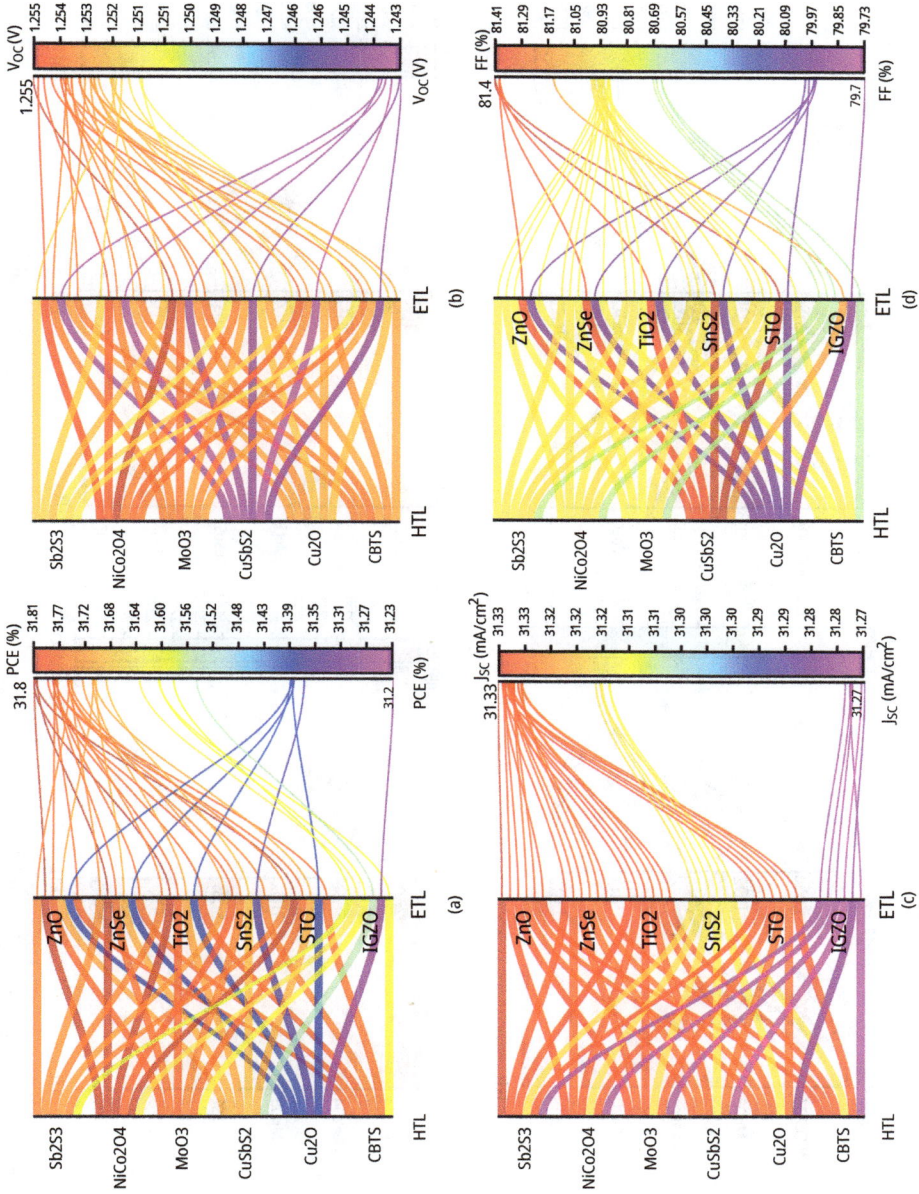

Figure 5.2: (i) and (ii) Transport layer alterations on key PV parameters: (a) PCE, (b) V_{OC}, (c) J_{SC}, and (d) FF.

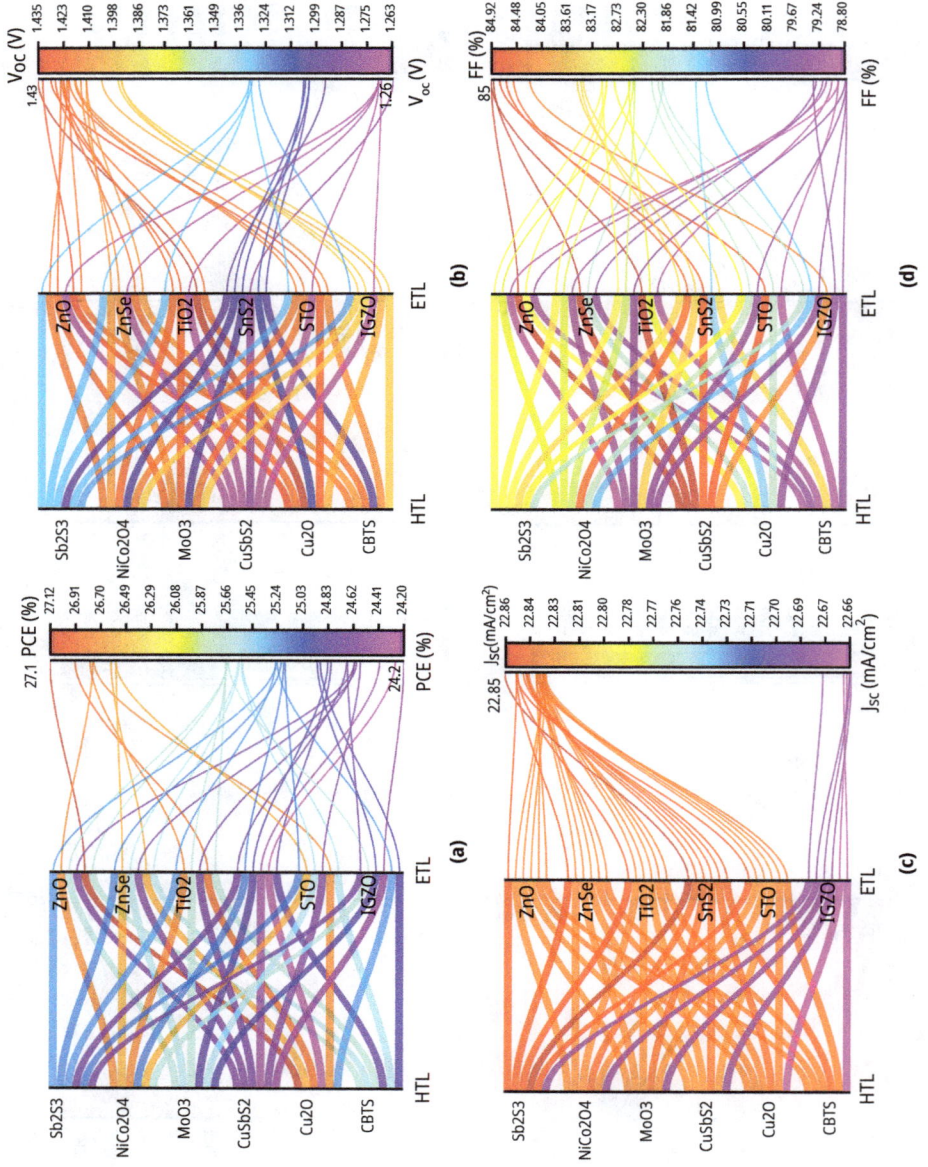

Figure 5.2 (continued)

1.25 V, J_{SC} of 31.33 mA/cm^2, FF at 80.92%). Making the surface conductive was important, so ZnO was picked because it is more conductive, has quicker electron movement, better fit for electron excitations, strong UV blocking, and is stable with different chemicals. Research in the lab also demonstrates that combining $NiCo_2O_4$ with common ETLs improves their results. During optimization, it was found that a thickness of 0.05 μm for ZnO and 0.07 μm for $NiCo_2O_4$ is best.

The same ETL and HTL materials were used in another MAPbI$_3$-based simulation study (Figure 5.2(ii) (a–d)), deciding that ZnO, together with Cu_2O, was most efficient for their device structure. This combination yielded a PCE of 27.12% (FF = 82.92%, J_{SC} = 22.83 mA/cm^2, and V_{OC} = 1.43 V) for a lead-based perovskite layer [25, 28, 30–32]. These results underscore that beyond efficiency gains, ETL/HTL optimization significantly enhances device stability and carrier transport by enabling customized energy-level alignment and thickness control, which are vital for PSC advancement.

5.3.2 Variation of thickness for MASnI$_3$ and MAPbI$_3$ layer

Efficient photovoltaic cells are achieved by setting the thickness of the active layer so that both photon absorption and movement of charges are equally great. Enlarging the absorber layer thickness significantly improves the way light is absorbed and produces more electron–hole pairs, taking the J_{SC} from 9.36 to 21.91 mA/cm^2. 3D absorber layer is the first variation introduced in this study. They looked at the results for a range of absorber thicknesses, from 0.1 to 1 μm. More carrier generation from the absorber layer causes both V_{OC} and PCE to grow until they reach peak values of 31.81% at 1 μm and bottom out at 19.68% for 0.05 μm thickness, as demonstrated in Figure 5.3(a). With absorbers that are very thin, oxidation does not make too much difference; once the thickness passes a point, oxidation decreases how far light is diffused, leading to a rise in recombination, fewer charges gathered at the circuit, and poorer PV performance. Raising the thickness of a device too many causes both FF to decline and the V_{OC} and PCE to fall. Figure 5.3(d) reveals that when the cell thickness is 1 μm, the FF of the thin-film solar cell lowers from 84.19% to 75.47% and V_{OC} drops from 1.26 to 1.24 V, as shown in Figure 5.3(b). Better photon absorption and avoiding unwanted recombination's led to reported device thickness values of 0.05–0.4 μm, depending on the chosen material and layout.

The combination of various absorber thicknesses shows that special attention must be given to cell thickness to ensure light harvesting, charge extraction, and low resistance. The best absorber layer thickness allows for adequate photon absorption but not powerful carrier diffusion, to guarantee that carriers can diffuse out quickly and not commonly be absorbed within the device. The best thickness of an active region depends on achieving maximum light capture while keeping recombination rates low, and results in different thicknesses, between 0.7m and 1 μm, based on device designs and material properties. The need to match absorber thickness to cell designs becomes essential for achieving efficient charge extraction and minimizing resistive losses, together with adequate light capture. The absorber layer should achieve adequate photon absorption, while remaining below the carrier diffusion length to prevent bulk recombination.

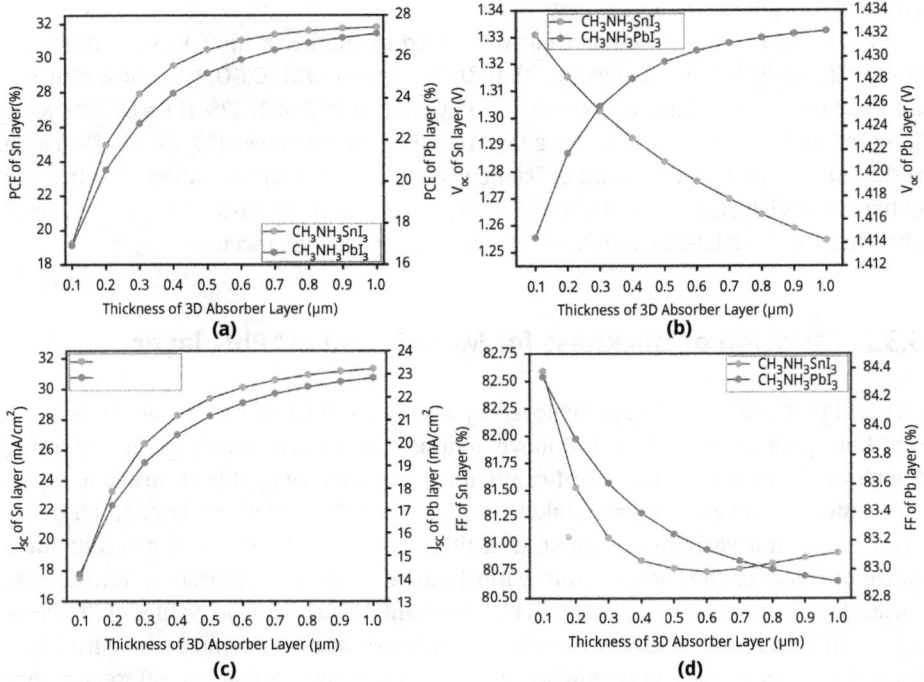

Figure 5.3: Impact of both 3D perovskite absorber layer (PAL) thickness on (a) PCE, (b) V_{OC}, (c) J_{SC}, and (d) FF.

The lead- and tin-based structures, with their J–V curves and EQE devices, with the thickness of their active layer ranging in thickness from as little as 0.2 to up to 1 μm (illustrated with "thickness increase is shown"). A graph of these variables, based on the arrow directions, is found in Figures 5.4(i) and (ii). How a solar cell's power varies with the amount of light that hits it is made by two input variables: applied voltage (V) and current density (J).

J–V curve: Graphs of the J–V curve are provided for five different thicknesses, and the worst is achieved for the lead-based device PSC, when reaching a minimum current density of 5.52 and a maximum of 22.83 mA/cm^2, which are measured at a uniform layer thickness of 1 μm, and for tin-based layer, a range from 4.03 to 31.33 mA/cm^2 is obtained at 1 μm thicknesses. Figure 5.4(i) (a) shows a J–V curve at different layer thickness, ranging from 0.2–1.0 μm, and Figure 5.4(ii) (a) also shows this value for lead-based device. The external quantum efficiency (EQE) curve is valuable

(Figure 5.4(b) (i–ii)) at different layer thicknesses, ranging from 0.2 to 1.0 μm. It allows researchers to find out how well a perovskite material absorbs photons. It also changes light from all parts of the solar spectrum. An optimal is one that almost every user should be able to activate EQE, by giving PSC an EQE score of nearly 100% throughout most of the sections of the visible spectrum [33]. We call this the EQE curve. It is developed for cases where the active layer thickness ranges between 0.2 and 1 μm. The EQE of the device is the highest in the range of 400–800 nm, a uniquely wide range, and in an extra step, as you can see in Figures 5.3(d) and 5.4(d), the process ended rapidly because perovskite can no longer react properly and absorb the photons.

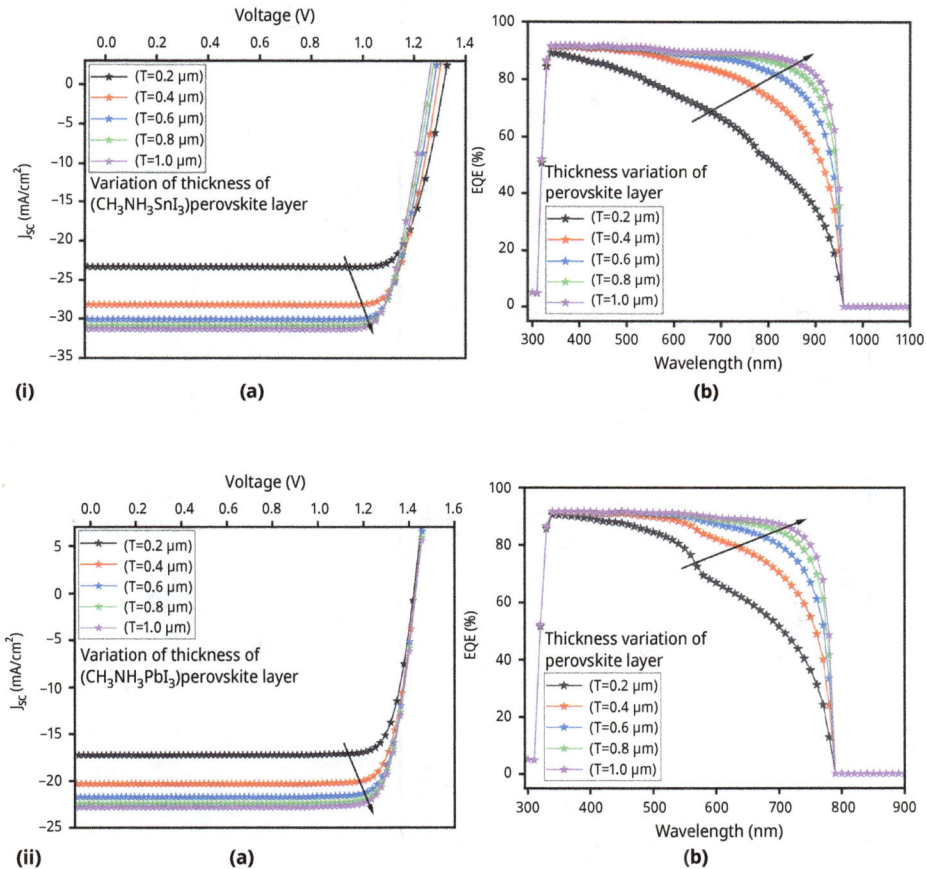

Figure 5.4: PSC characteristics affected by thickness alterations of **(i)** the MASnI$_3$ layer and **(ii)** the MAPbI$_3$ layer: (a) J–V curve and (b) EQE curve.

Table 5.2: Material attributes of HTLs and ETLs used in the present simulation work.

Parameter	HTL						ETL					
	Sb_2S_3	$CuSbS_2$	$NiCo_2O_4$	MoO_3	CBTS	Cu_2O	IGZO	ZnSe	ZnO	STO	SnS_2	TiO_2
Thickness (μm)	0.1	0.05	0.07	0.1	0.1	0.25	0.2	0.07	0.05	0.4	0.15	0.4
E_g (eV)	1.7	1.58	2.3	3	1.9	2.17	3.05	2.81	3.3	3.2	1.85	3.2
χ (eV)	3.8	4.2	3.48	2.5	3.6	3.2	4.16	4.09	4	4	4.26	4
ε (relative)	7	14.6	11.9	12.5	5.4	7.11	10	8.6	9	8.7	17.7	9
N_c (1/cm^3)	2.5×10^{19}	2×10^{18}	2.2×10^{18}	2.2×10^{18}	2.2×10^{18}	2.02×10^{17}	1×10^{18}	2.2×10^{18}	3.7×10^{18}	1.7×10^{19}	7.32×10^{18}	2×10^{18}
N_v (1/cm^3)	3.5×10^{19}	1×10^{19}	1×10^{19}	1.8×10^{19}	1.8×10^{19}	1.1×10^{19}	1×10^{18}	1.8×10^{18}	1.8×10^{19}	2×10^{20}	1×10^{19}	1.8×10^{19}
μ_e (cm^2/Vs)	7×10^{-2}	4.9×10^1	1.05	25	10	200	15	4×10^2	100	5.3×10^3	50	2×10^1
μ_h (cm^2/Vs)	2×10^{-2}	4.9×10^1	1.61	100	30	80	1×10^{-1}	1.1×10^2	2.5×10^1	6.6×10^2	25	1×10^7
V_e (cm/s)	1×10^7	1×10^7	1×10^7	1×10^7	1×10^7	1×10^7	1×10^7	1×10^7	1×10^7	1×10^7	1×10^7	1×10^7
V_h (cm/s)	1×10^7	1×10^7	1×10^7	1×10^7	1×10^7	1×10^7	1×10^7	1×10^7	1×10^7	1×10^7	1×10^7	1×10^7
N_D (1/cm^3)	0	0	0	0	0	0	1×10^{17}	1×10^{18}	1×10^{18}	2×10^{16}	9.85×10^{19}	1×10^{16}
N_A (1/cm^3)	1×10^{18}	1×10^{18}	1×10^{18}	1×10^{18}	1×10^{18}	1×10^{18}	0	0	0	0	0	0
N_t (1/cm^3)	1×10^{15}	1×10^{15}	1×10^{14}	1×10^{15}	1×10^{15}	1×10^{14}	1×10^{15}	1×10^{15}	1×10^{15}	1×10^{15}	1×10^{14}	1×10^{15}

5.3.3 Energy band graph of both MASnI$_3$ and MAPbI$_3$ DJ 2D-3D proposed structure

PSC performance is strongly affected by how the energies of electrons align in the two materials and by the rapid shifting of quasi-Fermi levels when a bias is applied (see Figure 5.5(a and b)). The alignment of E_v, E_c, F_p, and F_n helps maximize the way charges are transported in devices. Across all the tested DJ 2D films, the highest efficiency of up to 31.81% was measured using the FTO/ZnO/CH$_3$NH$_3$SnI$_3$/PeDAMA$_5$Pb$_6$I$_{19}$/NiCo$_2$O$_4$/Pt structure (Figure 5.5(a)), which uses ZnO as the ETL and NiCo$_2$O$_4$ as the HTL. Using this device structure, we achieve J_{SC} = 31.33 mA/cm^2 and V_{OC} = 1.25 V, while the lowest PCE of 27.12% is obtained using the lead-based (FTO/ZnO/CH$_3$NH$_3$PbI$_3$/PeDAMA$_5$Pb$_6$I$_{19}$/Cu$_2$O/Ni) structure. Fn and Ec form a neat alignment, whereas Fp matches with Ev to give the best separation between electrons and holes. The electrons created in the 2D-3D perovskite layer are brought by ZnO (ETL) to the electrodes, while letting holes pass to the electrodes via the NiCo$_2$O$_4$ (HTL). With a poor connection between HTL and Cu$_2$O, the device's efficiency is low – PCE = 27.12%, FF = 66.69% – caused by charge recombination and a clash of their energy levels (Figure 5.5(b)). An enhanced effectiveness of photocurrent occurs in the heterostructure ZnO/NiCo$_2$O$_4$ because the combined bands allow for better charge movement. The study demonstrates that getting ETL/HTL materials with appropriate band alignment near the Ec/Ev and Fn/Fp bands is essential to improve PSC light absorption and cut back on recombination.

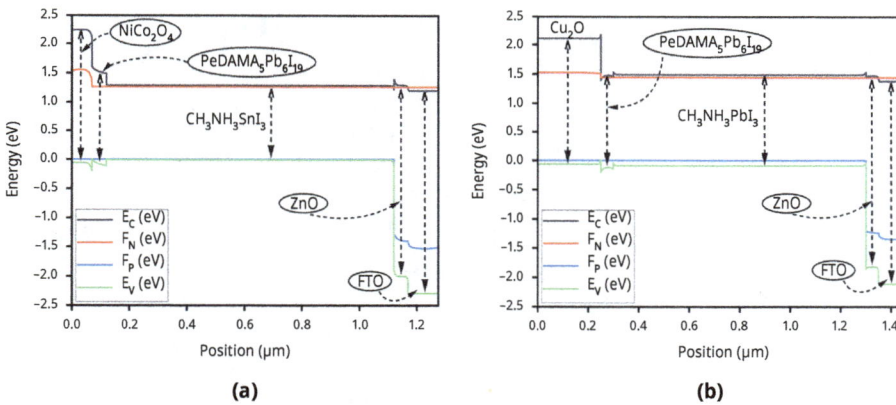

(a)   (b)

Figure 5.5: Energy band plot of solar cells indicating performance with different ETL and HTL with biasing (a) tin based and (b) lead based.

5.3.4 Effect of variations in temperature for both MASnI$_3$ and MAPbI$_3$ on solar cell PV parameters

The efficiency and stability of DJ 2D-3D PSCs are significantly influenced by their operational temperature conditions, as shown in Figure 5.6(a–b). Figure 5.6(b) presents how the V_{OC} falls from 1.25 V at 300 K to 1.17 V at 450 K. Higher temperatures cause the materials to release more carriers and help recombine them more quickly, lowering the V_{OC} [19]. As temperature goes from 300 to 450 K, the J_{SC} rises from 31.331 to 31.333 mA/cm^2, thanks to a slight improvement in carrier mobility (Figure 5.6(a)). At higher operating temperatures, dominant recombination effects lower the J_{SC}. FF declines from 80.92% to 75.77% (Figure 5.6(b)) due to growing heating, which increases the series resistance and leads to more recombination in the cells. While both combined effects operate, the PCE goes from 31.81% to 27.72% in the same temperature span (Figure 5.6(b)). Also, lead-based device in Figure 5.6(a) shows a declination in PCE from 27.12% to 21.93%. Because perovskites are thermally sensitive, their degradation is sped up by instability in charge movement and in their interfaces. Results from the study show that 300 K is best for performance because a stable environment at this temperature aids in retaining the cell's electronic functioning, reduces recombination, and cuts down on series resistance, to boost PCE and stability in DJ 2D-3D PSCs.

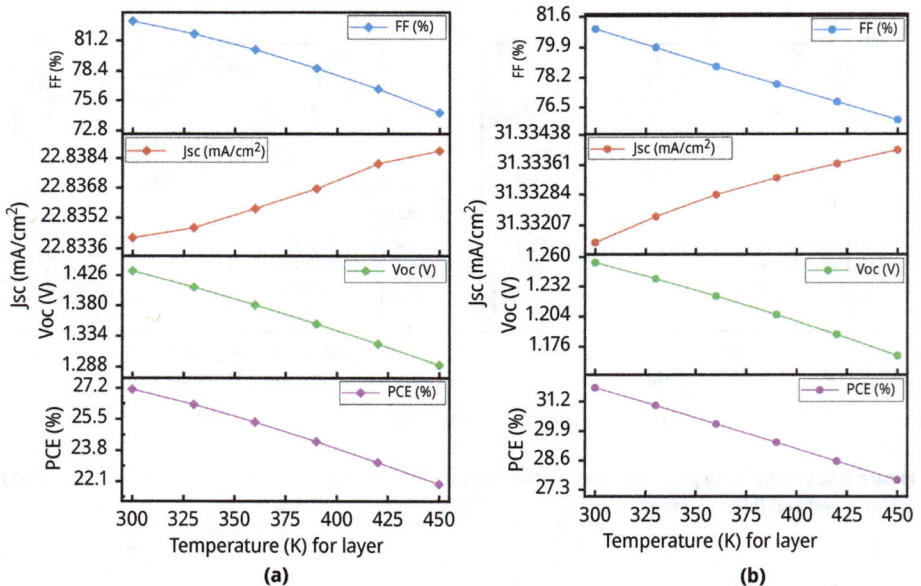

Figure 5.6: Impact of variations in temperature of PSC (DJ 2D-3D) structure on (a) MAPbI$_3$ based (b) MASnI$_3$ based.

5.3.5 Impact of variation in series resistance for both MASnI$_3$ and MAPbI$_3$ on solar cell PV parameters

The presence of series resistance in PSCs reduces their quality of operation; charge carriers cannot transport easily and some energy is wasted [19]. If there is no resistance, the ideal case is that $R_S = 0$, but in devices, resistance appears because of contacts, charge transport layers (ZnO with ETL and NiCo$_2$O$_4$ with HTL), and interfacial layers [34]. Rising R_S causes the photovoltaic properties to deteriorate quickly as PCE drops significantly with FF, though J_{SC} varies little and V_{OC} may either stay the same or present a minor increase. According to the results in Figure 5.7(a) and (b), when R_S changed from 3 to 8 $\Omega.cm^2$, PCE and FF both dropped, while V_{OC} slightly rose and J_{SC} stayed near the same level. The PCE decreased as R_S increased and FF also decreased, while the J_{SC} was steady. Also, V_{OC} remains nearly steady. It was shown through research that FF, along with PCE, showed the greatest degree of sensitivity to R_S, as a resistive effect can change the J-V curve shape, resulting in more resistance and thus more power dissipation. One should make certain that R_S is as small as possible, since minimizing resistive loss in the contact, interfaces, and transport layers contributes to better design and more stable operation.

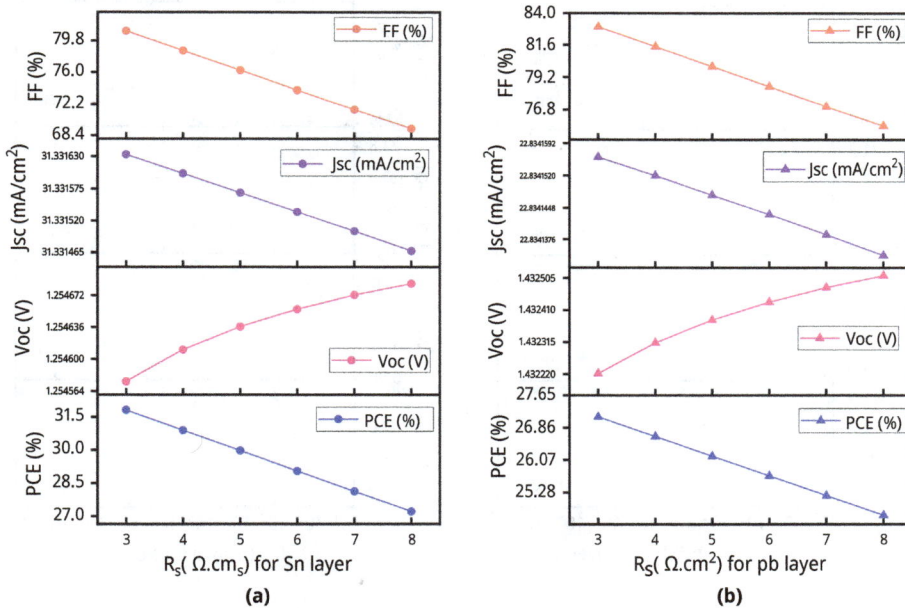

Figure 5.7: Impact of varying the R_S of the PSC (DJ 2D-3D) structure on (a) MASnI$_3$ based (b) MAPbI$_3$ based.

5.3.6 Effect of shunt resistance variation for both MASnI₃ and MAPbI₃ on solar cell PV parameters

The steadiness and performance of PSCs depend on the shunt resistance (R_{Sh}), as it changes both leakage and recombination of carriers within the device [19]. Parasitic currents form when R_{Sh} values are low, due to production issues such as pinholes or grain boundaries, instead of design errors, which results in several efficiency losses [1]. At low-light levels, the scarce number of photo-generated carriers results in more recombination losses. R_{Sh}, being in the range of 101–108 $\Omega.cm^2$, leads to better device metrics. The PCE increases from 1.88% to 31.81%, V_{OC} from 0.31 V to 1.27 V, and dark current goes down, as displayed in Figure 5.8(a), FF jumps from 24.90% to 80.29% and J_{SC} displays an upward trend, around 24.08 mA/cm² to 31.30 mA/cm². When a low-resistance collector is used, the shape of the J–V curve becomes straight and the MPP is reduced, while high resistance allows the curve to keep its original shape and increases both FF and stability. Low R_{Sh} readings result in the first increase in V_{OC}, but as R_{Sh} goes up, V_{OC} gets better because leakage currents go down. It is important for manufacturing methods to focus on reducing shunt loss, since this reduces both inefficiency and the risks of switching.

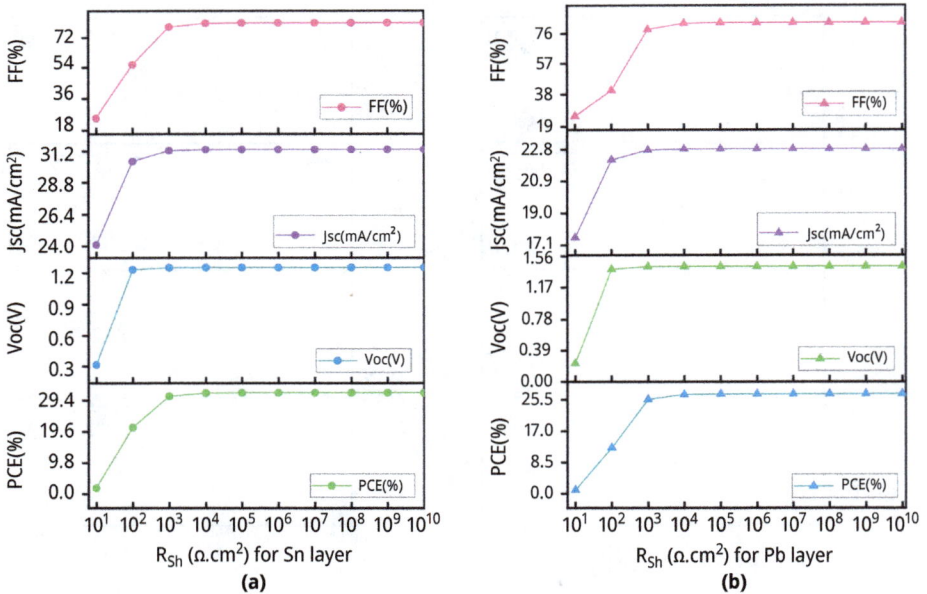

Figure 5.8: Effect of R_{Sh} on PSC structure on (a) MASnI₃ based and (b) MAPbI₃ based.

5.3.7 Effect of defect density and thickness in DJ 2D and 3D absorber layers on device performance

Solar cell performance of 2D Dion-Jacobson (DJ) PSC is very sensitive to the number of defects and how thin the absorber layer is. Researching active region thicknesses, ranging from 0.05 to 1 μm, with total defects of 1×10^{11} to 1×10^{18} cm^{-3}, found that performance is heavily dependent on the generation and collection of carriers as well as recombination patterns. A thicker absorber layer in PCE boosts its performance by improving light absorption and the creation of carriers. Rising with decreasing junction thickness until it reaches 0.3 μm, PCE achieves its maximum of 31.81% within a 0.05 μm thick layer, as the TDD increases from 0 to 1×10^{11} cm^{-3}. If the film is too thick and the TDD high, then reduced performance appears because carriers reach the base too quickly, causing increased recombination. Lower values of TDD (below 1×10^{12} cm^{-3}) prevent trap-assisted recombination and allow the V_{OC} and FF to be maintained. It follows that in Figure 5.9(b), the V_{OC} has a steady trend in most sample areas, even when the TDD is below 1×10^{15} cm^{-3}. Figure 5.9(d) shows that the FF does not decrease in its thickness stability for layers that contain up to 1×10^{15} cm^{-3} TDD. Additionally, Figure 5.9(c) indicates that the J_{SC} is highest for an 0.8 μm absorber thickness, since more light is absorbed, and reaches its maximum value at TDDs above 1×10^{15} cm^{-3}, since recombination plays a larger role than generation. At this point, so much defect action occurs that very few photons reach the active region, making recombination losses considerable. The combination of 0.05 μm thickness with 1×10^{11} cm^{-3} TDD creates the most efficient PCE result because it achieves maximum light absorption and lowest recombination levels. Strictly optimized performance occurs at 0.05 μm thickness, coupled with 1×10^{11} cm^{-3} TDD, because additional thickness or higher TDD intensifies the balance between carrier generation and recombination effects. The absorption thickness of the material must correspond to its defect-derived diffusion length so that collection of charges occurs before recombination becomes a significant issue [35].

A detailed analysis of the 3D perovskite absorber layer in Figure 5.10 demonstrates important trends in PV performance when thickness and total defect densities are adjusted. After making the straight TDD equal to 1×10^{11} cm^{-3} and reaching 1 μm thickness, the PCE of the DSSC can no longer be improved, as shown in Figure 5.10(a). When the carrier diffusion length goes beyond 1×10^{11} and 1×10^{14} cm^{-3}, the 3D structure can achieve 31.81% PCE. As soon as excess defect concentrations reach 1×10^{14} cm^{-3} in TDD, further increases in PCE are not possible, as these are the maximum defect concentrations available in these materials. Thickness ≤1 μm and TDD of 1×10^{11} cm^{-3} result in the best energy conversion efficiency based on the studies. Open-circuit voltages (V_{OC}) are stable across all thicknesses once defects fall below 1×10^{14} cm^{-3} (Figure 5.10b). Inversely, the current density at open-circuit gradually decreases as cells become thicker, from 0.8 μm to 200 μm, for different levels of defects, as shown in Figure 5.10(c). For TDDs greater than 1×10^{14} cm^{-3}, the FF declines, as shown in Figure 5.10(d), though it

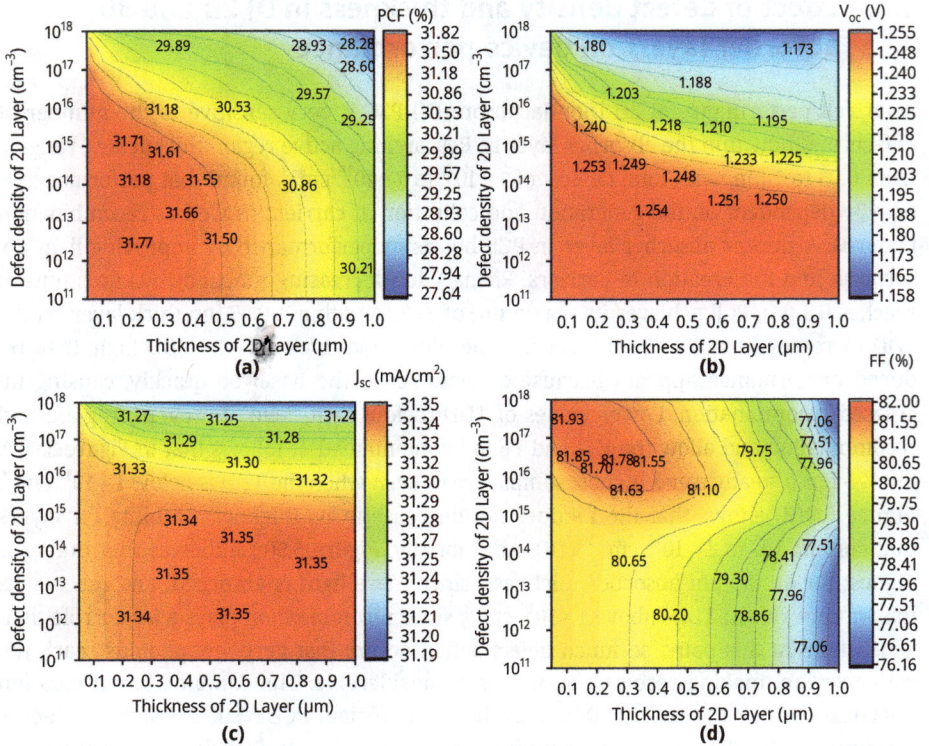

Figure 5.9: Impact of DJ 2D layer thickness and defect density on key PV parameters (a) PCE, (b) V_{OC}, (c) J_{SC}, and (d) FF.

remains unchanged for TDDs up to that level. Around 1 μm thickness and a 1×10^{11} cm^{-3} TDD, the cellular operation of 3D PSC is very efficient.

5.3.8 Combined effect of series and shunt resistance on MASnI$_3$-optimized structure

Photovoltaic device performance is greatly affected by the series and shunt resistors in the circuit. The resilience of the perovskite to electrical current, plus its HTL and ETL, along with the FTO and metal layer, is what results in R_S. The construction of the device is known as R_{Sh} and its performance is affected by pinholes, edge impacts, and recombination losses. Figure 5.11 illustrates the circuit for a PV cell that is equivalent to its actual one [6]. Computational analysis was performed to measure device performance as R_s values were set between 3 and 8 Ω.cm^2 and R_{Sh} levels altered between 10^1

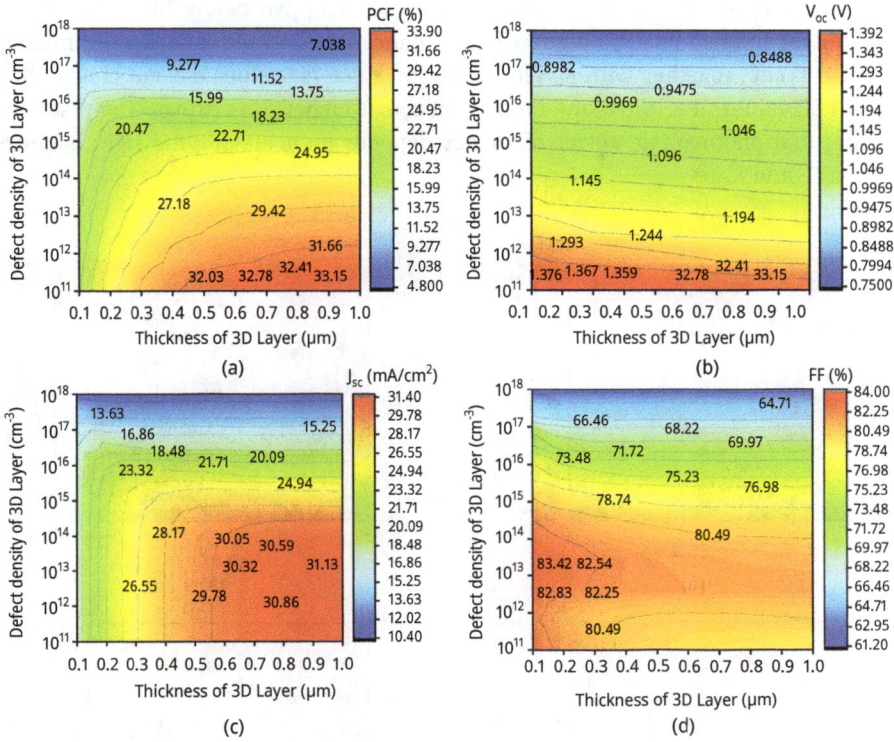

Figure 5.10: Impact of 3D layer thickness and defect density on key PV parameters: (a) PCE, (b) V_{OC}, (c) J_{SC}, and (d) FF.

and 10^8 $\Omega.cm^2$. When the sheet resistance rose from 3 to 8 $\Omega.cm^2$, the performance of the device really dropped, leading to an 80.29% drop in FF to 69.07%, a PCE of 1.36%, a J_{SC} of 17.41 mA/cm^2 and a V_{OC} remaining the same at 0.31 V, as shown in Figure 5.12.

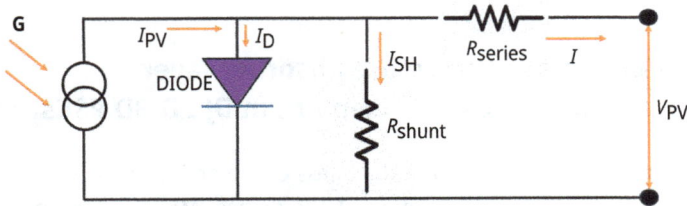

Figure 5.11: Equivalent circuit of solar cell.

The device efficiency diminished when R_{Sh} values dropped below 10^3 $\Omega.cm^2$, since power losses intensified. The highest performance level occurred when optimizing R_{Sh} to 10^8 $\Omega.cm^2$, together with R_S set at 3 $\Omega.cm^2$, since Jsc values did not change but FF and PCE values first lasted and then stabilized at higher R_{Sh} settings. The research indicates that photovoltaic device efficiency depends on achieving the right balance between R_S and R_{Sh}.

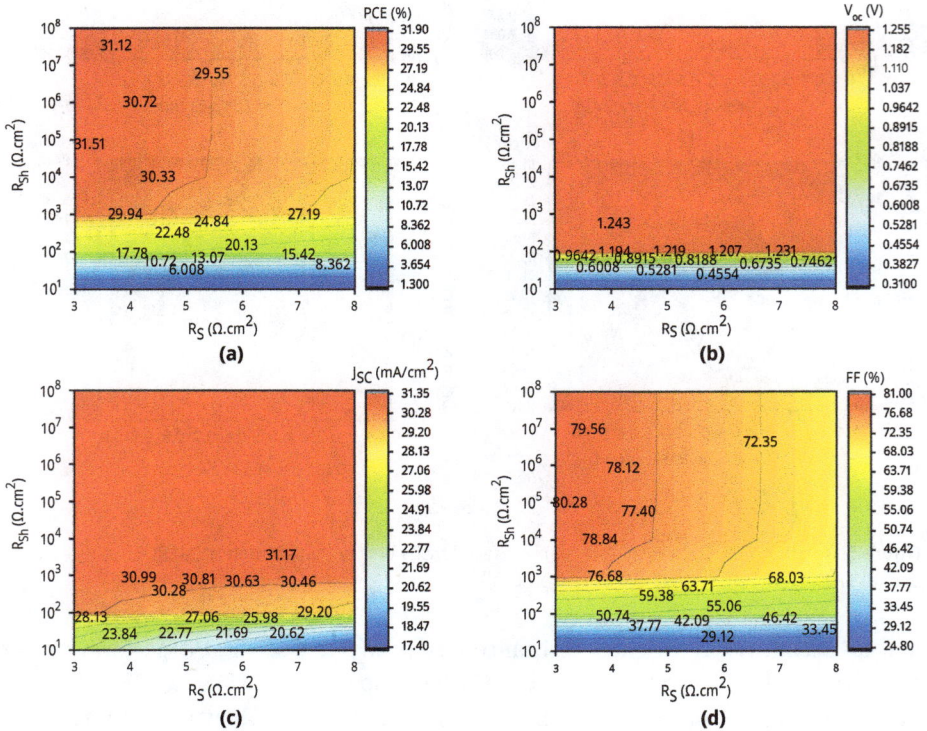

Figure 5.12: Impact of R_S and R_{Sh} on PSC structure on (a) PCE, (b) V_{OC}, (c) J_{SC}, and (d) FF.

5.3.9 Optimizing MASnI₃-based structure absorber layer thickness for performance enhancement in DJ 2D-3D PSCs

Performance in PSCs is mostly controlled by the thickness of the active layer, since it impacts both the stability and the overall function of the device. Making the active layer thicker helps absorb more photons and forms more electron–hole pairs which leads to better J_{SC} performance, as expected by simulation models (Figure 5.13(a–d)). Strongly thick active layers make the device less efficient, leading to more losses and reduced efficiency. With 1 μm in the 3D layer and 0.05 μm in the 2D layer, we were able to achieve the highest PCE (31.81%), highest J_{SC} (31.33 mA/cm^2), V_{OC} (1.25 V), and FF (80.92%) in DJ 2D-3D PSCs. The PCE lessened when 3D layers with thicknesses greater than 1.0 μm caused V_{OC} and FF to decline, because of recombination losses. Peak PCE of 31.81% was achieved with CH$_3$NH$_3$SnI$_3$ (3D), 1.0 μm and PeDAMA$_5$Pb$_6$I$_{19}$ (DJ 2D), 0.05 μm thickness in the simulation. Better efficiency at first results from layers that absorb more light, but after a certain point, the narrower bandgap leads to more recombination and then resistance problems (as shown in Figure 5.13(a)). The main goal for PSC development should be to optimize thickness, since this helps the absorber layer collect the most photons in spite of resistive and recombination effects [36, 19].

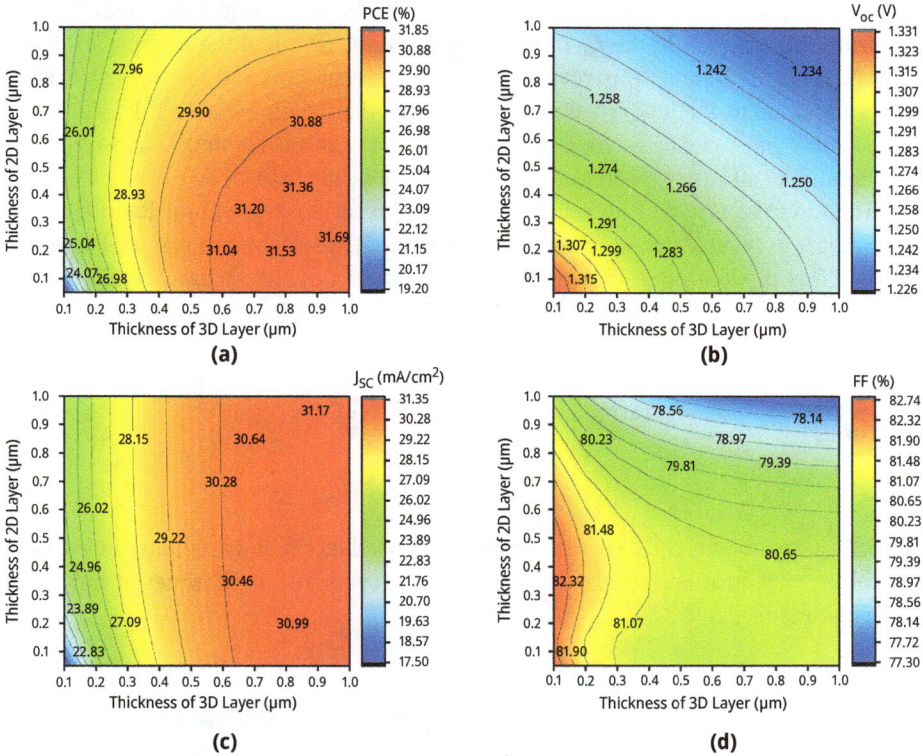

Figure 5.13: Influence of DJ 2D-3D perovskite absorber layer (PAL) thickness on (a) PCE, (b) V_{OC}, (c) J_{SC}, and (d) FF.

5.3.10 Impact of defect density of MASnI₃ in DJ 2D-3D PSCs

PSC performance relies a lot on defect density because it controls the speed at which charge carriers in the PSC recombine. The analysis reveals that V_{OC}, PCE, FF, and J_{SC} do not change until the defect density is less than 1×10^{13} cm⁻³, as random recombination is minimized at lower defect densities [19]. Both the photosensitive and J_{SC} properties of V_{OC}, PCE, and FF are constant until (N_t) is equal to 1×10^{12} cm⁻³, while J_{SC} stabilizes at (N_t) = 1×10^{13} cm⁻³. When the device reaches $N_t = 1 \times 10^{18}$ cm⁻³, its efficiency drops significantly; at point where J_{SC} hits 10.42 mA/cm² and PCE is 4.85%, the results in the graph show a clear drop (Figure 5.14). Because the defects act as places for charges to combine again, they lower both the photocurrent and efficiency. When hole concentration is reduced from 10^{18} to 10^{11} holes per cubic cm, the PCE increases from 4.85% to 31.81%, the V_{OC} from 0.47 V to 1.27 V, the J_{SC} from 1.35 to 31.30 mA cm⁻² and the FF from 26.39% to 80.29%. For both 2D and 3D perovskite structures, the best J_{SC} and V_{OC} are achieved at N_t values of about 1×10^{11} cm⁻³. The findings show that when the defect density is less than 1×10^{12} cm⁻³, performance is improved due to efficient recombination reduction and because of steadier transfer of electrical charge inside PSCs.

To study the effect of defect density in the absorber region and its interaction with adjacent layers, the Shockley–Read–Hall (SRH) recombination model was employed. This method offers a complete understanding of the recombination mechanisms influencing solar cell performance under varying defect conditions.

$$R_{SRH} = \frac{np - n_i^2}{\tau_p(n + n_i) + \tau_n(p + p_i)} \tag{5.4}$$

$$\tau_{n,p} = \frac{1}{\sigma v_{th,n,p} N_t} \tag{5.5}$$

Here, R_{SRH} represents the SRH recombination rate, where n and P denote the electron and hole concentrations, respectively. τ_p and τ_n represent the lifetimes of holes and electrons, respectively [25, 26].

5.3.11 Variation of work function for the back contact of MASnI₃

Good performance in solar cells relies largely on the work function, which tells how little energy is needed to take an electron from a surface. Efficiency becomes higher with high work functions since they lower the majority carrier barrier height and establish a cross section that supplies ohmic contact and improves the V_{OC} voltage as well as the total device performance. Efforts are made to find earth-abundant back-contact alternatives because platinum and gold are both plentiful and costly, with work functions of 5.7 and 5.1 eV, respectively [6]. Results showed the influence of back-contact materials with work functions ranging from 4.7 to 5.7 eV (see Table 5.3). Because its charge

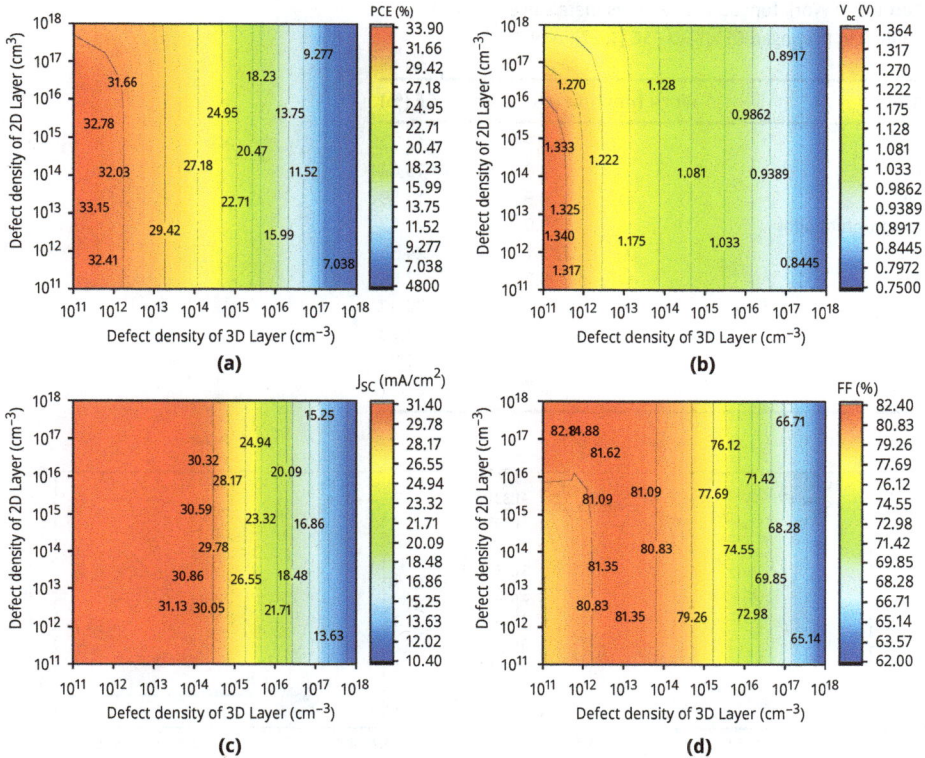

Figure 5.14: Influence of both perovskite absorber layer (DJ 2D-3D) defect density on (a) PCE, (b) V_{OC}, (c) J_{SC}, and (d) FF.

function was efficient and used the least amount of energy (Figure 5.15(a)), Pt was chosen as the best contact material. As a result of changing to the new device configuration, the FF shot up from 26.26% to 80.92% and V_{OC} decreased from 1.55 to 1.25 V, although J_{SC} improved by only a small amount, as shown in Figure 5.15(a) and (b). Because of band bending at the interface, although metal–semiconductor recombination decreases, contact is made more efficiently. This research has demonstrated that Pt is effective as an electrode for PSCs, while also pointing out that the larger-scale use of PSCs could be improved by finding low-cost alternatives with higher work function.

5.3.12 Influence of IDD of Sn-based DJ 2D-3D PSC on photovoltaic (PV) parameters of solar cell

Significant changes in device efficiency in DJ 2D-3D PSCs come from interface layers and their effect is affected in large part by interface defect density (IDD). The re-

Table 5.3: Work functions of various metals used as back contact in proposed DJ 2D-3D PSC [1, 37].

Metal	Work function (eV)	PCE (%)
Ag	4.7	12.72
Fe	4.8	15.64
Nb	4.9	18.59
Cu-graphite alloy	5.0	21.57
Au	5.1	24.58
Co	5.2	27.60
Ni	5.3	30.47
Ir	5.4	31.73
Pd	5.5	31.80
Pt	5.7	31.81

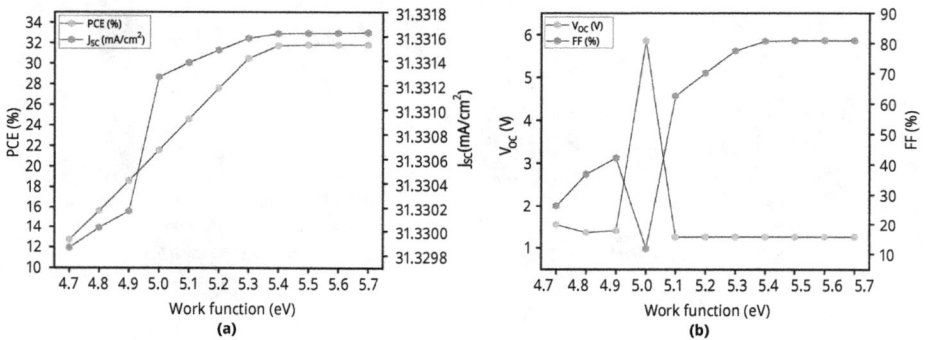

Figure 5.15: Influence of metal work function of DJ 2D-3D PSC on (a) PCE and J_{SC} and (b) V_{OC} and FF.

search explores changes in the diffusion rate at interface surfaces varying from 10^{10} to 10^{18} cm^{-2} and looks at the impact of these on major parameters (Figure 5.16(a–d)). Until 1×10^{12} cm^{-2}, both the interface between CH$_3$NH$_3$SnI$_3$/ZnO and ZnO/FTO are steady. The gap between the 2D and 3D interfaces leads to V_{OC} dropping from 1.25 to 0.94 V (Figure 5.16(b)). Short-circuit current density (J_{SC}) keeps its high level until IDD reaches a value of more than 1×10^{18} cm^{-2}. At the same time, FF declines throughout the IDD increase, as shown in Figures. Defects in the materials add more recombination processes and cause power efficiency to go down. A density of defects at 1×10^{18} cm^{-2} allows the cell to lose recombination particles, while still reaching a high-power voltage [33, 35].

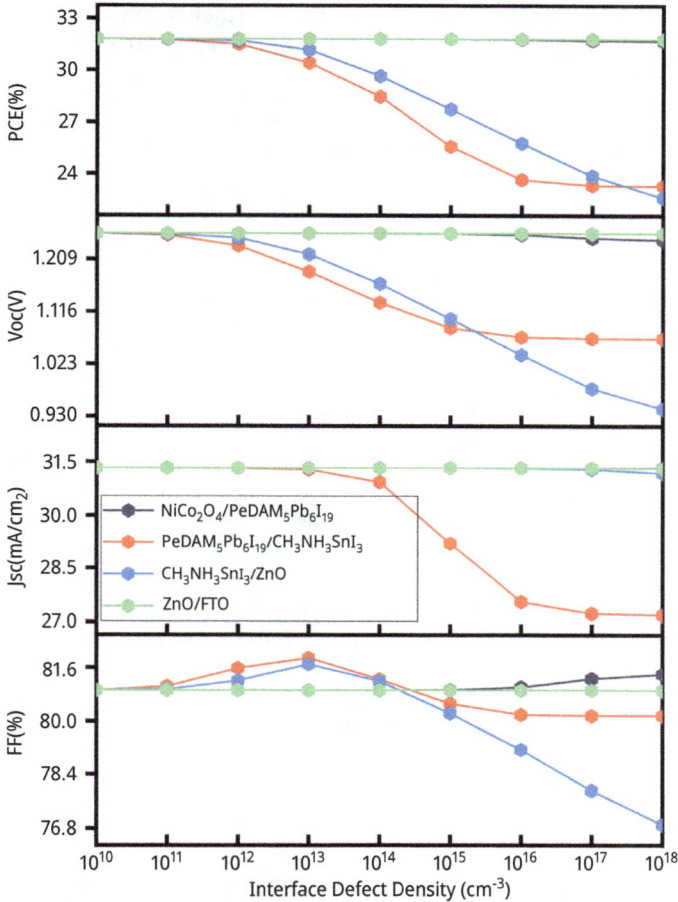

Figure 5.16: Influence of IDD on solar cell PV parameters: (a) PCE, (b) V_{OC}, (c) J_{SC}, and (d) FF.

5.3.13 Optimized J–V characteristics and external quantum efficiency (EQE) of MASnI$_3$ and MAPbI$_3$ layer

The thickness of the absorption layer in PSCs is most important for the power conversion efficiency, as well as current–voltage (J–V) and EQE [19]. A combination of 0.05 μm for DJ 2D layers and 1 μm for 3D layers gives strong photon absorption and lessens recombination losses for both layers (Sn and Pb), as depicted in Figure 5.17(a). The graph in Figure 5.16(b) shows that from 290 to 1,100 nm, response goes up, beginning at 350 nm, achieves an efficiency of 85–86% near 350–870 nm for the Sn layer, while Pb layer achieves it between 350 and 750 nm, and then drops sharply above 790 nm for lead and 900 nm for tin layer because the photon energy falls below the bandgap. If the photons have energies of less than 400 nm, thermalization causes the EQE to decrease. Problems with performance above 930 nm happen because the photons are blocked by the edge of the absorber's energy gap. To improve the performance of PSCs in different parts of the solar spectrum, both 3D and 2D layer thicknesses must be changed equally, and recombination rates must also be lowered with optimization.

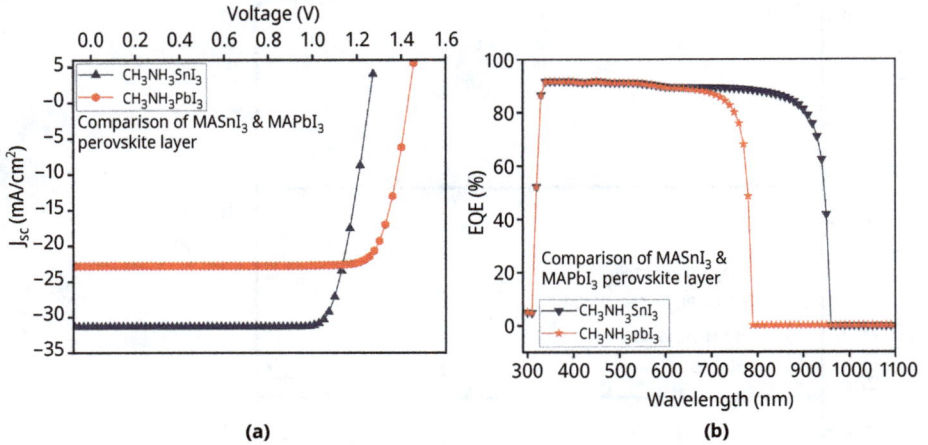

Figure 5.17: (a) J–V curve and (b) EQE curve of the MASnI$_3$- and MAPbI$_3$-based DJ 2D-3D PSC.

5.4 Conclusion

The action of PSCs depends primarily on their ETL and HTL, as these two types of materials are responsible for separating and carrying charges; their recombination; and the ultimate performance of the cell. The band alignment between the perovskite and the electrode layers allows efficient charging of carriers and leaves no room for opposite chargers to enter the device. Better use of materials and adjusting the layers boosts the PCE, FF, open-circuit voltage (V_{OC}), and short-circuit current density (J_{SC}). A potent ETL encourages electrons, minimizes energy loss, prevents electron and hole recombination, and consequently enhances the device's performance. Just like that, an effective HTL supports better movement of holes to the anode, cutting down resistance; poor hole transport harms efficiency and may cause charge buildup. Results from SCAPS-1D simulations confirm that ETL and HTL types as well as device parts greatly influence PTC performance. Researchers found that ZnO was the best candidate for pairing with NiCo$_2$O$_4$, when examined among ETLs (SnS$_2$, IGZO, STO, ZnSe, TiO$_2$, and ZnO) and HTLs (Cu$_2$O, CuSbS$_2$, NiCo$_2$O$_4$, MoO$_3$, CBTS, and Sb$_2$S$_3$). The photovoltaic pair of CH$_3$NH$_3$SnI$_3$ and PeDAMA$_5$Pb$_6$I$_{19}$ reached a PCE of 31.81% (for 1.25 V_{OC}

and 31.33 mA/cm^2 J_{SC}), with R_S of 3 Ω.cm^2 and R_{Sh} of 1×10^8 Ω.cm^2 when set up in a DJ-structured 2D-3D PSC, with layers of FTO/ZnO/CH$_3$NH$_3$SnI$_3$/PeDAMA$_5$Pb$_6$I$_{19}$/NiCo$_2$O$_4$/Pt. The device has a lead-free 3D perovskite known as CH$_3$NH$_3$SnI$_3$ and is combined with a 2D perovskite known as PeDAMA$_5$Pb$_6$I$_{19}$, whose organic and inorganic layers alternate efficiently. Major enhancements were made to the defect density, thickness of the active layer (0.05 μm for 2D and 1.0 μm for 3D) for both structures, low resistance in the series, high resistance in the shunt pathways, good generation and recombination rates, reduced number of defects at the interface, and stability at 300 K. Because of its higher conductivity, better electron movement, more suitable band arrangement, greater UV protection, and better stability, ZnO was preferred instead of TiO$_2$. In laboratory work, NiCo$_2$O$_4$, along with common ETLs, functions more efficiently than it does on its own. The results of a second lead-based (FTO/ZnO/CH$_3$NH$_3$PbI$_3$/PeDAMA$_5$Pb$_6$I$_{19}$/Cu$_2$O/Ni) simulation using the same above mentioned ETLs and HTLs found the combination of ZnO and Cu$_2$O to be the best for a lead-based PVK-layered structure, reaching a PCE of 27.12%. Thanks to the optimized DJ device's high performance and constant output at low fabrication and process costs, it is clear that PSCs could be used on a large scale to provide sustainable energy solutions. In addition to helping with efficiency, ETL/HTL and structural optimization improve device stability and how well carriers are transported by letting scientists customize energy positions, adjust film thicknesses, and reduce imperfections on the films.

References

[1] Islam, M.A., Paul RA., lead-free inorganic Cs2TiX6-based heterostructure perovskite solar cell design and performance evaluation, Optical and Quantum Electronics. 2023; 55: 957. https://doi.org/10.1007/s11082-023-05238-1.

[2] Kour R, Arya S, Verma S, Gupta J, Bandhoria P, Bharti V, et al. Potential substitutes for replacement of lead in perovskite solar cells: A review, Global Challenges. 2019; 3. https://doi.org/10.1002/gch2.201900050.

[3] Kojima A, Teshima K, Shirai Y, Miyasaka T. Organometal halide perovskites as visible-light sensitizers for photovoltaic cells, Journal of the American Chemical Society. 2009; 131: 6050–6051. https://doi.org/10.1021/ja809598r

[4] Anaya M, Lozano G, Calvo ME, Míguez H. ABX3 perovskites for tandem solar cells, Joule. 2017; 1: 769–793. https://doi.org/10.1016/j.joule.2017.09.017

[5] Kadem BY, Abbas EM. Optimization of several parameters towards 30 % efficiency perovskite based solar cell using SCAPS-1D software, Iraqi Journal of Physics University of Baghdad. 2024; 22: 117–129.

[6] Islam MS, Sobayel K, Al-Kahtani A, Islam MA, Muhammad G, Amin N, et al. Defect study and modelling of SnX3-based perovskite solar cells with SCAPS-1D, Nanomaterials. 2021; 11. https://doi.org/10.3390/nano11051218.

[7] Zhao J, Zhang Z, Li G, Aldamasy MH, Li M, Abate A. Dimensional tuning in lead-free tin halide perovskite for solar cells, Advanced Energy Materials. 2023; 13. https://doi.org/10.1002/aenm.202204233.

[8] Wang K, Zheng L, Hou Y, Nozariasbmarz A, Poudel B, Yoon J, et al. Overcoming Shockley-Queisser limit using halide perovskite platform?, Joule. 2022; 6: 756–771. https://doi.org/10.1016/j.joule.2022.01.009.

[9]   Dedecker K, Grancini G. Dealing with lead in hybrid perovskite: A challenge to tackle for a bright future of this technology?, Advanced Energy Materials. 2020; 10: 1–8. https://doi.org/10.1002/aenm. 202001471

[10]   Ou M, Qiu L, Ding C, Zhou W, Zheng C, Wu Y, et al. 2D Dion Jacobson/3D perovskite heterojunction solar cells without hole transport layer: Further optimize the performance by SCAPS-1D, Materials Today Communications. 2024; 40. https://doi.org/10.1016/j.mtcomm.2024.109575.

[11]   Chen HN, Wei ZH, He HX, Zheng XL, Wong KS, Yang SH. Solvent Engineering Boosts the Efficiency of Paintable Carbon-Based Perovskite Solar Cells to beyond 14%, Advanced Energy Materials. 2016; 6. https://doi.org/10.1002/aenm.201502087.

[12]   Nasti G, Abate A. Tin halide perovskite (ASnX3) solar cells: A comprehensive guide toward the highest power conversion efficiency, Advanced Energy Materials. 2020; 10. https://doi.org/10.1002/ aenm.201902467.

[13]   Filippetti A, Kahmann S, Caddeo C, Mattoni A, Saba M, Bosin A, et al. Fundamentals of tin iodide perovskites: A promising route to highly efficient, lead-free solar cells, Journal of Materials Chemistry A. 2021; 9: 11812–11826. https://doi.org/10.1039/d1ta01573g.

[14]   Yu BB, Chen Z, Zhu Y, Wang Y, Han B, Chen G, et al. Heterogeneous 2D/3D Tin-Halides perovskite solar cells with certified conversion efficiency breaking 14%, Advanced Materials. 2021; 33: 1–10. https://doi.org/10.1002/adma.202102055.

[15]   Zhao Z, Gu F, Li Y, Sun W, Ye S, Rao H, et al. Mixed-organic-cation tin iodide for lead-free perovskite solar cells with an efficiency of 8.12%, Advanced Science. 2017; 4. https://doi.org/10.1002/advs. 201700204.

[16]   Hao L, Li T, Ma X, Wu J, Qiao L, Wu X, et al. A tin-based perovskite solar cell with an inverted hole-free transport layer to achieve high energy conversion efficiency by SCAPS device simulation, Optical and Quantum Electronics. 2021; 53: 1–17. https://doi.org/10.1007/s11082-021-03175-5.

[17]   Kavitha MV, Anjali CK, Sudheer KS. Device simulation and optimization of HTL-free perovskite solar cell with CH3 NH3 SnBr3 as the absorber layer using solar cell capacitance simulator software, Journal of Ovonic Research. 2024; 20: 245–254. https://doi.org/10.15251/JOR.2024.202.245

[18]   Sunny A, Rahman S, Khatun MM, Ahmed SR A. Numerical study of high performance HTL-free CH3NH3SnI3-based perovskite solar cell by SCAPS-1D, AIP Advances. 2021; 11. https://doi.org/10. 1063/5.0049646.

[19]   Verma Akash A, Dwivedi DK, Lohia P, Singh PK, Yadav RK, Kumar M, et al. Achieving 31.16 % efficiency in perovskite solar cells via synergistic Dion-Jacobson 2D-3D layer design, Journal of Alloys and Compounds. 2025b; 1010: 177882. https://doi.org/10.1016/j.jallcom.2024.177882.

[20]   Zeng F, Kong W, Liang Y, Li F, Lvtao Y, Su Z, et al. Highly stable and efficient formamidinium-based 2D Ruddlesden–Popper perovskite solar cells via lattice manipulation, Advanced Materials. 2023; 35. https://doi.org/10.1002/adma.202306051.

[21]   Blancon JC, Even J, Stoumpos CC, Kanatzidis MG, Mohite AD. Semiconductor physics of organic–inorganic 2D halide perovskites, Nat Nanotechnol. 2020; 15: 969–985. https://doi.org/10. 1038/s41565-020-00811-1

[22]   Chaurasia S, Lohia P, Dwivedi DK, Pandey R, Madan J, Agarwal S, et al. Enhancing perovskite solar cell efficiency to 28.17% by Integrating Dion-Jacobson 2D and 3D phase perovskite absorbers, Inorganic Chemistry Communications. 2024a; 170: 113140. https://doi.org/https://doi.org/10.1016/j. inoche.2024.113140.

[23]   Mao L, Ke W, Pedesseau L, Wu Y, Katan C, Even J, et al. Hybrid Dion-Jacobson 2D lead iodide perovskites to cite this version : HAL Id : Hal-01714840 Hybrid Dion-Jacobson 2D lead iodide perovskites Optical and Quantum Electronics. 2018; 140 (10): 3775-3783.

[24]   Mohammed MKA, Al-Mousoi AK, Kumar A, Sabugaa MM, Seemaladinne R, Pandey R, et al. Harnessing the potential of Dion-Jacobson perovskite solar cells: Insights from SCAPS simulation

techniques, Journal of Alloys and Compounds. 2023; 963: 171246. https://doi.org/10.1016/j.jallcom. 2023.171246.

[25] Chaurasia S, Lohia P, Dwivedi DK, Pandey R, Madan J, Yadav S, et al. Highly efficient and stable Dion–Jacobson(DJ) 2D-3D perovskite solar cells with 26 % conversion efficiency: A SCAPS-1D study, Journal of Physics and Chemistry of Solids. 2024b; 191: 112038. https://doi.org/10.1016/j.jpcs.2024. 112038.

[26] Katiyar P, Dwivedi DK, Lohia P, Pandey R, Madan J, Anand A, et al. Journal of physics and chemistry of solids synergistic combination of DJ 2D-3D layers : Achieving 30 . 75 % perovskite solar cell efficiency, Journal of Physics and Chemistry of Solids. 2025; 207: 112877. https://doi.org/10.1016/j. jpcs.2025.112877.

[27] Pathak G, Dwivedi DK, Lohia P, Singh YK, Pandey R, Madan J, et al. Optimizing the filtered spectrum and various transport layers of synergistic Dion–Jacobson 2D–3D perovskite tandem solar cell: Achieving 32.11% efficiency, Journal of Inorganic and Organometallic Polymers and Materials. 2025. https://doi.org/10.1007/s10904-025-03863-9

[28] Islam A, Muhammad G. Optik photoelectric performance of environmentally benign Cs 2 TiBr 6 -based perovskite solar cell using spinel NiCo 2 O 4 as HTL, Optik (Stuttg). 2023; 272: 170232. https://doi.org/10.1016/j.ijleo.2022.170232

[29] Patel PK. Device simulation of highly efficient eco-friendly CH3NH3SnI3 perovskite solar cell, Scientific Reports. 2021; 11: 1–11. https://doi.org/10.1038/s41598-021-82817-w

[30] Singh YK, Dwivedi DK, Lohia P, Pandey R, Madan J, Hossain MK, et al. Filtered spectrum modeling of high-performance perovskite tandem solar cells: Tailoring absorber properties and electron/hole transport layers for 31.55 % efficiency, Journal of Physics and Chemistry of Solids. 2024; 192: 112096. https://doi.org/10.1016/j.jpcs.2024.112096.

[31] Hossain MK, Islam MA, Uddin MS, Paramasivam P, Hamid JA, Alshgari RA, et al. Design and simulation of CsPb.625Zn.375IBr2-based perovskite solar cells with different charge transport layers for efficiency enhancement, Scientific Reports. 2024a; 14: 1–22. https://doi.org/10.1038/s41598-024-81797-x.

[32] Hossain MK, Islam S, Sakib MN, Uddin MS, Toki GFI, Rubel MHK, et al. Exploring the optoelectronic and photovoltaic characteristics of lead-free Cs2TiBr6 double perovskite solar cells: A DFT and SCAPS-1D Investigations, Advanced Electronic Materials. 2024b; 2400348: 1–18. https://doi.org/10. 1002/aelm.202400348.

[33] Verma Akash A, Dwivedi DK, Lohia P, Agarwal S, Kulshrestha U, Kumar M, et al. Enhancing efficiency of lead-free Cs2TiIxBr6-x perovskite solar cells through linear and parabolic grading strategies: Toward 31.18% efficiency, Progress in Photovoltaics: Research and Applications. 2025; 1–17. https://doi.org/10.1002/pip.3895.

[34] Verma Akash A, Dwivedi DK, Lohia P, Pandey R, Madan J, Agarwal S, et al. Innovative design strategies for solar cells: Theoretical examination of linearly graded perovskite solar cell with PTAA as HTL, Journal of Physics and Chemistry of Solids. 2025a; 196: 112401. https://doi.org/10.1016/j.jpcs. 2024.112401.

[35] Singh YK, Dwivedi DK, Lohia P, Pandey R, Madan J, Agarwal S, et al. Current matching and filtered spectrum analysis of wide-bandgap/narrow-bandgap perovskite/CIGS tandem solar cells: A numerical study of 34.52 % efficiency potential, Journal of Physics and Chemistry of Solids. 2025; 196: 112300. https://doi.org/10.1016/j.jpcs.2024.112300.

[36] Kumar A, Singh S, Mohammed MKA, Esmail Shalan A. Effect of 2D perovskite layer and multivalent defect on the performance of 3D/2D bilayered perovskite solar cells through computational simulation studies, Solar Energy. 2021; 223: 193–201. https://doi.org/10.1016/j.solener.2021.05.042

[37] Rumble J CRC Handbook of Chemistry and Physics version 2008 2008.

[38] Best Research-Cell Efficiency Chart | Photovoltaic Research | NREL. n.d. https://www.nrel.gov/pv/ cell-efficiency.html (accessed March 27, 2025).

Sunita Kumari Sah, Pooja Lohia, and R.K. Chauhan

Chapter 6
Simulation tools for perovskite solar cell design and optimization

Abstract: This article highlights improved stability and superior optical performance achieved through the utilization of amalgamated-cation lead amalgamated-halide perovskites, specially $[FA_yCs_{1-y}Pb(I_yBr_{1-y})3]_y = 0.85$, as the functional layer. The transport layer for electrons (ETL) consists of Nb_2O_5, while the transport layer for holes (HTL) employs Spiro-MeOTAD. The cell configuration $FTO/Nb_2O_5/[FA_yCs_{1-y}Pb(I_yBr_{1-y})3]_y = 0.85/Spiro-MeOTAD/Au$ is analyzed using SCAPS-1D software, making the efficiency of power conversion (PCE) of 17.71%. Additionally, the analysis outcome into the optimization by hole transport layer materials, namely $CuAlO_2$, PTAA, Cu_2O, $SrCu_2O_2$, CNTS, MoO_3, and Spiro-MeOTAD. It is found that the hole transport layer of MoO_3, with a carrier mobility of approximately 100 cm^2/Vs, has an increased diffusion length of carrier and its conductance can extend up to a few micrometers. Consequently, MoO_3 achieves the highest efficiency of power conversion of 23.13%, while $CuAlO_2$ exhibits the least PCE of 16.63%. The solar cell structure with Nb_2O_5 as ETL, MoO_3 as HTL and $[FA_yCs_{1-y}Pb(I_yBr_{1-y})3]_y = 0.85$ absorber layer is investigated for different device constraints such as bandgap, thickness, and defect density present in the absorber layer, the effect of interface defect density at the region of ETL/absorber, absorber/HTL interfaces, as well as internal series-shunt resistances. It is also analyzed that for the optimal perovskite solar cell performance, Au is used as the rear end contact material. Furthermore, the study explores the alternative transport HTL materials to Spiro-MeOTAD. The temperature settings were adjusted and meticulously calculated to achieve their optimal values. Simulations were carried out by taking the optimized parameters and the obtained outcomes shows a short circuit current density (J_{sc}) of 30.3319 mA cm^{-2}, a fill factor of 84.66%, an open circuit voltage (V_{oc}) of 1.0553 V and a PCE of 27.09%. Hence, the PCE of 27.09% is gained for the refined solar cell configuration: $FTO/Nb_2O_5/[FA_yCs_{1-y}Pb(I_yBr_{1-y})3]_y = 0.85/MoO_3/Au$. These results offer valuable insights for designing and constructing highly effective solar cell technology.

Keywords: ETL, HTL, PCE, perovskite, SCAPS-1D

Acknowledgment: The authors thank Dr. Marc Burgelman and his team at the University of Gent, Belgium, for granting access to the SCAPS-1D simulator.

Sunita Kumari Sah, Department of Electronics and Communication Engineering, Madan Mohan Malviya University of Technology, Gorakhpur 273010, India, e-mail: sunitasah2k7@gmail.com
Pooja Lohia, and R.K. Chauhan, Department of Electronics and Communication Engineering, Madan Mohan Malviya University of Technology, Gorakhpur 273010, India

https://doi.org/10.1515/9783111726847-006

6.1 Introduction

The perovskite solar cells (PSCs) are made by the composition of organic and inorganic material such as lead-metal and halide has been of great concern since they have the highest power conversion efficiency (PCE), they are cheap to make and utilize common elements on earth hence driving the photovoltaic (PV) technology. While perovskite material is named after the Russian mineralogist and crystallographer Lev Perovski, is a mineral type that was identified two centuries ago. It gained recognition for its intriguing magnetic and electrical properties which exhibit distinctive characteristics, including customizable bandgap, extensive diffusion length, minimal exciting binding energy, and exceptional absorption capabilities [1–3].

The perovskite structure is characterized by the common compound composition ABX_3. Here, A denotes a cation of organic–inorganic material which includes species like Rb^{2+}, $CH_3NH_3^+$ (methylammonium), FA^+ ($CH_3(NH_2)_2^+$ (formamidinium), and Cs^{2+}, while B corresponds to a cation of an inorganic material such as Co^{2+}, Sn^{2+}, Fe^{2+}, Pb^{2+}, and Cu^{2+}. X denotes the halide ion, encompassing species like I^-, Br^-, and Cl^- [4]. Forming thin film perovskite of superior quality involves in managing the crystallization process, often employing a combination of formamidinium an organic material, cesium an inorganic material and lead perovskite. Encapsulated PSCs demonstrated exceptional thermal endurance, maintaining efficiency without any signs of degradation even after enduring 3,300 h at 65 °C [5]. The incorporation of a small quantity of Cs^+ ions into the basic structure of $FAPbI_3$ (forming $FA_{1-y}Cs_yPbI_3$) significantly enhances the stability of the phase, enduring above 100 h of exposure to light irradiation in contrast to the unaltered $FAPbI_3$. Additionally, this modified compound demonstrates notable resilience to moisture, withstanding up to 90% humidity for approximately 4 h, a substantial improvement over the pure material [6]. These perovskites have numerous advantageous and intriguing characteristics, such as affordability and straight forward synthesis methods, highly tuneable, large absorption coefficients, vibrant photoluminescence, elevated dielectric constants, and inherent ferroelectric polarization leading to high conversion rate of power more than 22% [7, 8].

Spiro-MeOTAD stands out as an exceptional small molecule HTL, providing increased solubility and eliminating the need for heat treatment after fabrication. Its elevated melting point and amorphous nature, coupled with enhanced conductivity through dopants, make it a perfect preference for the HTL "transport layer for holes" in PSCs. Moreover, the matched bandgap of this HTL with perovskite facilitates efficient charge transport. Introducing niobium pentoxide (Nb_2O_5) into PSCs offers various innovative benefits, such as more stable, higher mobility of electron, appropriate band alliance, and broad bandgap characteristics. Additionally, it holds promise as an apparent conductive oxide. These advantages collectively add to the advancement of more proficient and steady perovskite based solar cell technologies, crucial for the ongoing progress for solutions of renewable energy [9].

In this study, the most popular PSC simulation tools are discussed with their capabilities and uses in designing and optimization. The simulation of cesium-based PSC was conducted using SCAPS-1D simulation software. Nb_2O_5 was chosen as a beneficial material to make this solar cell more stable. Moreover, the forbidden energy region of Nb_2O_5 has the potential to increase the cell's open-circuit voltage (Voc). However, intrinsic Nb_2O_5 exhibits low electron mobility when utilized as an oxide. One approach to improve the conductivity of Nb_2O_5 is doping it in electron transport layers (ETLs). However, precise control over deposition parameters is necessary. By altering oxygen flow during deposition, conductivity can be enhanced without introducing structural defects. Oxygen vacancies are formed by decreasing the rate of flow of oxygen, which consecutively increases film conductivity, thereby contributing to the efficiency improvement of the solar cells. The cell configuration $FTO/Nb_2O_5/[FA_yCs_{1-y}Pb(I_yBr_{1-y})_3]_{y=0.85}/$ Spiro-MeOTAD/Au is analyzed using the software named SCAPS-1D, resulting in 17.71% conversion efficiency of power. At the interface of spiro-MeOTAD and perovskite, there exists an energy gap of 1.8 eV. This gap effectively restricts the crossing of photoelectrons from the active area (perovskite) to the HTL (spiro-MeOTAD) area. Correspondingly, at the Nb_2O_5/perovskite interface, there exists an energy gap of 1.10 eV, which obstructs the flow of hole from the active area to the electron transport area. Additionally, the mismatch in the conductions bands between Nb_2O_5 and perovskite inhibits electron transfer from Nb_2O_5 to the active area. These obstacles play a crucial role in the setup of a proper n(electron)–i(intrinsic)–p(holes) hetero-junction, thereby improving the overall functionality of the device [10–12].

To achieve well-organized structures and enhance stability of photocurrent, researchers often combine various frequencies and compositions of ABX_3 perovskites. Here, "A" represents organic-inorganic cations such as $CH(NH_2)^{2+}$ (formamidinium [FA]), Cs^+, and $CH_3NH_3^+$ (methylammonium [MA]). "B" denotes inorganic metallic cations like Ge_2^+, Pb_2^+, and Sn_2^+, while "X" signifies mono-valence anions such as I^-, Br^-, and Cl^-. This diverse combination allows for tailored properties and improved performance in perovskite-based devices [12–14]. Shivani et al. utilized various HTLs in the $FTO/SnO_2/[FA_xCs_{1-x}Pb(I_xBr_{1-x})_3]x = 0.85/HTL/Au$ configuration. They stated power conversion efficiencies (PCE) of approximately 7.5% for CIS, 8.9% for P_3HT, 14.2% for CuI, 14.3% for PEDOT: PSS, 14.8% for $CuSbS_2$, 18.2% for CuSCN, and 18.7% for Cu_2O. These findings demonstrate the effectiveness of different HTL materials in improving the functionality of perovskite-based solar cells [15, 16]. In a further replication, Vaibhava et al. investigated different types of transport HTLs within the $FTO/SnO_2/[FA_xCs_{1-x}Pb(I_xBr_{1-x})_3]_x = 0.85/HTLs/Au$ solar cell configuration design. Their study revealed PCE of approximately 20.59% for $SrCu_2O_2$, 17.58% for $CuAlO_2$, 17.53% for PTAA, 13.99% for CNTS, 20.86% for MoO_3, 19.46% for D-PBTTT-14, and 18.65% for Cu_2O. These results highlight the significant influence of various HTL materials on enhancing the functionality of perovskite-based solar cells [17].

Table 6.1: Parameters used for the FTO layer, perovskite layer, HTL, and ETL [17–22].

Parameter	FTO	Nb$_2$O$_5$	[FA$_y$Cs$_{1-y}$Pb(I$_y$Br$_{1-y}$)3]$_y$ =0.85	Spiro-MeOTAD
Material thickness (nm)	500	50	350	50
Electron affinity χ, EA (eV)	4.0	3.9	4.09	2.20
ε (relative)	9.0	9.0	6.6	3.0
Bandgap, E_g (eV)	3.5	3.4	1.59	2.90
Effective DOS of CB, N_c (cm^{-3})	2.2×10^{18}	3.7×10^{18}	2.0×10^{19}	2.2×10^{18}
Effective DOS of VB, N_V (cm^{-3})	2.2×10^{18}	1.8×10^{19}	2.0×10^{18}	2.2×10^{18}
Mobility, μ_e (cm^2/Vs)	20	100	8.16	1.0×10^{-4}
Mobility, μ_h (cm^2/Vs)	10	25	2.0	1.0×10^{-4}
Density of donor, N$_D$ (cm^{-3})	1.0×10^{15}	1.0×10^{18}	1.3×10^{16}	–
Density of acceptor, N$_A$ (cm^{-3})	–	–	1.3×10^{16}	1.3×10^{18}
Density of defect, N$_t$ (cm^{-3})	1.0×10^{18}	1.0×10^{15}	1.0×10^{15}	1.0×10^{15}

Table 6.2: Defect density at the interface [6, 15, 23].

Interface	(n, p) (cm^2)-capture cross section	Defect type	Distribution of energy	EV (eV)-energy level	Density (cm^{-2})
Hole transport layer/absorber layer	1.0×10^{-19}	Neutral	Single	0.60	1.0×10^{10}
	1.0×10^{-19}				
Absorber layer/electron transport layer	1.0×10^{-19}	Neutral	Single	0.60	1.0×10^{10}
	1.0×10^{-19}				

6.2 Device architecture and simulation tools

Designing and optimizing PSCs requires simulation software capable of modeling optical, electrical, and material properties of the cells [24, 25]. Here are some commonly used tools for PSC simulation:

- **COMSOL multiphysics:** It is an advanced simulation software for designing and optimizing PSCs. It utilizes the finite element method) to accurately model complex optical, electrical, and thermal processes in PSCs. COMSOL provides various physics modules tailored for different aspects of PSC simulation. These include wave optics module for optical simulation, semiconductor module for electrical simulation, heat transfer module for thermal simulation and structural mechanics module for mechanical simulation. It has a multiphysics coupling capability of combining all these modules in a single framework. Extensive coding expertise is not required for simulation but scripting is available through MATLAB/Livelink.

Allows parameter sweeps for optimization of materials, structures, and operating conditions with full three dimensional (3D) modeling. This simulation tool enhances PSC efficiency by evaluating optical absorption, charge transport, and thermal stability.

- **General-purpose PV device model:** PSC simulation is an open-source simulation software specifically designed for modeling PV devices, including PSCs. It can simulate optical, electrical, and thermal properties to optimize solar cell performance. It supports one dimension (1D) and two dimension device simulation to models thin-film and planar heterojunction structures. It uses a transfer matrix method for light absorption modeling and solves drift-diffusion equations for charge transport in the PSC. It is capable to simulate transient response and hysteresis effects and models temperature-dependent stability and degradation of the PSC and organic solar cell. Comparably these tools are faster than COMSOL.
- **Automat for simulation of heterojunction solar cells:** It is a simulation tool specifically designed for heterojunction solar cells, including PSCs and silicon-perovskite tandem cells. It solves optoelectronic transport equations in 1D and provides detailed analysis of carrier transport and recombination. It is a 1D-simulation software that simulations of energy band diagrams for different material interfaces. Capable of modeling multijunction perovskite/silicon structures with accurate electrical transport and interface defect parameters of the cell.
- **Quokka3:** Fast 3D simulation of PSCs and silicon solar cells, particularly silicon-based and tandem structures (e.g., perovskite-silicon tandem cells). It is known for its high computational speed and accuracy in solving semiconductor transport equations using the quasi-3D finite-volume method. It is faster than full 3D finite element solvers. It analyze interface recombination and band alignment for planar and textured solar cells. It is capable to support realistic textured silicon surface with balance speed and accuracy. Ideal for perovskite-Si tandem design and optimization simulation.
- **The SCAPS-1D** stands as a pivotal simulation software utilized for assessing the performance as quantum efficiency (QE) of PSC device structures. Electronics and information systems have created this package at the Ghent University in Belgium. It serves as a vital link bridging various experimental and theoretical approaches in fabricating device structures. This software is widely adopted for its multifaceted significance and its accessibility as a freely available tool. Primarily grounded on fundamental applications such as the Poisson calculations and continuity calculations for both holes and electrons, it serves as a robust simulator in the arena of perovskite based solar cell exploration.

The equation of Poisson is as follows:

$$\frac{\partial}{\partial x}\left(\varepsilon(x)\frac{d\Psi}{dx}\right) = -e\left[p(x) - n(x) + N_D^+(x) - N_A^-(x) + p_t(x) - n_t(x)\right] \tag{6.1}$$

$$\frac{\partial n}{\partial t} = G_n(x) + \frac{1}{e}\frac{\partial J_n}{\partial x} - U_n(x, t) \tag{6.2}$$

$$\frac{\partial p}{\partial t} = G_p(x) + \frac{1}{e}\frac{\partial J_p}{\partial x} - U_p(x, t) \tag{6.3}$$

In the solar cell context, x represents positional coordinates, e is the electronic charge, ψ signifies electrostatic potential, n as electron concentrations and p as hole concentrations, p_t and n_t represent trapped hole and electron distributions, N_A^- and N_D^+ denote the uniform shallow acceptor and donor concentrations, $G(x)$ represents the ratio of carrier generated, U_n and U_p denotes the recombination rates of charge carriers respectively. J_n denotes the current absorption of electrons and J_p as the current absorption of holes [26]. The thermal speed across all layers is considered as 1.00×10^7 cm/s. Structure of the solar cell was illuminated by the air mass of 1.50 having sunlight spectrum with solar energy of 1×10^3 W/m^2 at 300 K temperature. This structure of solar cell necessitates a low (R_s) series resistance of 1 Ω cm^2 and a high (Rsh) shunt resistance of 1,000 Ω cm^2 to optimize performance. The input values used for numerical analysis are detailed in Tables 6.1 and 6.2.

6.3 Results and discussion

Figure 6.1(a) illustrates energy band diagram of the PSC along with the suggested device configuration. Near the junction of hole transport material and the active layer of perovskite, there exists a barrier potential of 1.90 eV which prevents the crossing of electrons from the active absorber toward hole transport material. Similarly, a barrier potential of 1.81 eV near the electron transport material and absorber material obstructs the movement of holes from the active absorber material to ETL. As a result, the presence of an energy barrier between their conduction bands restricts the drive of electrons from ETL to the absorber material. Consequently, these junctions are pivotal in forming an appropriate n-i-p junction and improving the electrical properties of the solar cell.

Here, Figure 6.2(a) shows the achieved spectral response curve and Figure 6.2(b) illustrates the current density–voltage (J–V) characteristics curve. Analysis of these figures reveals a simulated QE of 86.0% across the electromagnetic spectrum, alongside a resulting "short-circuit current density – J_{sc}" of 20.36 mA/cm^2.

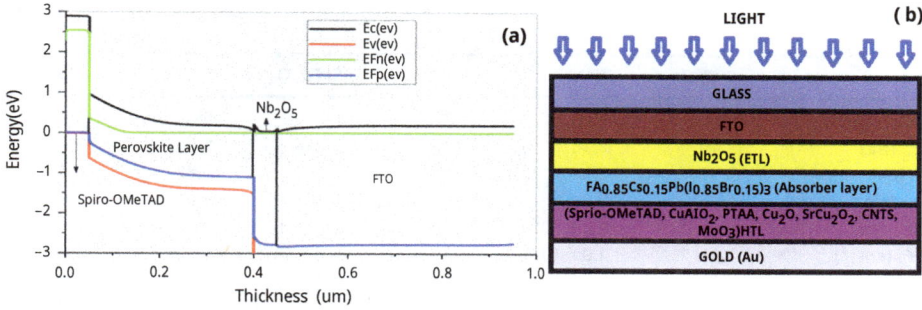

Figure 6.1: (a) Energy band diagram and (b) configuration of the suggested device.

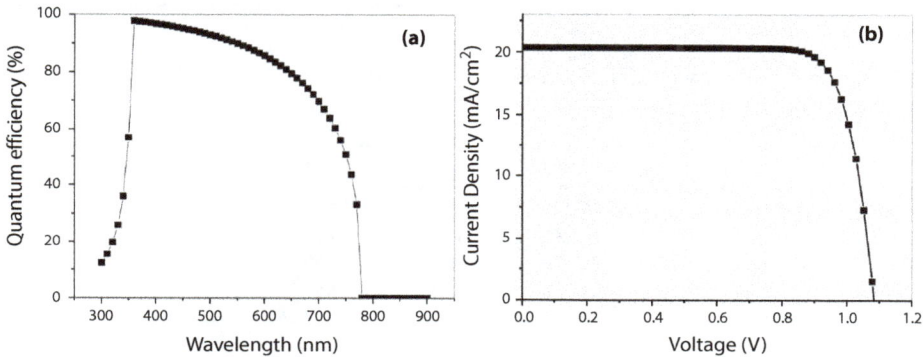

Figure 6.2: (a) Quantum efficiency of the suggested device, (b) JV curve of the suggested device.

6.3.1 Impact of the hole transport layer

The preceding segment has elaboration on the initial architecture of the PSC. This section underlines the significance of the hole transport layer corresponding to various organic and inorganic materials. Among them, Nb_2O_5 has been chosen as the electron transport layer. The investigated hole transport layer materials, namely $CuAlO_2$, PTAA, Cu_2O, $SrCu_2O_2$, CNTS, MoO_3, and Spiro-MeOTAD, have been identified through extensive literature review and references. Detailed material parameters for these substances are detailed in Table 6.3.

All HTL layers were set with a uniform thickness of 50 nm. Subsequent measurements were conducted to obtain the spectral response and the JV curve of the devices. JV characteristics indicate an excellent design of the perovskite-based solar cell with *n-i-p* configuration. Table 6.4 displays the response of PSCs with various types of HTLs. In *n-i-p* structure, light enters through the ETL, Nb_2O_5, which has an energy gap of 3.4 eV, enabling it to absorb higher-energy photons and allowing MoO_3 to facilitate

Table 6.3: Different HTLs used for the proposed device structure [8, 10, 12, 27, 28].

Parameters	CuAlO$_2$	PTAA	Cu$_2$O	SrCu$_2$O$_2$	CNTS	MoO$_3$
Thickness (nm)	50	50	50	50	50	50
E$_g$ (eV)	3.46	2.96	2.17	3.3	1.74	3.0
χ (eV)	2.5	2.3	3.20	2.2	3.87	2.5
ε (relative)	60	9.0	7.11	9.77	9.0	12.5
N$_c$ (cm^{-3})	2.2 × 10^{18}	1.0 × 10^{21}	2.02 × 10^{17}	2.2 × 10^{18}	2.2 × 10^{18}	2.2 × 10^{18}
N$_v$ (cm^{-3})	1.8 × 10^{19}	1.0 × 10^{21}	1.01 × 10^{19}	1.8 × 10^{19}	1.8 × 10^{19}	1.8 × 10^{19}
μ_e (cm^2 Vs)	2.0	1	200	0.1	11	25
μ_h (cm^2 Vs)	8.6	40	80	0.46	11	100
N$_D$ (cm^{-3})	–	–	–	–	–	–
N$_A$ (cm^{-3})	3.6 × 10^{18}	1.0 × 10^{20}	1.0 × 10^{17}	6.1 × 10^{17}	1.0 × 10^{19}	1.0 × 10^{18}
N$_t$ (cm^{-3})	1.0 × 10^{15}	1.0 × 10^{15}	1.0 × 10^{14}	1.0 × 10^{15}	1.0 × 10^{15}	1.0 × 10^{15}

Table 6.4: Performance parameters for different HTLs in FTO/Nb$_2$O$_5$/[FA$_y$Cs$_{1-y}$Pb(I$_y$Br$_{1-y}$)$_3$]$_y$=0.85/HTL/Au PSCs.

HTLs	V_{oc} (V)	J_{sc} (mA/cm^2)	FF (%)	η (%)
SrCu$_2$O$_2$	1.297	20.361	86.87	22.96
CuAlO$_2$	1.214	20.360	67.23	16.63
PTAA	1.111	20.359	86.15	19.50
CNTS	1.296	20.362	87.18	23.01
MoO$_3$	1.298	20.361	87.53	23.13
Cu$_2$O	1.277	20.362	83.85	21.80

efficient charge transport. This results in high shunt resistance (R_{sh}) and low series resistance (R_s), enhancing the generation of charge carriers in PSCs. MoO$_3$, with a carrier mobility of approximately 100 cm^2/Vs, has an increased diffusion length of carrier and its conductance can extend up to a few micrometers. Consequently, MoO$_3$ achieves the highest efficiency of power conversion of 23.13%, while CuAlO$_2$ exhibits the least PCE of 16.63%.

6.3.2 Effect of heat on PSCs

As the heat increases, the speed of charged particles increases and hence, the recombination rate of electrons and holes also increases, leading to a decrease in the count of charge carriers which are free. This change in heat affects the overall functionality of the solar cell. To analyze the effect of working point temperatures, ranging from 280 to 420 K, on the PV constraints of solar cells are obtained, focusing on the optimal absorber thickness.

As per the results shown in Figure 6.3(a–d), we see that with increasing temperature, the Voc, the efficiency of power conversion and the fill factor (FF) of all type of solar cells decreases, while the only factor, which remains constant across all solar cell configurations, is the short-circuited current density. This behavior occurs because the bandgap, concentration of carriers, mobility of charge carrier, and material resistance changes at higher temperatures. Additionally, the electrons gaining more energy at higher temperatures becomes more unstable, causing to increase the recombination rate and ensuing in lower PCE. To achieve optimal response, 300 K was chosen as the temperature of working point [29]. Voc remarkably drops with the temperature rising from 1.32 to 1.16 V, FF from 88.11% to 83.51% and the efficiency of power conversion 23.75% to 20.14%.

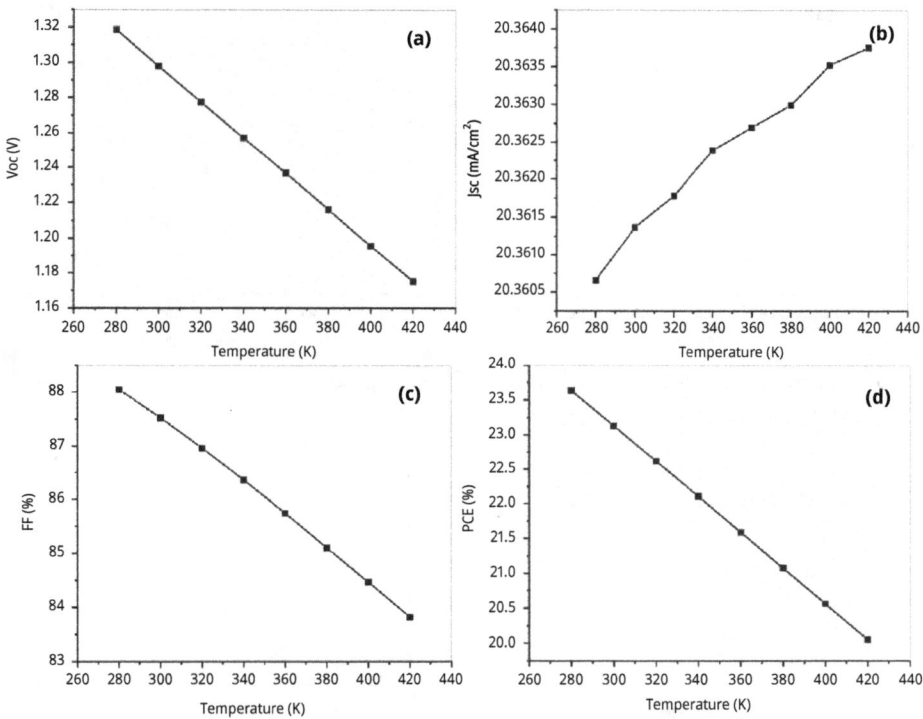

Figure 6.3: Performance parameters with varying temperatures of photovoltaic cell.

6.3.3 Impact of bandgap of the active absorber layer

To analyze the effect of bandgap for the active layer is changed from 1.3 eV to 2.0 eV as shown in Figure 6.4(a–d). To achieve a desirable PCE it is required to have an optimal bandgap for the HTL, absorber layer, and ETL. The results of simulation show that as the bandgap increases, all performance parameters except for the V_{oc} decrease. This occurs because V_{oc} is directly proportional to the bandgap. A wider bandgap results in a higher Voc due to the segregation of excitons during charge carrier generation. However, as the bandgap widens, the light photons absorption diminishes, leading to the reduction in the Jsc. An increase in bandgap also decreases FF, primarily due to mismatch of the lattice between the HTL and the active absorber layer. This mismatch adds to the lower presentation of the device configuration at higher active layer bandgaps [30]. Therefore, to attain the efficiency of 25% for the perovskite based on FA material need to have bandgap of 1.40 eV for the absorber layer.

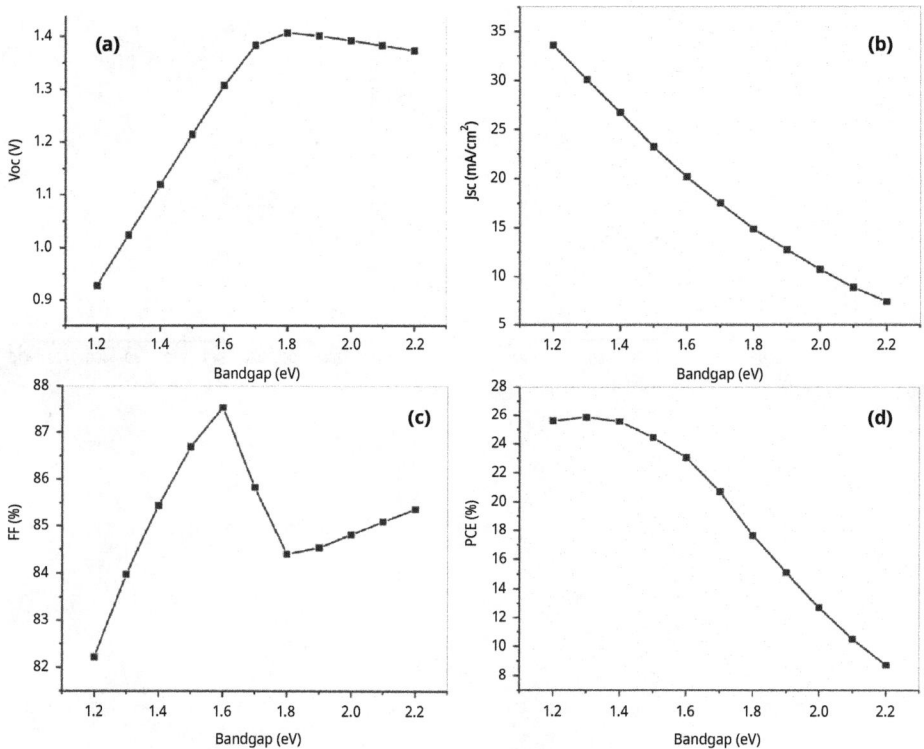

Figure 6.4: Performance parameters with varying bandgap of absorber layer of photovoltaic cell.

6.3.4 Impact of width of the absorber layer

To improve the functionality and stability of the solar cell, active layer width optimization is very essential. This layer absorbs light and produces carriers generated by the photons. The width of the active absorber is changed from 100 to 1,000 nm within the cell arrangement, the simulation results are examined to analyze the working of the solar cell, while all other values are kept unchanged.

As revealed in Figure 6.5(b), the short-circuit current density increases due to the increase in the width of the absorber region. This increase in Jsc is accredited to the creation of increased hole and electron pairs with a thicker absorber layer. Short-circuit current density increases quickly up to a thickness of 410 nm, beyond which it continues to increase, but at a much slower rate. This phenomenon occurs, as a thick absorber region does not produce a significantly higher number of carriers.

Again Figure 6.5(c) shows the FF changes with the change in the absorber layer width. The FF starts to decrease as the width of the region increases. This is also because of the increased power losses in the solar cell [21, 31]. The expression for V_{OC} is:

$$V_{oc} = \frac{nKT}{q} \ln\left(1 + \frac{I_{sc}}{I_0}\right) \tag{6.4}$$

Here, n is the ideality factor, $\frac{KT}{q}$ is the thermal equivalent of voltage, I_{sc} is the current under illuminance, and I_0 is the dark current.

Electron–hole recombination is reduced when the absorber layer is very thin; this results in lowering the I_0. Conversely, the increase in thickness makes the additional carrier concentration higher, and due to which short circuit current increases and consequently the open circuit voltage decrease Figure 6.5(a). Hence, efficiency of power conversion (26.45%), current density (24.126 mA/cm^2), FF (87.15%), and open circuit voltage (1.272 V) are achieved when keeping the width of 900 nm for absorber of the solar cell.

6.3.5 Impact of work function of the electrode

Electrodes of metal are used to make the solar cells stable, which significantly impact device performance. Typically, elements like Ni, Pt, C, Ag, Pd, Se, Fe, and Au serve as terminal materials. In statistical simulations, the anode terminal material used is Au. As illustrated in Figure 6.6(a–d), contact electrodes with lower work function results in the decrease in built-in potential, leading to reduced collection of photo-generated carriers. Consequently, the Jsc as well as the V_{oc} decrease. This is due to the formation of Schottky barrier at the interface of perovskite material and rear contact material, when the value is below 5 eV, significantly reducing the functionality of the perovskite-based solar cells. FF and the efficiency of power conversion also decreases due to the hole collection at the rear side contact of the device which is due to the effect of increased height of Schottky barrier [14, 30]. The response of perovskite-based solar cell improves with increasing metal work function up to approximately 5.1 eV, beyond which no further enhancement is observed. For optimal PSC performance, Au is used as the rear end contact material.

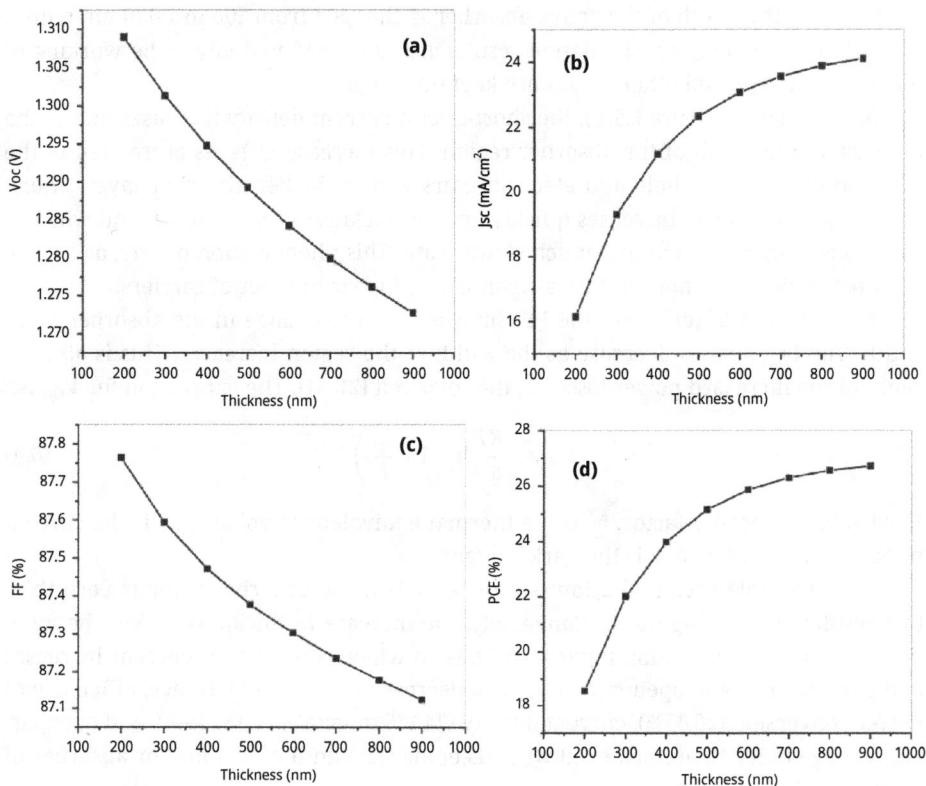

Figure 6.5: Performance factors with varying width of absorber layer of photovoltaic cell.

6.3.6 Effect of defect concentration for perovskite layer of PV cell

The functionality of a cell largely involves the good quality material with acceptable defects for the perovskite layer which possesses key features such as photo generation of electrons, transportation of carrier, and recombination. Consequently, studying defects is crucial for enhancing the properties of photovoltaic cells. Point defects, including Schottky defects, vacancies, anti-site defects and interstitials, can either favorably or unfavorably affect the conductivity of charge carriers within the device. Defects such as acceptor defect introduce holes into the valence band, while donor defects contribute electrons to the conduction band. The Shockley–Read–Hall (SRH) recombination model is useful for describing the impact of densities of different defects on the device response [32]:

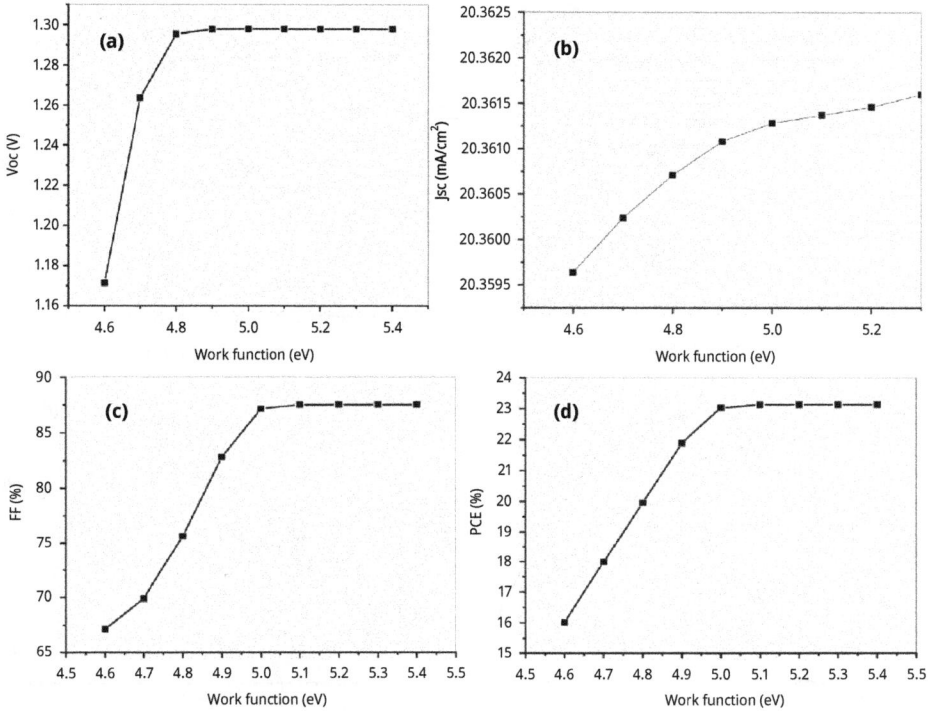

Figure 6.6: Performance parameters with varying metal electrode work function of photovoltaic cell.

$$R = \frac{np - n_i^2}{\tau_p(n+n_1) + \tau_{n(p+p_1)}} \tag{6.5}$$

Here, n_i is the built-in concentration of carrier, n_1 and p_1 represent the SRH densities, τ_n signifies the electron's lifetime, τ_p signifies the holes' lifetime, and p and n denote the concentrations of holes and electrons at nonequilibrium. The change in the solar cell parameters by varying the defect immersion is illustrated in Figure 6.7(a–d). It is observed that as the defect concentration in the perovskite based solar cell layer increases from 10^{11} cm^{-3} to 10^{16} cm^{-3}, the efficiency of power conversion reaches 15.54%, the FF is approximately 42.86%, the open circuit voltage is around 1.12 V, and the "short circuit current density" is about 16.82 mA/cm^2. The rapid decline in solar cell features with increasing defect immersion indicates that higher defect densities act as recombination centers, reducing carrier lifetimes.

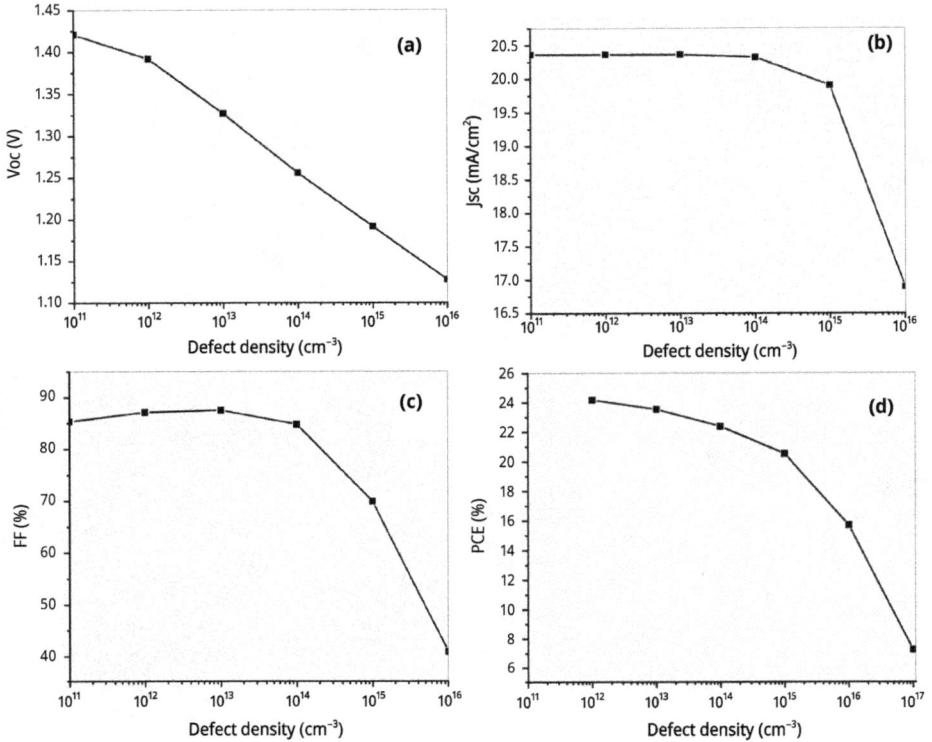

Figure 6.7: Performance parameters with varying defect concentration of photovoltaic cell.

6.3.7 Effect of defect concentration for interfacing layer of PV cell

The density of interface defects increases due to exposure to light, oxygen, humidity, and elevated operating temperatures, leading to the degradation of device performance. Therefore, investigating the interfacing layer defect concentration is crucial. Here, the interface defect density was changed from 10^9 cm^{-2} to 10^{17} cm^{-2} between the $Nb_2O_5/[FA_yCs_{1-y}Pb(I_yBr_{1-y})_3]_{y\,=0.85}$ and $[FA_yCs_{1-y}Pb(I_yBr_{1-y})_3]_{y\,=0.85}/MoO_3$.

As shown in Figure 6.8(a) and (b), there is a rise in traps and recombination centers with the increase in defect concentration at the ETL region and the active region interface. And so, the PCE drops from 23.30% to 20.51% with the increased defect concentration from 10^9 to 10^{17} cm^{-2} at this interface increases. It is evident that the reduction in PCE due to the rise in defect concentration at ETL region and the active region interface is significantly greater than the reduction caused by an increase in defect density at the active region and the HTL region interface. This has a great impact on the device performance. At 10^9 cm^{-2} for the electron transport layer and the absorber layer interface Jsc of around 20.516 mA/cm², an Voc of approximately 1.29 V, PCE of 23.30% and FF of approximately 87.56%.

The cell performance is also degraded by the increased defect density at the absorber layer and the hole transport layer interface as shown in Figure 6.8(c and d). When the defect level rises from 10^9 cm^{-2} to 10^{17} cm^{-2} at this interface, the PCE decreases from 23.35% to 15.52%, indicating a diminished effect of defect density as it increases. And when it is less than 10^9 cm^{-2}, results can be comparatively good, given that defects inherently exist at the interface. Both the V_{oc} and the J_{sc} decrease as the defect level of the active region and the HTL region interface increases. Consequently, a defect level of 10^{13} cm^{-2} for the active region and the HTL region interface was selected, resulting in a PCE of around 21.51%, V_{oc} of 1.5 V, FF of about 86.52%, and a J_{sc} of approximately 23.52 mA/cm².

Under light illumination, there is a tenfold increase in the accumulation of hole-electron pairs at the ETL region and the active region interface compared to the active region and the HTL region interface. The excess ETL region and the active region interface tend to raise the rate of recombination [30]. The results obtained for the two different interfaces show better consistency compared to previous studies. A high performing device can be achieved when the concentration of defect is below 10^9 cm^{-2}, taking into account the actual interface defects.

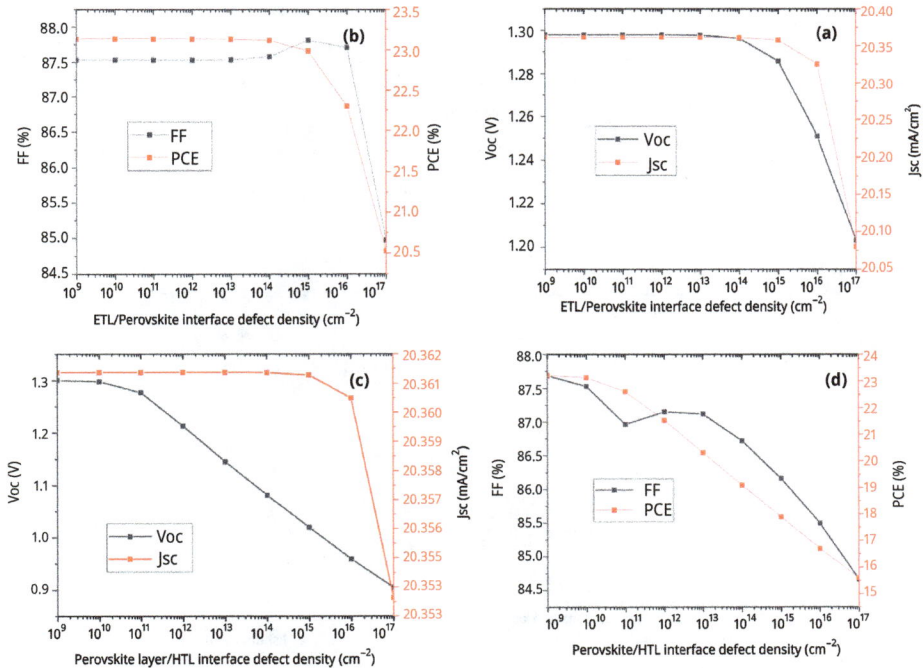

Figure 6.8: Performance parameters with varying interface defect density of photovoltaic cell.

6.3.8 Effect of resistance of PV cell

Figure 6.9(a–d) illustrates the simulated changes in performance characteristics based on the device resistance in series (R_s) and shunt (R_{sh}). Here R_s arises from internal resistance sources such as contacts of metallic electrodes at the front and rear, and the base and emitter current flow. In contrast, R_{sh} is due to mechanical losses like industrial defects and leak currents across junctions. The series resistance predominantly influences the PCE, the J_{sc}, and the FF of the device, whereas the shunt resistance influences the V_{oc}, the PCE, and the FF. The analysis involves varying R_s from 1 to 35 $\Omega \cdot cm^2$ and R_{sh} from 1,000 to 9,000 $\Omega \cdot cm^2$.

All the performance constraints except for V_{oc} are adversely affected as the value of series resistance increases. A slight increase in R_s significantly impacts the FF of the PV cell, which in turn decreases functionality of the device. This reduction in J_{sc} is due to transmission losses and FF is reduced due to degradation in soldering. Conversely, as shunt resistance is increased all parameters are improved except for J_{sc}. The FF is negatively impacted in the same manner as described above when J_{sc} remains stable despite changes in R_{sh}, which consequently deteriorates the cell's PCE. Thus, it can be concluded that the FF of a PSC is **persuaded mutually** by series and shunt resistance. Taking an R_s of 1 $\Omega \cdot cm^2$ and an R_{sh} of 103 $\Omega \cdot cm^2$, the device parameters are FF (84.66%), V_{oc} (1.0553 V), PCE (27.10%), and J_{sc} (30.3319 mA/cm^2).

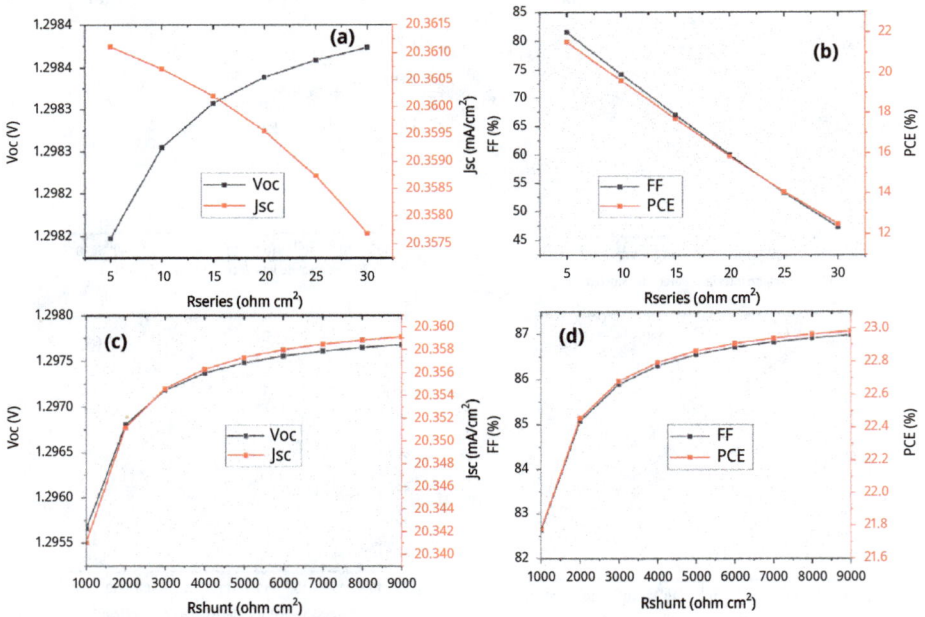

Figure 6.9: Impact of series and shunt resistance on performance of photovoltaic cell.

6.4 Optimized final result

To upgrade the working of PSCs, several aspects such as defect concentration, active layer thickness, shunt and series resistance, interface defects at HTL region/active region and active region/ETL region junctions, operating temperature, absorber layer bandgap, and the rear contact work function have been optimized. Simulations were carried out by taking the optimized parameters and the obtained outcomes shows J_{sc} of 30.3319 mA/cm^2, FF of 84.66%, V_{oc} of 1.0553 V, and PCE of 27.09%. The J–V characteristics curve and QE are detailed in Figure 6.10(a and b), **demonstrating** the enhanced solar cell performance. Table 6.5 presents the comparative representation of solar cell parameters for normal and optimized devices.

Table 6.5: Performance parameters for proposed and optimized PSCs.

Devices	V_{oc} (V)	J_{sc} (mA/cm^2)	FF (%)	η (%)
Normal	1.297	20.361	87.53	23.13
Optimized	1.055	30.331	84.65	27.09

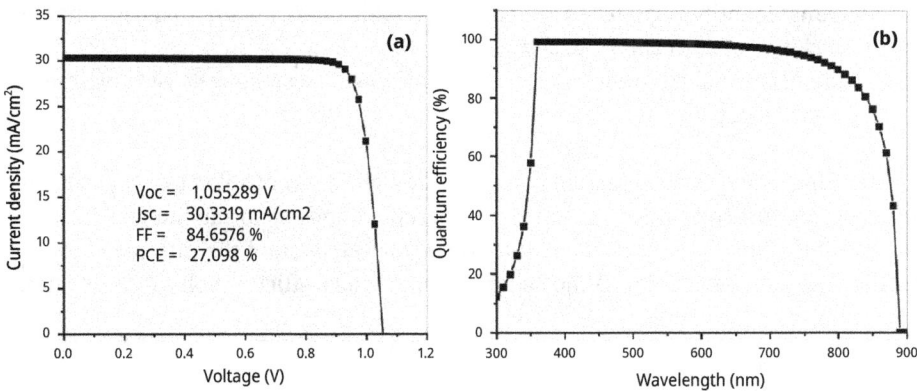

Figure 6.10: (a) J–V curve, and (b) Q–E curve of optimized photovoltaic cell.

A contour plot is created as shown in Figure 6.11 to visualize the effect of changing the width of the perovskite layer and the acceptor density on the efficiency of a solar cell that can be highly insightful. As analyzed in results section, efficiency of solar cell is influenced by numerous factors, and the absorber layer width and acceptor density are two critical parameters. This plot shows the effect of absorber layer width and acceptor density on the efficiency of a solar cell and is interpreted as follows:
– **Contours:** Each contour line represents a constant efficiency value. The closer the lines, the steeper the change in efficiency.

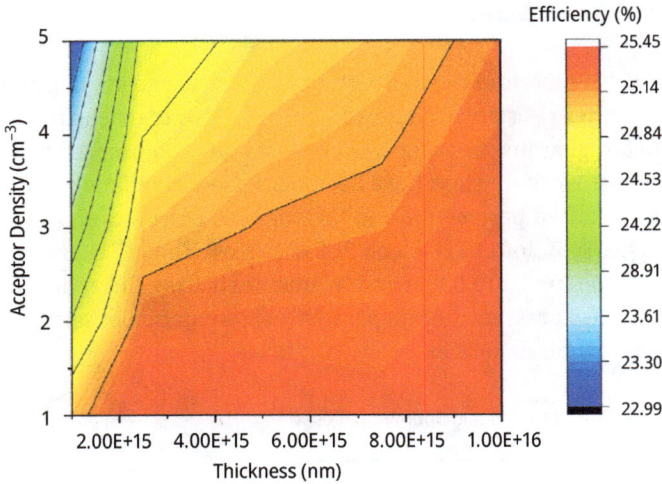

Figure 6.11: A contour plot by varying the width of the (active) absorber layer and the acceptor density on the efficiency of a solar cell.

- **Regions:** Identify regions of high efficiency, which indicate optimal combinations of thickness and acceptor density.
- **Trade-offs:** Analyze trade-offs between different parameters to achieve the desired efficiency.

By analyzing the contour plot, we can optimize the design of solar cells by selecting appropriate absorber layer thickness and acceptor density to maximize efficiency. This method allows for a visual and quantitative assessment of the interplay between different parameters, aiding in the development of more efficient solar technologies.

6.5 Conclusion

In this study, an organic layer that is lead based is used for the design of PSC using the SCAPS-1D simulation software. The device arrangement FTO/SnO$_2$/[FA$_y$Cs$_{1-y}$Pb(I$_y$Br$_{1-y}$)$_3$]$_{y=0.85}$/Spiro-MeOTAD/Au gained the PCE of 15.36%, closely matching experimental proved results. To make the device structure more stable and functional, a mixture of organic-inorganic materials for the HTLs were tested to replace Spiro-OMeTAD, which is prone to oxidation and causes instability. Among the tested HTLs, molybdenum trioxide (MoO$_3$) demonstrated the best performance.

Key parameters affecting device performance were studied, including thickness, bandgap, defect concentration, functional high temperature and defect concentration of the absorber layer. Additionally, the effects of interface defect density at the region of ETL/absorber, absorber/HTL interfaces, as well as internal series-shunt resistances were analyzed. The enhanced device configuration achieved efficiency of power conversion of 27.09%, a QE of 98.84%, a short circuit current density of 30.3319 mA/cm^2, an open circuit voltage of 1.055 V and a FF of 84.65% with illumination in the visible region. This work delivers valued perceptions for the future challenge of highly functional and steady PSCs.

Data availability: On request the data will made available.

Competing Interests: The authors declare that they have no competing interests.

Funding: The authors declare that no funds, grants, or other support were received during the preparation of this manuscript.

Authorship contribution statement credit:
Sunita Kumari Sah: Conceived and designed, formal analysis, writing original draft.
Pooja Lohia and R.K. Chauhan: Supervision. All authors read and agreed to the final versions of manuscript.

References

[1] Srivastava P, Sadanand RS, Lohia P, Dk D, Qasem H, et al. Theoretical study of perovskite solar cell for enhancement of device performance using SCAPS-1D, Physica Scripta [Internet]. 2022; 97(12): 125004. Available from: https://dx.doi.org/10.1088/1402-4896/ac9dc5.

[2] Deo M, Chauhan RK. Tweaking the performance of thin film CIGS solar cell using InP as buffer layer, Optik (Stuttg) [Internet]. 2023; 273: 170357. Available from: https://www.sciencedirect.com/science/article/pii/S0030402622016151.

[3] Gupta A, Srivastava V, Yadav S, Lohia P, Dwivedi DK, Umar A, et al. Performance enhancement of perovskite solar cell using SrTiO 3 as electron transport layer, Journal of Nanoelectronics and Optoelectronics. 2023 Apr 1; 18: 452–458.

[4] Chaurasia S, Lohia P, Dwivedi DK, Pandey R, Madan J, Yadav S, et al. Highly efficient and stable Dion-Jacobson(DJ) 2D-3D perovskite solar cells with 26 % conversion efficiency: A SCAPS-1D study, Journal of Physics and Chemistry of Solids [Internet]. 2024; 191 March: 112038. Available from: https://doi.org/10.1016/j.jpcs.2024.112038.

[5] Srivastava V, Chauhan RK, Lohia P, Yadav S. Achieving above 25 % efficiency from FA0.85Cs0.15Pb (I0.85Br0.15)3 perovskite solar cell through harnessing the potential of absorber and charge transport layers, Micro and Nanostructures. 2023 August; 184: 20769.

[6] Zhao J, Fürer SO, McMeekin DP, Lin Q, Lv P, Ma J, et al. Efficient and stable formamidinium–caesium perovskite solar cells and modules from lead acetate-based precursors, Energy & Environmental Science [Internet]. 2023; 16(1): 138–147. Available from: http://dx.doi.org/10.1039/D2EE01634F.

[7] Gupta A, Yadav S, Srivastava V, Dwivedi DK, Lohia P, Umar A, et al. Simulation of carbon-based perovskite solar cell using PBS-TBAI as a Hole Transport Layer (HTL), Science of Advanced Materials. 2023; 15(5): 655–661.

[8] Bhattarai S, Pandey R, Madan J, Sahoo GS, Hossain I, Wabaidur SM, et al. Numerical investigation of toxic free perovskite solar cells for achieving high efficiency, Materials Today Communications

[Internet]. 2023; 35: 105893. Available from: https://www.sciencedirect.com/science/article/pii/
S2352492823005846.

[9]   Srivastava V, Chauhan RK, Lohia P, Yadav S. Investigation of eco-friendly perovskite solar cell
      employing niobium pentoxide as electron transport material using SCAPS-1D, Transactions on
      Electrical and Electronic Materials. 2024; 25(3): 294–303.

[10] Yadav S, Lohia P, Sahu A. Enhanced performance of double perovskite solar cell using WO3 as an
      electron transport material, Journal of Optics [Internet]. 2023; 52(2): 776–782. Available from:
      https://doi.org/10.1007/s12596-022-01035-3.

[11] Yadav S, Lohia P, Sahu A, Chaudhary AK. Design and optimization of (FA)2BiCuI6-based double
      perovskite solar cells using kesterite CBTS as hole transport layer for high power conversion and
      quantum efficiency, Physica Scripta [Internet]. 2024; 99(9): 95516. Available from: http://dx.doi.org/
      10.1088/1402-4896/ad69cb.

[12] Yadav S, Lohia P, Sahu A. Impact of generation recombination rate in STO-enabled (FA)2BiCuI6-
      based double perovskite solar cell without HTL, Journal of Optics [Internet]. 2023; Available from:
      https://doi.org/10.1007/s12596-023-01535-w.

[13] Kumar S, Chauhan RK. Performance up-gradation of CIGS solar cell using Ag2S quantum dot as
      buffer layer, Journal of Materials Research [Internet]. 2023; 38(10): 2689–2700. Available from:
      https://doi.org/10.1557/s43578-023-00992-0.

[14] Yadav S, Lohia P, Sahu A. Design insights into (FA) 2 BiCuI 6 based double perovskite solar cells
      employing different charge transport layers, Optical and Quantum Electronics [Internet]. 2024;
      56(10): 1–22. Available from: https://doi.org/10.1007/s11082-024-07487-0.

[15] Srivastava V, Chauhan R, Lohia P. Investigating the performance of lead-free perovskite solar cells
      using various hole transport material by numerical simulation, Transactions on Electrical and
      Electronic Materials. 2022 Aug 4; 24 (184).

[16] Yadav S, Lohia P, Sahu A. Modeling and enhancement of double perovskite solar cell using WO3 as
      electron transporting material and SrCu2O2 as hole transporting material, Materials Today:
      Proceedings [Internet]. 2024 (February); 2–6. Available from: https://doi.org/10.1016/j.matpr.2024.
      02.033.

[17] Eperon GE, Stranks SD, Menelaou C, Johnston MB, Herz LM, Snaith HJ. Formamidinium lead
      trihalide: A broadly tunable perovskite for efficient planar heterojunction solar cells, Energy &
      Environmental Science. 2014; 7(3): 982–988.

[18] Kim EB, Akhtar MS, Liu C, Wang Y, Ameen S. Possibility of highly efficient 2D-3D perovskite/CIGS
      tandem solar cells with over 30% efficiency, Journal of Materials Chemistry A. 2024; 12(20):
      12262–12273.

[19] Etgar L, Gao P, Xue Z, Peng Q, Chandiran AK, Liu B, et al. Mesoscopic CH3NH3PbI3/TiO2
      heterojunction solar cells, Journal of the American Chemical Society. 2012 Oct; 134(42): 17396–17399.

[20] Haque MD, Ali MH, Rahman MF, Azmt I. Numerical analysis for the efficiency enhancement of MoS2
      solar cell: A simulation approach by SCAP-1D, Optical Materials (Amsterdam). 2022 Sep; 131.

[21] Hossain MK, Toki GFI, Kuddus A, Rubel MHK, Hossain MM, Bencherif H, et al. An extensive study on
      multiple ETL and HTL layers to design and simulation of high-performance lead-free CsSnCl3-based
      perovskite solar cells. Scientific Reports, 2023 Dec; 13(1): 2521.

[22] Jošt M, Köhnen E, Al-Ashouri A, Bertram T, Tomšič Š, Magomedov A, et al. Perovskite/CIGS Tandem
      Solar Cells: From Certified 24.2% toward 30% and beyond, ACS Energy Letters. 2022 Apr; 7(4):
      1298–1307.

[23] Srivastava V, Chauhan R, Lohia P. Lead free perovskite solar cell using TiO2 as an electron transport
      materials and Cu2O as a hole transport materials, In. 2022; (911):305–311.

[24] Kowsar A, Debnath SC, Shafayet-Ul-Islam M, Mj H, Hossain M, Chowdhury AK, et al. An overview of
      solar cell simulation tools, Solar Energy Advances [Internet]. 2025; 5 October 2024: 100077. Available
      from: https://doi.org/10.1016/j.seja.2024.100077.

[25] Tala-Ighil R, Oudjehani C, Tighilt K. SCAPS Simulation for Perovskite Solar Cell, Journal of Solar Energy Research Updates. 2021 Apr 30; 8: 21–26.

[26] Jayan KD, Sebastian V. Comparative study on the performance of different lead-based and lead-free perovskite solar cells, Advanced Theory and Simulations [Internet]. 2021 May 1; 4(5): 2100027. Available from: https://doi.org/10.1002/adts.202100027.

[27] Dong H, Ran C, Gao W, Li M, Xia Y, Huang W. Metal Halide Perovskite for next-generation optoelectronics: Progresses and prospects, eLight. 2023; 3: 3.

[28] Dikhit A, Beriha S, Kumar TS. Improving the efficiency and stability in a MAPbI$_3$ perovskite solar cell using dimensionally engineered PbS quantum dot passivation layer, Optional Practical Training [Internet]. 2023; 12: 100461. Available from: https://www.sciencedirect.com/science/article/pii/S266695012300113X.

[29] Patel AK, Mishra R, Soni SK. Performance enhancement of CIGS solar cell with two dimensional MoS2 hole transport layer, Micro and Nanostructures. 2022 May; 165.

[30] Chaudhary AK, Verma S, Chauhan R~K. Design of a low-cost, environment friendly perovskite solar cell with synergic effect of graphene oxide-based HTL and CH$_{3}$NH$_{3}$GeI$_{3}$ as ETL, Engineering Research Express. 2023 Sep; 5(3): 35039.

[31] Shrivastav N, Madan J, Pandey R. A short study on recently developed tandem solar cells, Materials Today: Proceedings. 2023; In Press(2): 30.370.

[32] Yadav S, Lohia P, Sahu A, Dwivedi DK. Design insights into (FA)2BiCuI6 based double perovskite solar cells employing different charge transport layers, Optical and Quantum Electronics [Internet]. 2024; 56(10): 1628. Available from: https://doi.org/10.1007/s11082-024-07487-0.

Gauri Pathak, Yashwant Kumar Singh, D.K. Dwivedi*,
and Pravin Kumar Singh

Chapter 7
Tandem solar cells: design and performance enhancement

Abstract: Tandem solar cells (TSC) provide a higher efficiency than the single-junction cells used in the past. In this study, a dual-layer structure is simulated, where the top cell (T_{CELL}) and bottom cell (B_{CELL}) are modeled and their working is mostly controlled by the absorber material's bandgap energies. A Dion–Jacobson (DJ) material has been used in the T_{CELL} in which a 2D perovskite layer with composition $PeDAMA_4Pb_5I_{16}$ is set on top of a 3D perovskite layer $FAPbI_3$. This structure enhances the properties of 3D perovskite with added theory-based efficiency along with its stability. In addition, the B_{CELL} has a copper indium gallium selenide (CIGS) absorber (with a bandgap of 1.27 eV) which is known for its ability to absorb more light at low temperature, and it is less toxic than CdTe. The simulations were run using SCAPS-1D which simulates solar cells under AM1.5 spectrum at 1,000 W/m² and 300 K. In standalone configuration, the T_{CELL} delivered a PCE of 23.18% (V_{OC} = 1.37 V, J_{SC} = 20.96 mA/cm², FF = 80.39%), while the B_{CELL} showed a higher PCE of 26.32% (V_{OC} = 0.89 V, J_{SC} = 33.95 mA/cm², FF = 86.72%). Further optimization has been performed by adjusted the thickness of the perovskite (T_{CELL}) and CIGS (B_{CELL}) layers and the number of defects in them. Using spectral filtering and current matching, the tandem perovskite-CIGS cell (PCTSC) demonstrated outstanding results: 35.63% PCE, 2.19 V V_{OC}, 19.0 mA/cm² J_{SC}, and 85.63% FF. The detailed study gives useful results for enhancing the efficiency and sustainability of TSC.

Keywords: Tandem, solar cell, spectral filtering, perovskite, current matching

*Corresponding author: D.K. Dwivedi, Photonics and Photovoltaic Research Lab (PPRL), Department of Physics and Material Science, Madan Mohan Malaviya University of Technology, Gorakhpur 273010, Uttar Pradesh, India, e-mail: todkdwivedi@gmail.com
Gauri Pathak, Yashwant Kumar Singh, Photonics and Photovoltaic Research Lab (PPRL), Department of Physics and Material Science, Madan Mohan Malaviya University of Technology, Gorakhpur 273010, India
Pravin Kumar Singh, Institute of Advanced Materials, IAAM, Gammalkilsvägen 18, 59053 Ulrika, Sweden

https://doi.org/10.1515/9783111726847-007

7.1 Introduction

A solar cell (SC) uses the photovoltaic (PV) effect to convert sunlight into electricity [1, 2]. As light photons hit the p-n junction in a semiconductor, they generate electron–hole pairs, and the built-in electric field separates them to start a current [3]. A fraction of solar rays is absorbed, bounced or scattered by Earth's atmosphere, but much of the light and heat from the sun reaches the ground and can be easily captured by PV panels because of energy losses that cannot be avoided [4–6]. The maximum potential efficiency of a single-junction solar cell is limited to almost 33% by the Shockley–Queisser limit (SQL)) but its limitations are: The photons that do not fit into the material's bandgap range, photon energy that is lost as heat when electric circuits are used, pairwise recombination between electrons and holes, and the inefficiency at the voltage contacts causes voltage losses [7–10]. The multijunction design is what sets tandem solar cells (TSC) above traditional SQLs [11, 12].

The top cell in these modules has a higher bandgap for short-wavelength response, but the bottom cell has a lower bandgap so it is sensitive to longer-wavelength photocarriers. As a result of this design, thermalization and transmission losses are reduced, so the efficiency of PV cells can theoretically reach 47% under common lighting conditions. With recent improvements in arrays, for example, in perovskite-silicon and perovskite-CIGS systems, high efficiencies of more than 33% have been demonstrated, suggesting they will likely revolutionize solar energy [13, 14]. Besides saving energy, tandem technologies also help reduce global warming by increasing electrical power from less ground, consuming less space and speeding the move to power sources that are friendly to the environment [15]. These experiments include efforts to achieve gap alignment, modify the boundaries between materials and manufacture these gadgets on a larger scale for commercial use.

The quick rise in energy required around the world because of population and technology has worsened climate issues such as global warming [16]. As a result, PV systems and other renewable energy technology are becoming important ways to produce affordable and clean energy. Around the world, countries are making green energy changes to meet their goals for a sustainable future [17]. An example is India's Panchamrit plan which aims to reduce carbon intensity by 45% by 2030 using more clean energy and says it will achieve net-zero emissions by 2070 [18, 19]. India is focused on development as well as reducing its carbon footprint by making its net-zero commitment for the year 2070 [20]. With 7% of the world's CO_2 emissions coming from India's towns and cities, it believes the Panchamrit Initiative will help meet its goals of renewable energy and fossil-free targets [21]. By comparison, China (27% of emissions) has vowed to reach net zero by 2060 and is supported by its rise in solar manufacturing and a plan for 1,200 GW of renewable power; the United States (11%) and the EU (7%) both aim for net-zero by 2050, compared to the G7's 2040 target and this is made possible by polices like the Inflation Reduction Act and the Fit for 55 packages shown in Figure 7.1(a) [22]. There are regions that, collectively, account for about half the world's emissions, with the other half coming from other nations with different plans (2050–2070) [23]. If we present this data as a pie chart, China stands out as having the largest emissions, followed by the United States, India, the EU and other countries, each with annotations stating when zero emissions are aimed and main policies such as India's PLI plan or the EU's carbon border tax shown in Figure 7.1(b) [24].

Perovskite-based TSC have made significant impact in the solar field. By combining elements of 2D and 3D perovskites, the hybrid design offers more stability without reducing efficiency which is a challenge in conventional perovskite solar cells [25–27]. Usually, 3D materials such as methylammonium lead halide has octahedral structure that share corners, allowing these perovskites to absorb light well and outperform other materials, boosting their PCE above 25%. However, the 3D structure is prone to instability and breaks down fast due to heat, moisture, and ion movement. Alternatively, 2D perovskite materials are stabilized by adding thick organic cations, like phenethyl-ammonium, which divide the 3D structure into separate quantum well layers [28, 29]. With this structure, ion movement and moisture are kept at bay, but it causes difficulty in charging due to less mobile carriers and higher-energy bound excitons.

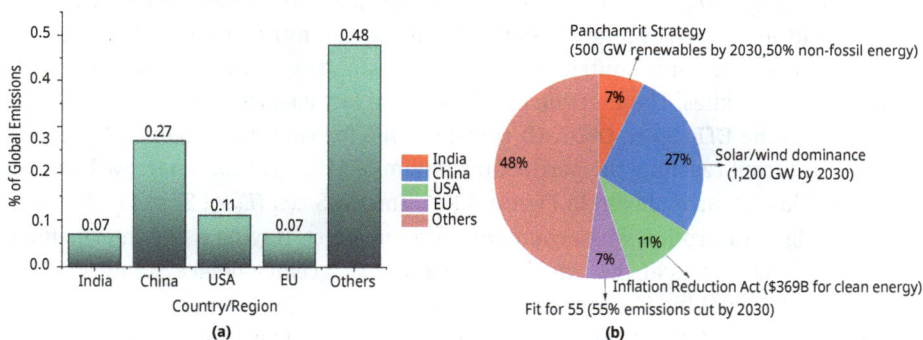

Figure 7.1: (a) Graph between different countries versus global emission in percentage, **(b)** pie chart of different countries into groups based on their target years.

In 2014, copper-zinc-tin-sulfide (CZTS) was presented for the first time as a tandem ingredient. Now, much research centres are working on perovskite-based TSCs which are now considered as strong competition to familiar single-junction technologies like c-Si and CIGS, as well as narrow-bandgap perovskites [30]. While perovskite/c-Si tandem cells are expected to lead in the PV market thanks to their efficiency, their adoption is being slowed by the high cost of making them.

On the other hand, perovskite–CIGS composites are a leading choice because they mix cost-effectiveness, a light build, the ability to bend and high efficiency with the best placement of bandgaps for broad-spectrum collecting. Improving NiO_x with

4PADCB increases charge extraction and cuts down on losses to reach 22.8% efficiency and 1.82 V for flexible PVK-CIGS tandem cells [31]. The presence of an IZO protective coating prevents the AZO layer from being damaged, so flexible perovskite/CIGS TSCs can use heat-resistant materials and maintain a 21.24% efficiency [32]. Building on improved NiO_x (22.8%) and IZO-protected CIGS (21.24%), a simulated perovskite-CIGS tandem cell achieved 35.63% efficiency and 2.19 V V_{OC}.

The present work extends above findings by simulating a $PeDAMA_4Pb_5I_{16}$ (2D) – $FAPBI_3$ (3D)-based mixed perovskite T_{CELL} and a CIGS bottom cell. Each cell has been optimized thoroughly via variation of thickness, defect density, transport layers, temperature, and eventually utilizing filtered spectrum and current matching technique to obtain a tandem architecture.

7.2 Architecture and modeling tools for devices

In this work, 2D-3D mixed perovskite hybrid architecture is used in the top cell (T_{CELL}) (Figure 7.2(a)). $PeDAMA_4Pb_5I_{16}$ is a 2D perovskite that adopts the DJ-layered structure and has a bandgap (E_g) of 1.65 eV, yielding efficient photoluminescence quantum efficiency. Lead iodide perovskite ($FAPbI_3$, energy gap 1.6 eV) is used as the main absorber in this structure, contributing to better stability than methylammonium-containing perovskites. The function of TiO_2 is to extract electrons efficiently from the perovskite as the ETL. Spiro-OMeTAD is responsible for collecting holes and preventing moisture from reaching the perovskite. Bottom cell (B_{CELL}) uses a narrow-bandgap CIGS-based layout, as is shown in Figure 7.2(b). The CIGS cell (E_g = 1.27 eV) is designed for strong light absorption when conditions are indistinct. Having a CdS intermediate layer diminishes losses at the interface and supports greater device stability. i-ZnO ranks as the ETL and buffer, enhancing the ease of electron movement. Adding aluminum to zinc oxide (ZnO:Al) controls electric conductivity, turning the oxide transparent for collecting current. According to Figure 7.2(c), the tandem configuration splits the wavelengths of light, letting the T_{CELL} absorb only the high-energy AM1.5 spectrum and the B_{CELL} filter low-energy photons from the AM1.5 light that the B_{CELL} receives.

In Figure 7.2(d), we can see the absorption spectrum versus wavelength plot, a tool vital for tuning TSC performance. With this graph, the spectral response of T_{CELL} and B_{CELL} can be set accurately so that there is little overlap and the top cell and bottom cell receive almost all available solar radiation. The diagrams in Figure 7.2(e) represent the energy bands of layered structures:

- T_{CELL} developed: Tin oxide (FTO) covered with TiO_2/3D perovskite absorber $FAPbI_3$/2D perovskite layer $PeDAMA_4Pb_5I_{16}$/hole transport material Spiro-OMeTAD. Due to the staggered band positions, the charge carriers are easily transported and separated.

B_{CELL} has a CIGS layer, CdS buffer layer, an i-ZnO layer and a ZnO:Al transparent electrode. The infrared photon absorption is improved by the gradient bandgap in CIGS.

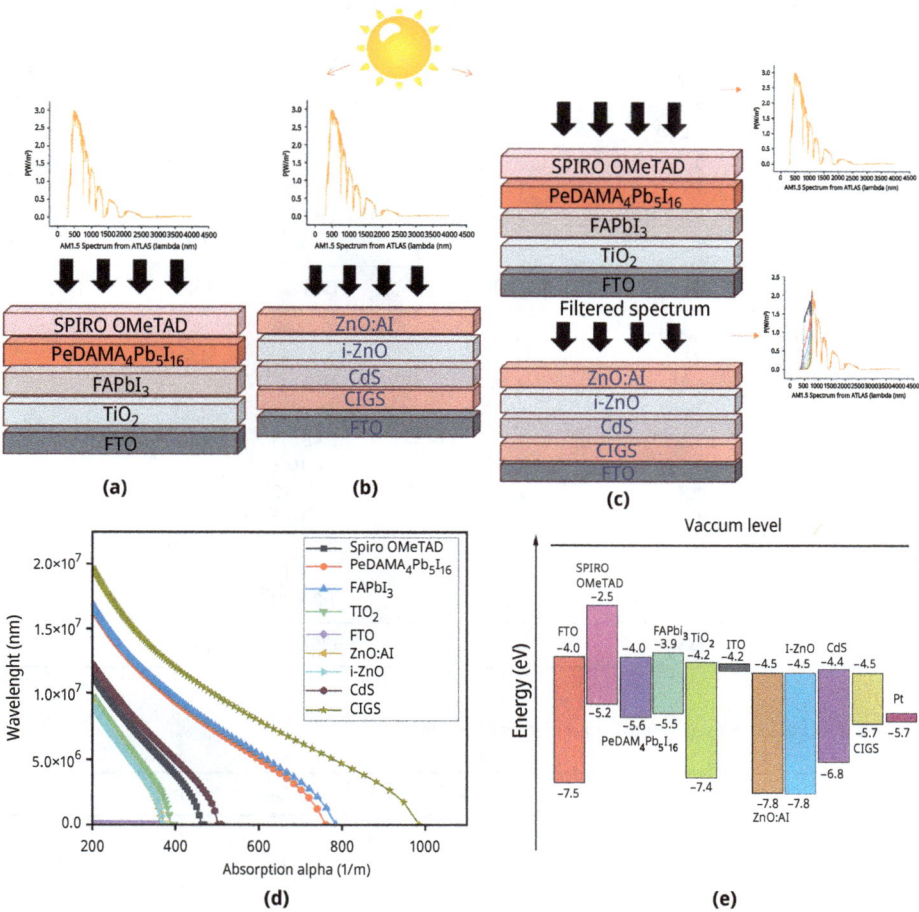

Figure 7.2: (a) and (b) Individual designs of the upper and lower sub cells operating on their own. **(c)** Interconnected parasite particles arranged one behind the other. **(d)** Comparison of how much a material absorbs light of various colors to its wavelengths. **(e)** Schematic of how the energy in each sub cell is aligned for the dual system.

Table 7.1: During simulation electric parameters of T_{CELL} and B_{CELL}.

Parameter of T_{CELL}	Spiro OMeTAD	PeDAMA$_4$Pb$_5$I$_{16}$	FAPbI$_3$	TiO$_2$	FTO
χ (eV)	2.5	4.0	3.9	4.26	4
ε (relative)	3	25	6.5	50	9
Thickness (nm)	300	200	500	100	100
E_g (eV)	2.7	1.6	1.6	3.2	3.5
μ_e (cm^2/Vs)	2×10^{-2}	1.4	2	2×10^4	2×10^1
N_D (1/cm^3)	0	1×10^{15}	0	6×10^{19}	2×10^{19}
N_A (1/cm^3)	1×10^6	1×10^{15}	1×10^{13}	0	0
N_c (1/cm^3)	2.5×10^{18}	7.5×10^{17}	1×10^{17}	2×10^{18}	2.2×10^{18}
N_v (1/cm^3)	1.8×10^{18}	1.8×10^{18}	1×10^{17}	1.8×10^{19}	1.8×10^{19}
References	[33]	[34]	[33]	[33]	[33]
Parameter of B_{CELL}	ZnO:Al	i-ZnO	CdS	CIGS	
χ (eV)	4.5	4.5	4.4	4.5	
ε (relative)	8.49	8.49	10	13.6	
Thickness (nm)	150	50	40	2,500	
E_g (eV)	3.37	3.37	2.48	1.27	
μ_e (cm^2/Vs)	1×10^2	1.1×10^2	1×10^2	1×10^2	
μ_h (cm^2/Vs)	1.3×10^1	2.5×10^1	2.5×10^1	2.5×10^1	
N_D (1/cm^3)	1×10^{18}	0	2×10^{18}	0	
N_A (1/cm^3)	0	0	0	1×10^{18}	
N_c (1/cm^3)	2.2×10^{18}	2.2×10^{18}	2.2×10^{18}	2.2×10^{18}	
N_v (1/cm^3)	1.8×10^{19}	1.8×10^{19}	1.8×10^{19}	1.8×10^{19}	
Reference	[35]	[35]	[35]	[35]	

Table 7.2: During simulation electric parameters ETLs and HTLs.

Parameter	Cu$_2$O	CuO	V$_2$O$_5$	CuSbS$_2$	CuI	C$_{60}$	SnO$_2$	WS$_2$
E_g (eV)	2.2	1.51	2.20	1.58	3.1	1.7	3.6	1.8
χ (eV)	3.4	4.07	4	4.2	2.1	4.5	4	3.95
ε (relative)	7.5	18.1	10	14.6	6.5	10	9	13.6
N_c (1/cm^3)	2×10^{19}	2.2×10^{19}	9.2×10^{17}	2×10^{18}	2.8×10^{19}	2.2×10^{18}	2.2×10^{18}	1×10^{18}
N_v (1/cm^3)	1×10^{19}	5.5×10^{20}	5×10^{18}	1×10^{19}	1×10^{20}	1.8×10^{19}	1.8×10^{19}	2.4×10^{19}
μ_e (cm^2/Vs)	200	100	3.2×10^2	49	100	0.1	100	100
μ_h (cm^2/Vs)	8,600	0.1	40	49	43.9	0.1	25	100
N_A (1/cm^3)	1×10^{18}	1×10^{18}	1×10^{18}	1×10^{18}	1×10^{18}	0	0	0
N_D (1/cm^3)	0	0	0	0	0	1×10^{18}	1×10^{17}	1×10^{18}
N_t (1/cm^3)	1×10^{15}	1×10^{15}	1×10^{15}	1×10^{15}	1×10^{15}	1×10^{15}	1×10^{15}	1×10^{15}
References	[2, 35, 36]	[2, 35, 36]	[2, 35, 36]	[2, 35, 36]	[2, 35, 36]	[2, 35, 36]	[2, 35, 36]	[2, 35, 36]

7.3 Outcomes and analysis

7.3.1 A configuration with an energy level of 1.65 eV assigned to the T_{CELL} and 1.2 eV to the B_{CELL}

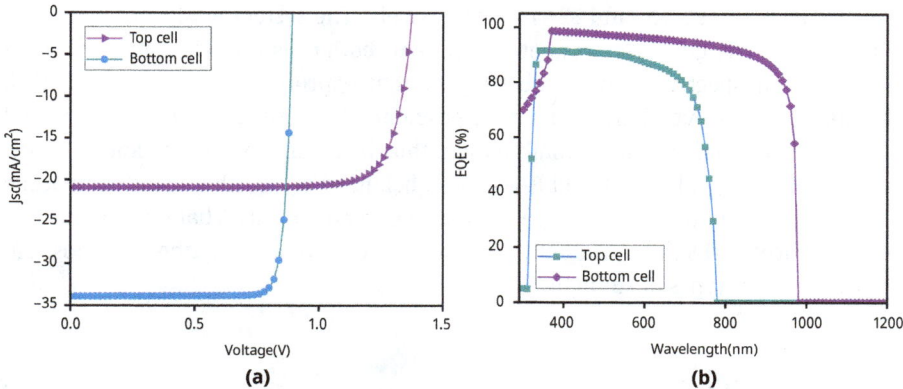

Figure 7.3: (a) Variation of short-circuit current density (J_{sc}) with open-circuit voltage (V_{oc}), and **(b)** variation of external quantum efficiency (EQE) across different wavelengths.

The section presents the J-V characteristics (mA/cm^2) for both T_{CELL} and B_{CELL} as well as the quantum efficiency spectrum (EQE %), both simulated using SCAPS-1D software. The output current per square area is labeled as current density and EQE represents the percentage of incoming photons that are transformed into charge carriers in the cell. Figure 7.3(a) shows the J-V curve, where T_{CELL} manages 20.96 mA/cm^2 at 1.46 V and B_{CELL} achieves 33.95 mA/cm^2 at 0.9 V. Figure 7.3(b) illustrates EQE, with T_{CELL} having a highest EQE of 91.49% at 350 nm and a very fast decrease beyond 770 nm, whereas B_{CELL} reaches its best EQE of by combining the results of these analyses, we stress the role of electrical properties and clarity of the solar glass in ensuring that these tandem cells perform as well as possible shown in Table 7.1.

7.3.2 Analysis of different configurations of electron transport layers (ETLs) and hole transport layers (HTLs)

The success of a solar cell depends largely on the ETL and HTL. The ETL is built to assist electrons in moving freely, remove charge recombination and tightly connect to the part absorbing light energy, reducing energy losses. In contrast, the HTL helps extract holes effectively, decreases recombination and produces a strong interface with the active part of the cell to make charge separation work efficiently.

In this work, the performance of several different ETL and HTL materials was assessed by simulating T_{CELL} and B_{CELL} for a TSC in SCAPS-1D. For T_{CELL} simulation, HTL consisted of Spiro-OMeTAD and the ETL was TiO_2 the device reached a record PCE of 25.9% (Figure 7.4(a–d)) with V_{OC} at 1.4 V and J_{SC} and FF of 21.10 mA/cm^2 and 84.60%, respectively. For B_{CELL} simulation, HTL consisted of CuSbS$_2$ and the ETL was CuI, the device reached a record PCE of 11.22% (Figure 7.5(a–d)) with V_{OC} at 1.14 V and J_{SC} and FF of 11.85 mA/cm^2 and 82.94%, respectively. The system works because B_{CELL} can absorb wavelengths T_{CELL} cannot which helps both cells provide complete coverage of the solar spectrum. The result suggests that appropriately choosing ETL/HTL materials can improve charge transport efficiency by cutting down on interfacial flaws that lead to defects and recombination, thus increasing durability and stability. It is clear from T_{CELL}'s huge FF and B_{CELL}'s well-balanced J_{SC}-V_{OC} that using compatible materials is important for the best performance of tandem cells. That means effective use of advanced ETLs and HTLs may lead to solar cells that can gather the most energy from various types of sunlight.

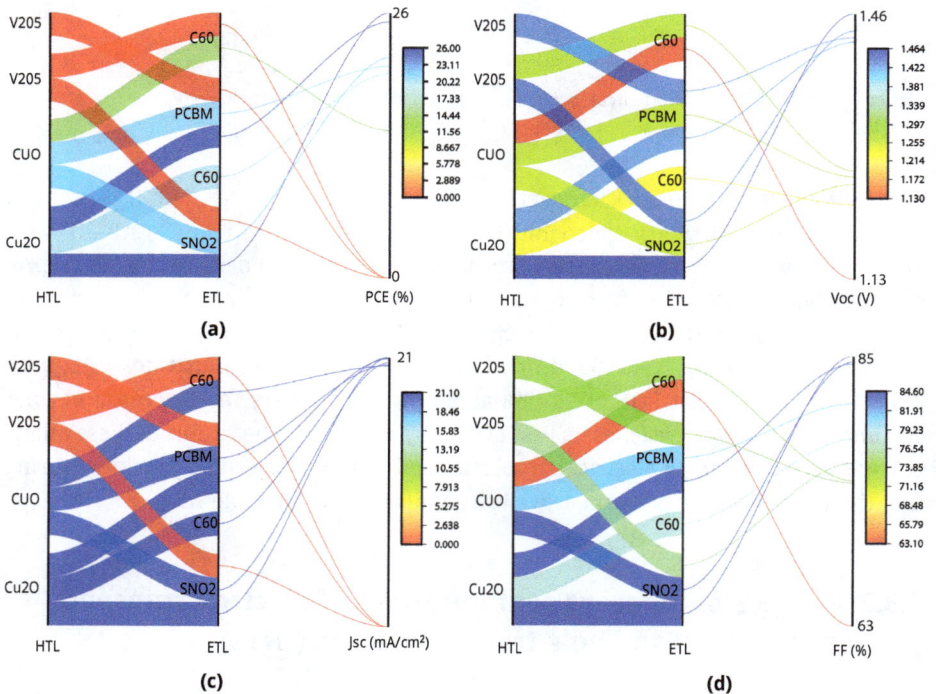

Figure 7.4: Effect of various electron transport layers (ETLs) and hole transport layers (HTLs) in the T_{CELL} on: **(a)** power conversion efficiency (PCE, %), **(b)** open-circuit voltage (V_{OC}, V), **(c)** short-circuit current density (J_{SC}, mA/cm^2), and **(d)** fill factor (FF, %).

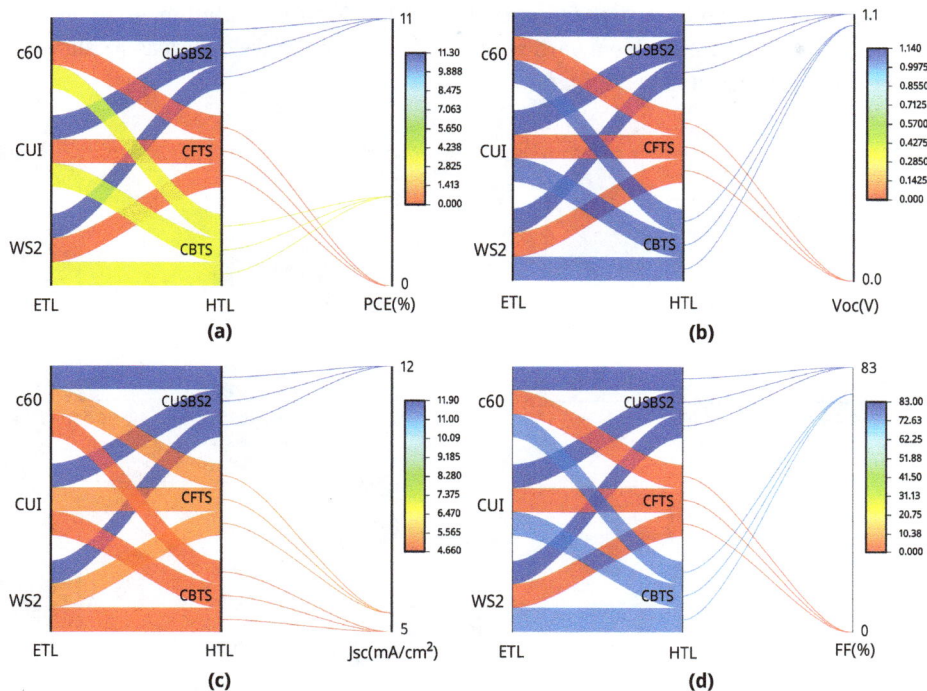

Figure 7.5: Effect of various electron transport layers (ETLs) and hole transport layers (HTLs) in the B_{CELL} on: **(a)** power conversion efficiency (PCE, %), **(b)** open-circuit voltage (V_{OC}, V), **(c)** short-circuit current density (J_{SC}, mA/cm^2), and **(d)** fill factor (FF, %).

7.3.3 Influence of thermal conditions on PV parameters

PV devices must be optimally heated or cooled to function well. Simulations were performed for our proposed solar cell over a range of temperatures from 300 K to 400 K, since solar cells placed in desert areas regularly experience these temperatures.

The top cell faces high-temperature conditions in direct sunlight. Because the material's bandgap is smaller due to the temperature rise, both its efficiency and the likelihood of thermal losses from recombination are increased. Instead, the B_{CELL} receives sunlight that was filtered earlier by the T_{CELL} which helps avoid heat stress and boosts its efficiency by decreasing heat losses. Figure 7.6(a–d) illustrates the way cell parameters change with temperature. $T_{CELL's}$ PCE maximizes at 300 K with 23.32% and its value decreases steadily until it reaches low points when the temperature reaches 340 K. Similarly, the V_{OC} and J_{SC} both reduce as temperature rises. Yet, the fill factor (FF) rises up to 80.60% at 340 k before slowly falling to 78.43% when the temperature reaches 460 K. As shown in Figure 7.6(c) and (d), the relative efficiency of B_{CELL} also falls with rising temperature, from 27.59% at 280 K to a minimum of 16.08% at 460 K. V_{OC} and J_{SC} trends downward with a rise in temperature, whereas FF shows a decrease regardless of how hot it gets.

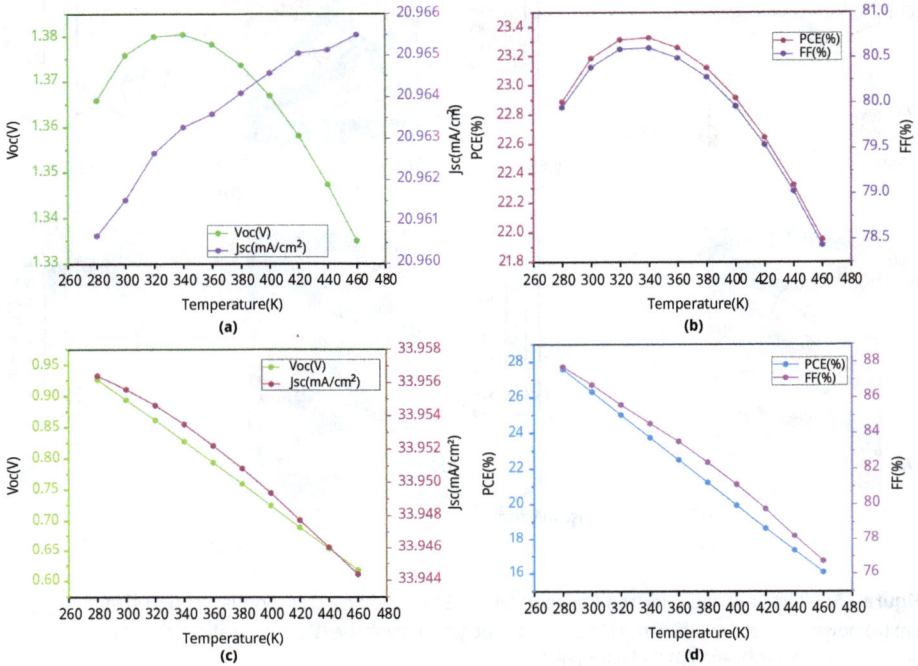

Figure 7.6: (a–b) Influence of temperature on the performance of the top cell, and **(c–d)** corresponding effects on the bottom cell.

7.3.4 Influence of absorber layer thickness on the properties of DJ 2D-3D and CIGS

Using SCAPS-1D software at AM 1.5G, 300 K, with no interfacial reflection and resistive losses. Solar cell efficiency and performance largely depend on the thickness of the absorber layer, but this effect is more important in 2D materials than in 3D materials.

Top choice is a DJ 2D (PeDAMA$_4$Pb$_5$I$_{16}$)-3D (FAPbI$_3$) hybrid perovskite material. The 2D perovskite layer (PeDAMA$_4$Pb$_5$I$_{16}$) prevents damage to the 3D perovskite (FAPbI$_3$) by using its bandgap of 1.65 eV and special Dion-Jacobson structure. The high quantum efficiency of the 2D layer keeps most recombination nonradiative, whereas the strong light absorption of the 3D layer increases the rate of photocurrent generation. Figure 7.7(a and b) shows 2D layer thickness falls between 0.1 and 1.0 μm. Increasing the thickness of the 2D material from 0.1 to 1.0 μm boosts the short-circuit current (J_{sc}) from 20.6 to 21.8 mA/cm^2 as a result of better light harvesting. Yet, the FF gradually decreases from 81% to 73%, probably because of either resistive losses or problems at the interface in thicker films. Because of this, η reaches a maximum of 23.2% with particles between 0.2 and 0.3 μm but drops to 21.8% when they are 1.0 μm wide. The measurement of V_{oc} at ~1.37 to 1.38 V indicates that recombination has had little effect. Most often, this layer is between 0.1 and 1.0 μm. Figure 7.7(c and d) shows that a 10-times thicker 3D layer in J_{SC} increases current from 16.6 mA/cm^2 (0.1 μm) to 22.1 mA/cm^2 (1.0 μm) due to more photon capturing. The efficiency increases from 18.5% to 23.8%, even though the FF falls from 81.5% to 78.1% due to bulk recombination. Stable V_{oc} (~1.37–1.38 V) is present, likely the effective charge extraction by TiO₂ in the ETL and Spiro-OMeTAD in the HTL. The CIGS layer (E_g = 1.27 eV) is paired with a CdS buffer layer and i-ZnO/ZnO:Al ETL to optimize infrared absorption and charge transport.

CIGS thickness (0.1–1.0 μm): Increasing thickness dramatically improves all parameters (Figure 7.7(e and f)):
-   J_{sc} triples from 13.2 to 30.7 mA/cm^2 due to enhanced infrared photon absorption.
-   V_{oc} rises from 0.78 to 0.87 V, reflecting reduced recombination at the CdS/CIGS interface.
-   FF improves from 79.8% to 86%, indicating better charge collection. Efficiency surges from 8.2% (0.1 μm) to 22.9% (1.0 μm), demonstrating the necessity of thicker CIGS for high performance.

7.3.5 Influence of defect density on the properties of DJ 2D-3D and CIGS

In this portion, the density of defects on a 2D layer is measured in the range of 1×10^{10} to 1×10^{18} cm^{-3}. A 2D layer is mainly used to keep out moisture and protect the roof. However, the problem of recombination at the interface means defects here lower FF and V_{oc}, but not J_{sc} because the 2D layers are not as intense. Having a defect density of more than 1×10^{14} cm^{-3} in crystalline silicon creates trap sites that make it challenging to speedily extract the charge. From Figure 7.8(a and b), it can be seen that material has a PCE of ~23.4% for all samples with ≤1×10^{12} cm^{-3} defects. The PCE drops fast tp 13.1% at 1×10^{16}cm^{-3} and only rises slightly (~13%) after it. V_{oc} drops from 1.38 to 0.96 V around 1×10^{16} cm^{-3} and maintains this value afterwards. Light absorption is only little affected because the J_{sc} stays stable (~21 mA/cm^2). At high defect densities, the pairwise coverage with FF drops from 81% to 67%. Then, 3D layer defect density can range from 1×10^{10} cm^{-3} to 1×10^{18} cm^{-3}. From Figure 7.8(c and d), it is known that the defect density with PCE is lowest at 23.7% when defects are present at fewer than 1 million per square cm. Kindly change this line to: "The PCE goes to 1.32% of its original value at 1×10^{18} cm^{-3} after recombination." Initially, the voltage is stable at ~1.38 V, but it lowers to 1.19 V at a density of 1×10^{16} cm^{-3} before becoming much higher at 1×10^{18} cm^{-3} (this increase is

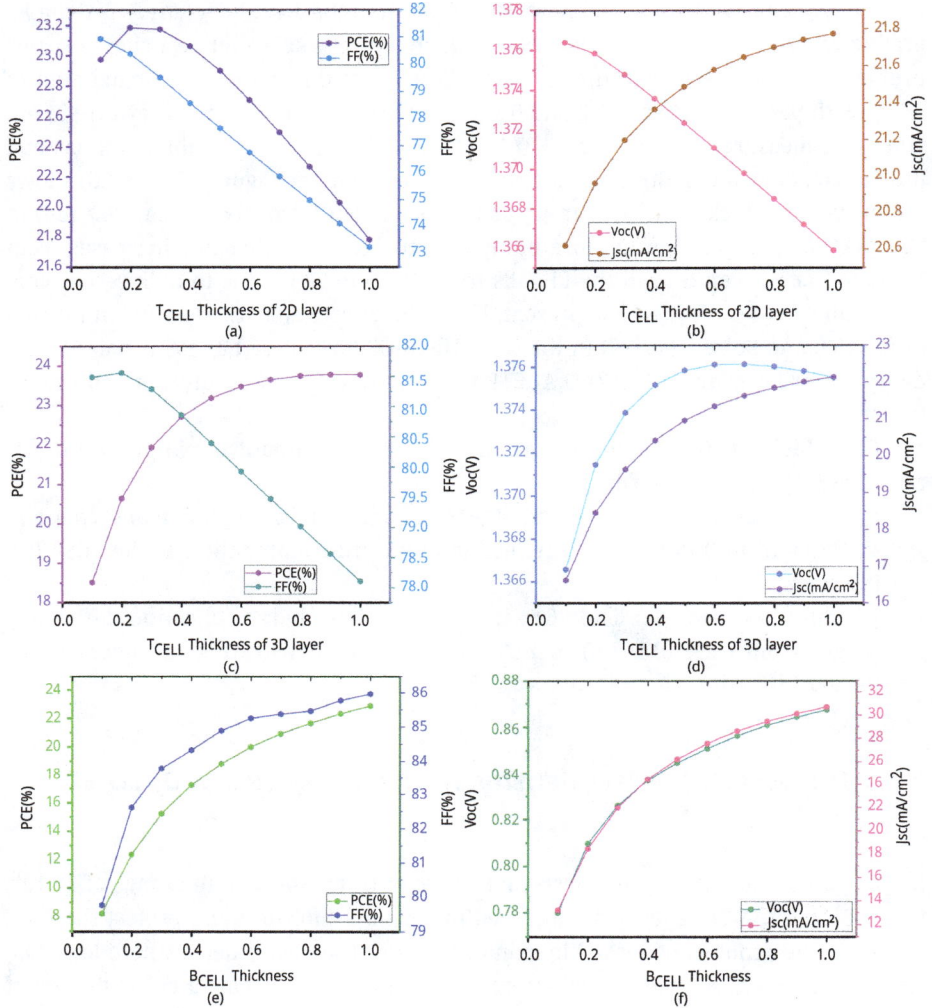

Figure 7.7: (a and b) Top cell 2D layer thickness with all four parameters, **(c and d)** top cell 3D layer thickness with all four parameters, **(e and f)** bottom cell thickness with all four parameters.

probably due to Fermi level pinning or trap-assisted tunneling). The current density stays steady (~21 mA/cm^2) up to 1×10^{16} cm^{-3}, from which it drops abruptly to 5.66 mA/cm^2 as it reaches 1×10^{18} cm^{-3}. The charge transport decreases by almost two-thirds, at 68%. The main role of the 3D layer (FAPbI$_3$) is absorption, so it is very sensitive to errors. From Figure 7.8(e and f), it is known that the detector has a defect density of ≤1 × 10^{14} cm^{-3}. The efficiency is at ~26.4% and has remained constant without losing V_{oc}, J_{sc}, or FF. Based on its flexibility, the variation in CIGS is suppressed by the recombination lower layer of CdS. High defect density is a situation where, at a minimum, 1 in every

100,000,000,000 atoms is defective. Defects near the middle of the band gap leads to additional Shockley–Read–Hall recombination. Carriers only have a short time to be collected because severe defects cause a big drop in carrier lifetime. When FF goes down (from 86.8% to 72.7%), it is because resistive losses and bad quality of junctions are now getting in the way of efficient charge flow.

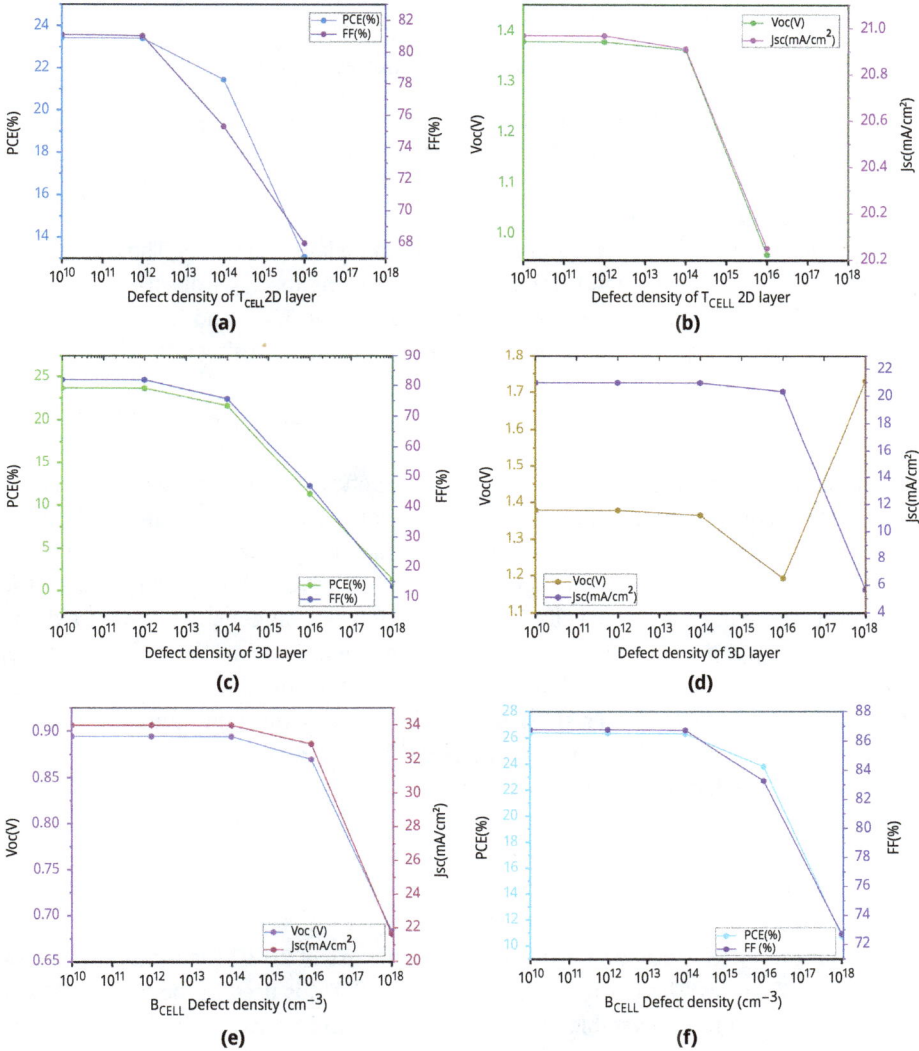

Figure 7.8: (a and b) Top cell 2D layer defect density with all four parameters, **(c and d)** top cell 3D layer defect density with all four parameters, and **(e and f)** bottom cell defect density with all four parameters.

7.3.6 Multijunction cell configuration

The T_{CELL} in TSCs blocks out certain light wavelengths and allows certain others to reach the B_{CELL}. T_{CELLs} mostly receive the strongest photons and B_{CELLs} gather those that get through to the T_{CELL}. The way a layer absorbs energy is dictated by the absorption equation.

It is calculated as follows:

$$= \text{Incident photon} - \text{transmitted photon}$$

$$= I(\lambda) - I(\lambda)exp(-\propto (\lambda) \times d))$$

$$= I(\lambda)(1 - exp(^{-\propto}(\lambda) \times d))$$

Here, $I(\lambda)$ is incident photon.

For each layer, $\propto$ refers to absorption and d to the layer thickness. The transmitted spectrum for B_{CELL} illumination is improved by optimizing the T_{CELL} absorber thickness, as per Figure 7.9(a). Matching the photocurrents from the T_{CELL} and B_{CELL} plays a key role in improving the performance of a TSC. This is reflected by using this equation:

$$J = q \cdot \int \text{Ifiltered}(\lambda) \cdot \propto (\lambda).\phi(\lambda)d\lambda$$

In this case, $\phi(\lambda)$ represents the quantum efficiency and q the electron charge. Figure 7.9(b) suggests that a maximum current density of 14.18 mA/cm^2 is obtained with 600 nm for the B_{CELL} active layer and T_{CELL} thickness of 0.2 μm.

Additional investigations in Figure 7.9(c and d) reveal that the tandem cell is more efficient than both single cells. When operating as a tandem, the structure delivers more than 2.19 V of combined voltage, more than what one of the single structures offers. As a result of this voltage boost, the efficiency improves. The representation in Figure 7.9(d) suggests that TSCs show marginally higher external quantum efficiency and nearly perfect power conversion efficiency, indicating that they offer useful options for the growth of solar cell technology.

7.4 Conclusion

In this study, a PCTSC structure is examined, making use of 2D-3D Dion-Jacobson perovskite and CIGS layers. The DJ 2D material (PeDAMA$_4$Pb$_5$I$_{16}$) is used in the top cell (T_{CELL}) of the device, owing to its stable and wide bandgap and a CIGS layer with a smaller bandgap (1.27 eV) is inserted below to cover a wide spectrum of photons in the bottom cell (B_{CELL}). The unique structure reduces flaws at the interface and allows charge to flow smoothly, while the CIGS film provides high performance in converting low-energy photons into electricity. This fusion configuration is best for reaching high current matching and covering most wavelengths efficiently. The device was improved through

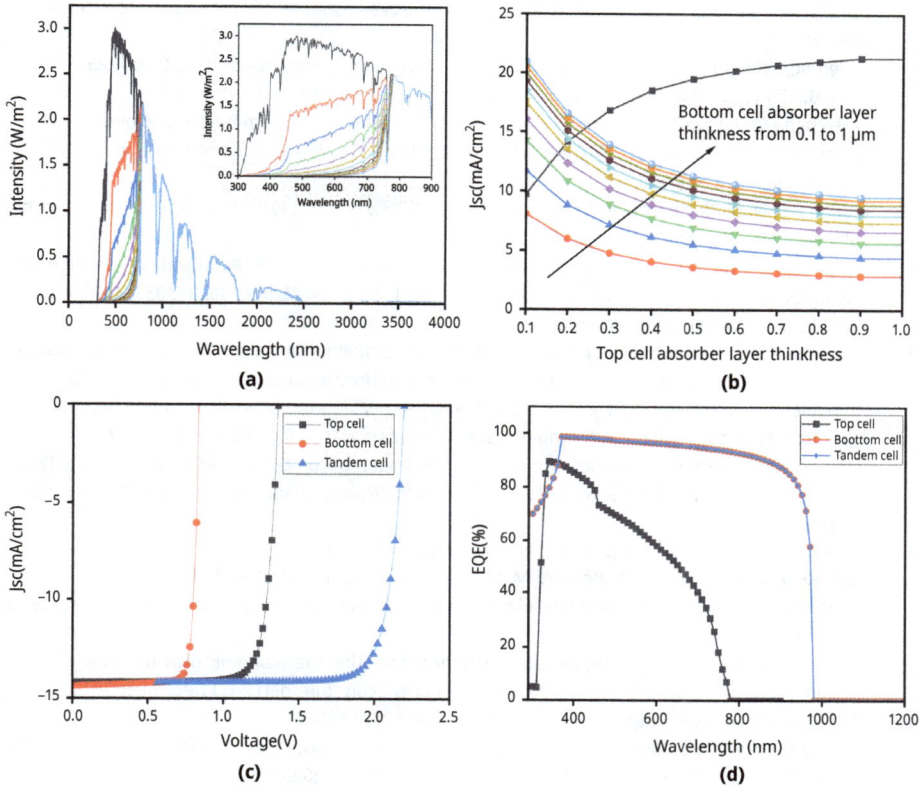

Figure 7.9: (a) Filtered light spectrum, **(b)** current density matching curve, and **(c and d)** J_{SC} versus voltage and EQE versus wavelength plots for tandem device simulation.

systematic simulations that checked whether different thicknesses and types of ETL/HTL or defects density affected how the device worked at different temperatures. The highly optimized tandem cell showed excellent results, including an efficiency of 35.63% in converting light into power, with a V_{OC} of 2.19 V, a J_{SC} of 19.0 mA/cm^2 and an FF of 85.63%.

References

[1] Solanki CS, Sangani CS, Gunashekar D, Antony G. Enhanced heat dissipation of V-trough PV modules for better performance, Solar Energy Materials and Solar Cells. 2008; 92(12): 1634–1638.

[2] Singh YK, Dwivedi DK, Lohia P, Pandey R, Madan J, Hossain MK, et al. Filtered spectrum modeling of high-performance perovskite tandem solar cells: Tailoring absorber properties and electron/hole transport layers for 31.55 % efficiency, Journal of Physics and Chemistry of Solids. 2024 Sep 1; 192.

[3] Razeghi M. Semiconductor p-n and metal-semiconductor junctions, Fundamentals of Solid State Engineering. 2009; 1–56.

[4] Hakuba M. Solar absorption in the atmosphere – Improved estimates from surface and satellite observations. 2015.

[5] Green AC, Whiteman DC. Solar Radiation, Cancer Epidemiology and Prevention. 2009; 1–25.

[6] Cook G, Billman L, Adcock R. Photovoltaic Fundamental. 1995;1–68.

[7] Li H, Zhang W. Perovskite tandem solar cells: From fundamentals to commercial deployment, Chemical Reviews [Internet]. 2020 Sep 23; 120(18): 9835–9950. Available from: https://doi.org/10.1021/acs.chemrev.9b00780.

[8] Madan J, Shivani PR, Sharma R. Device simulation of 17.3% efficient lead-free all-perovskite tandem solar cell. Solar Energy, 2020 Feb 1; 197:212–221.

[9] Ehrler B, Alarcón-Lladó E, Tabernig SW, Veeken T, Garnett EC, Polman A. Photovoltaics reaching for the Shockley–Queisser limit, ACS Energy Lett [Internet]. 2020 Sep 11; 5(9): 3029–3033. Available from: https://doi.org/10.1021/acsenergylett.0c01790.

[10] Ašmontas S, Mujahid M. Recent progress in perovskite tandem solar cells. 2023; Nanomaterials, 13.

[11] Beckedahl J Record-breaking tandem solar cell now with precise scientific explanations. 2023; Available from: https://phys.org/news/2023-07-record-breaking-tandem-solar-cell-precise.html

[12] Mariotti S, Köhnen E, Scheler F, Sveinbjörnsson K, Zimmermann L, Piot M, et al. Interface engineering for high-performance, triple-halide perovskite–silicon tandem solar cells, Science (80-) [Internet]. 2023; 381(6653): 63–69. Available from: https://www.science.org/doi/abs/10.1126/science.adf5872.

[13] CIGS and perovskite solar cells – an overview, 2020; 9: 812–824.

[14] Bush KA, Palmstrom AF, Yu ZJ, Boccard M, Cheacharoen R, Mailoa JP, et al. 23.6%-efficient monolithic perovskite/silicon tandem solar cells with improved stability, Nature Energy. 2017 Mar 13; 2(4): 17009.

[15] Acheampong AO, Boateng E, Amponsah M, Dzator J. Revisiting the economic growth–energy consumption nexus: Does globalization matter? Energy Econ [Internet] 2021 December 2020; 102: 105472. Available from: https://doi.org/10.1016/j.eneco.2021.105472.

[16] Lindsey R, Dahlman L. Climate change: Global temperature. https://WwwClimateGov/News-Features/Understanding-Climate/Climate-Change-Global-Temperature. 2023;1–5.

[17] www-globalgoals-org-goals-7-affordable-and-clean-energy-.pdf.

[18] MNRE. solar power in india [Internet]. Wikipedia. 2016; 1–9. Available from: https://en.wikipedia.org/wiki/Solar_power_in_India

[19] lack-insurance-solar-modules-holding-indian-modules 3.

[20] India's commitment to achieving net-zero emissions by 2070 India's Carbon Footprint - The Commitment to Net Zero by 2070 Decarbonization Goals : A Strategic Roadmap. 2024.

[21] Samprikta B. Panchamrit – A striding step towards achieving India's goal of net zero by 2070. 2023.

[22] www.iea.org. 947360.

[23] World Economic Forum. These 8 countries have already achieved net-zero emissions, Energy Transit. 2022.

[24] Bo P. Together, China, the US, and the EU emit nearly half of the world s greenhouse gases. 2023.

[25] Chaurasia S, Lohia P, Dwivedi DK, Pandey R, Madan J, Agarwal S, et al. Enhancing perovskite solar cell efficiency to 28.17% by integrating Dion-Jacobson 2D and 3D phase perovskite absorbers, Inorganic Chemistry Communications. 2024 Dec 1; 170.

[26] Kim EB, Akhtar MS, Shin HS, Ameen S, Nazeeruddin MK. A review on two-dimensional (2D) and 2D-3D multidimensional perovskite solar cells: Perovskites structures, stability, and photovoltaic performances. Journal of Photochemistry and Photobiology C: Photochemistry Reviews. 2021, 48: 100405. Available from: https://www.sciencedirect.com/science/article/pii/S1389556721000046.

[27] Rai N, Rai S, Singh PK, Lohia P, Dwivedi DK. Analysis of various ETL materials for an efficient perovskite solar cell by numerical simulation, Journal of Materials Science: Materials in Electronics [Internet]. 2020; 31(19): 16269–16280. Available from: https://doi.org/10.1007/s10854-020-04175-z.

[28] Pathak G, Pooja DKD, Yashwant L, Singh K, Pandey R, Madan J. Optimizing the filtered spectrum and various transport layers of synergistic Dion – Jacobson 2D – 3D perovskite tandem solar cell. Journal of Inorganic and Organometallic Polymers and Materials. 2025.

[29] Verma AA, Dwivedi DK, Lohia P, Singh PK, Yadav RK, Kumar M, et al. Achieving 31.16 % efficiency in perovskite solar cells via synergistic Dion-Jacobson 2D-3D layer design, Journal of Alloys and Compounds [Internet]. 2025 Jan 5; cited 2025 May 29 1010: 177882. Available from: https://www.sci encedirect.com/science/article/pii/S0925838824044700.

[30] Alve HR, Dey M, Rahman H, Ahsan E. Design and Optimization of Copper Zinc Tin Sulfide (CZTS) thin film solar cell refbacks, International Journal of Smart grid. 2024; 12(1): 1–3.

[31] Advanced Energy Materials – 2025 – Tang – Record-Efficient Flexible Monolithic Perovskite CIGS Tandem Solar Cell with VOC.pdf.

[32] Tian W, Yao L, Bi E, Zhang S, Tian Y, Zhou J, et al. Inert interlayer secures flexible monolithic perovskite/CIGS tandem solar cells with efficiency beyond 21%, ACS Energy Letters. 2025; 10(1): 562–568.

[33] Chaurasia S, Lohia P, Dwivedi DK, Pandey R, Madan J, Yadav S, et al. Highly efficient and stable Dion–Jacobson(DJ) 2D-3D perovskite solar cells with 26 % conversion efficiency: A SCAPS-1D study, Journal of Physics and Chemistry of Solids. 2024 Aug 1; 191.

[34] Mohammed MKA, Al-Mousoi AK, Kumar A, Sabugaa MM, Seemaladinne R, Pandey R, et al. Harnessing the potential of Dion-Jacobson perovskite solar cells: Insights from SCAPS simulation techniques, Journal of Alloys and Compounds [Internet]. 2023 March; 963: 171246. Available from: https://doi.org/10.1016/j.jallcom.2023.171246.

[35] Singh YK, Dwivedi DK, Lohia P, Pandey R, Madan J, Agarwal S, et al. Current matching and filtered spectrum analysis of wide-bandgap/narrow-bandgap perovskite/CIGS tandem solar cells: A numerical study of 34.52 % efficiency potential, Journal of Physics and Chemistry of Solids. 2025 Jan 1; 196.

[36] Hossain MK, Toki GFI, Kuddus A, Rubel MHK, Hossain MM, Bencherif H, et al. An extensive study on multiple ETL and HTL layers to design and simulation of high-performance lead-free CsSnCl3-based perovskite solar cells, Scientific Reports. 2023 Dec 1; 13(1): 2521.

Sadanand Maurya* and Mrinmoy Kumar Chini

Chapter 8
Hybrid systems incorporating perovskite solar cells

Abstract: The advent of hybrid perovskite solar cells (HPSCs) has been nothing short of a game-changer for the field of photovoltaics, balancing high efficiency with low-cost manufacturing and adaptable optoelectronic properties. Attention has focused on HPSCs because these materials built with an ABX3 perovskite structure exhibit outstanding light filtering capabilities, as well as long carrier diffusion lengths and the ability to be processed from a solution. The increase in efficiency of hybrid perovskites from 3.8% in 2009 to over 23% in just under a decade fervently accelerated the development of HPSCs. Still, issues like ion migration, charge recombination, and the influence of microstructure on the device remain significant obstacles to stability and performance. On the one hand, the recombination of charge at a defect site decreases efficiency, and on the other hand, ion migration within the perovskite lattice leads to operational instability. Furthermore, microstructural changes in crystallinity versus grain size and interface porosity also impact long-term durability and charge transport. Regardless of the hurdles, relentless progress in device optimization and material engineering helps to create more efficient and stable HPSCs. Once addressing the fundamental limitations is met, HPSCs stand to transform solar energy technologies in comparison to conventional photovoltaics.

Keywords: HPSCs, photovoltaic, optoelectronics, efficiency, stability

8.1 Solar energy

Global energy consumption is constantly rising. According to the Energy Information Administration's prediction, there will be a rise of about 48% in consumption by 2040 [1]. At present, more than two-thirds of the energy required for consumption is produced by conventional nonrenewable energy sources such as oil, gas, and coal. A real problem has developed with the increase in pollution and climate change; therefore, we need clean and affordable energy more than ever, especially as we immerse ourselves in a sea of electronic devices. The ultimate solution is the Sun, which provides

*Corresponding author: Sadanand Maurya, Department of Applied Sciences, Galgotias College of Engineering and Technology, Greater Noida 201310, Uttar Pradesh, India
Mrinmoy Kumar Chini, Department of Applied Sciences, Galgotias College of Engineering and Technology, Greater Noida 201310, Uttar Pradesh, India

https://doi.org/10.1515/9783111726847-008

enough energy in about 2 h to meet the world's yearly energy demand. Solar cell devices can harvest energy from the Sun and convert it into usable electrical energy, offering a promising solution to meet humankind's ever-increasing energy demands. Inorganic solar cells such as GaAs, CdTe, and CIGS have been extensively studied over the past decades to achieve high-power conversion efficiencies of over 25% in a single-junction structure [1, 2]. These solar cell materials provide a benchmark for comparing the efficiency of new and upcoming solar cell materials to determine if they can be as successful as their predecessors. The production cost and the material cost are equally important as the power conversion efficiency (PCE) of the solar cells. It is also necessary that new solar cell materials have the potential to be upscaled using simple processing techniques required by both industry and academia [3].

Hybrid perovskite solar cells (HPSCs) have emerged as a transformative technology in photovoltaics. They offer high PCE, low-cost fabrication, and tunable optoelectronic properties. In a very short time, the PCE of HPSCs has increased from 3.8% in 2009 to 22.1% in 2016, making them the fastest-advancing solar cell technology to date [4].

Although there has been remarkable growth in HPSCs, the working device physics of these solar cells is still poorly understood. The operation of these solar cells is unique and requires new physical models for optimization and characterization. To improve the performance and stability of these devices, it is essential to understand the role of all physical processes that govern their operation. Figure 8.1 illustrates the progressive improvement in the efficiencies of different solar cell technologies over time, based on data from the National Renewable Energy Laboratory (NREL) [5].

8.2 Hybrid perovskites

The term "perovskite" originated from the mineral $CaTiO_3$ and now refers to a group of compounds with the general formula ABX_3, where A and B are cations and X is an anion as shown in Figure 8.2. Perovskite solar cells are primarily used as the absorber layer in photovoltaic devices and are typically referred to as hybrid because they consist of both organic and inorganic compounds. In hybrid perovskites, the A moiety is a monovalent organic cation (e.g., methylammonium (MA) and formamidinium (FA)), B is a divalent metal cation (commonly lead (Pb) or tin (Sn)), and X is a halide anion (chlorine (Cl), bromine (Br), or iodine (I)) [6]. Due to their hybrid nature, hybrid perovskites exhibit strong light absorption, long carrier diffusion lengths, and low-exciton binding energies, making them promising candidates for high-efficiency solar cells [7]. Its structure consists of a three-dimensional framework in which the **B** cation is surrounded by six **X** anions, forming a BX_6 octahedron.

The A-site cation influences the material's overall stability and electrical characteristics by fitting into the cavities of the perovskite framework (Weber, 1978). One of the most important properties of HPSCs is their high absorption coefficient, which al-

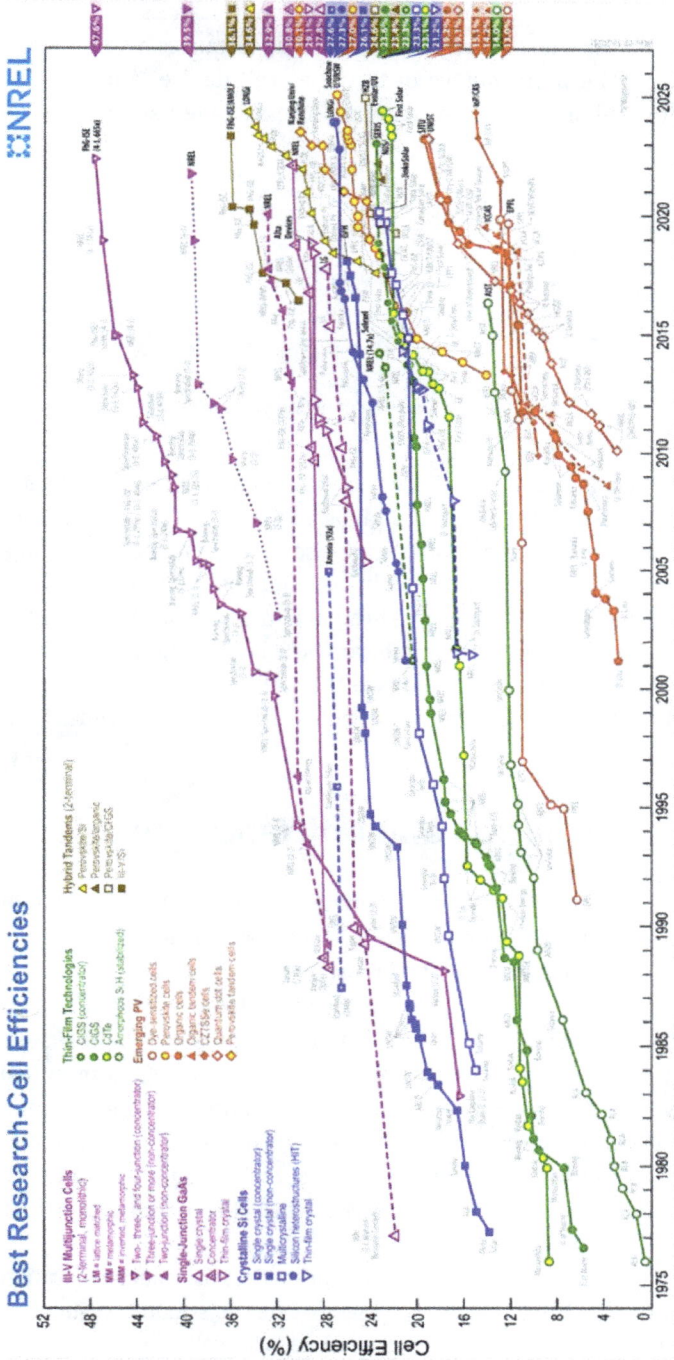

Figure 8.1: Efficiency versus time plot for various solar cell technologies including crystalline silicon, CdTe, DSSC, and organic solar cells. Perovskite solar cells have shown a remarkable increase in power conversion efficiency, surpassing other photovoltaic technologies at an unprecedented rate (ref. taken from NREL chart) [5].

lows them to efficiently absorb sunlight with minimal material thickness [8]. The long carrier diffusion length enables charge carriers to travel long distances before recombining, thereby improving the performance of the solar cell device. Additionally, HPSCs can be easily fabricated using low-cost solution processing methods, unlike traditional silicon-based photovoltaic solar cells [9, 10].

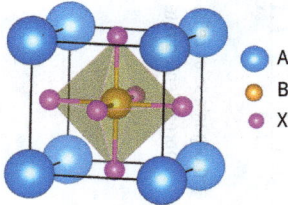

Figure 8.2: Crystal structure of hybrid perovskite compound.

The HPSCs offer several advantages in comparison to the tradition silicon-based solar cells. In very short spam of the time the PCE reach to the above 25%. The low-cost fabrication enables large-scale, cost-effective manufacturing also by varying the halide composition (Br, I, and Cl), and hybrid perovskite can be used for different optoelectronic applications [11, 12]. Even though hybrid perovskites have many promising qualities, they also have drawbacks such as lead toxicity, moisture sensitivity, and thermal instability. Thus, studies are being conducted to enhance the stability and environmental compatibility of HPSCs through material engineering and encapsulation techniques [13].

8.3 Performance evolution and efficiency growth

The first perovskite solar cell was developed by Kojima et al. in 2009, with a PCE of 3.8% [6]. However, due to extensive research, materials engineering, and interface optimization, the efficiency has surpassed 25% within just a decade. This remarkable improvement is attributed to the superior light absorption, long carrier diffusion lengths, and high defect tolerance of perovskite materials. The key advancements driving this efficiency surge also include (i) the introduction of FA and cesium (Cs) cations, which improve thermal and phase stability; (ii) interface optimization through passivation of grain boundaries, which reduces charge carrier recombination, leading to enhanced stability and higher open-circuit voltage; and (iii) the development of tandem structures by combining perovskite with silicon and other solar materials, pushing the efficiency beyond 29% [14].

Although perovskite solar cells exhibit exceptional properties as a next-generation photovoltaic technology, they face significant challenges related to scalability, stability, and environmental impact. The main issues with perovskite solar cells are (i) their high sensitivity to moisture – exposure to humidity leads to degradation of the thin film [15];

(ii) ion migration within the lattice, which affects long-term stability and causes hysteresis [16]; and (iii) the toxicity of lead in high-efficiency perovskite solar cells, raising serious environmental concerns.

In the current ongoing research goal is to develop more stable and efficient materials, scalable fabrication methods, and encapsulation strategies to make it commercially available for the near future. Perovskite solar cells have numerous advantages and have great potential to revolutionize the solar energy industry, providing a sustainable and incredibly effective substitute for traditional photovoltaics.

8.4 Charge recombination: a major challenge in hybrid perovskite solar cells

HPSCs have shown outstanding efficiencies, but one of the key factors which limit their performance is charge recombination. This process, where photogenerated electrons and holes recombine before contributing to electrical power, significantly impacts the efficiency and stability of perovskite devices. Understanding the origins of recombination and developing strategies to mitigate its effects are crucial for advancing this technology toward commercial viability. Perovskite materials are known for their long carrier diffusion lengths and defect tolerance, which theoretically reduce recombination losses [7]. However, in real-world devices, recombination occurs due to imperfections within the perovskite film, grain boundaries, and interfaces with adjacent layers. One of the most detrimental types of recombination is trap-assisted (nonradiative) recombination, which takes place at defect sites where charge carriers lose their energy without generating electricity [17]. This issue is particularly pronounced in perovskite films with poor crystallinity, leading to lower open-circuit voltage (V_{OC}) and reduced overall efficiency.

Another major concern is interfacial recombination, which occurs at the boundaries between the perovskite absorber layer and charge transport materials [18, 19]. If these interfaces are not well engineered, energy mismatches and defect states can create pathways for recombination, reducing charge extraction efficiency. Researchers have tackled this problem through interface passivation techniques, where materials like self-assembled monolayers (SAMs), 2D perovskite layers, and fullerene derivatives are used to reduce recombination at interfaces [13]. Material engineering has played a vital role in suppressing recombination losses. The introduction of mixed-cation perovskites (such as FA-Cs perovskites) has improved structural stability while minimizing defects [20]. Additionally, optimizing the perovskite crystallization process through advanced deposition techniques – such as antisolvent dripping, hot-casting, and vapor-assisted growth – has led to smoother films with fewer trap states, ultimately reducing recombination losses [21].

Apart from material engineering, device architecture modifications have proven effective in mitigating recombination effects. Tandem solar cells, where perovskites are paired with silicon or other materials, help maximize light absorption and charge carrier extraction, improving efficiency beyond 29% [14]. Additionally, using buffer layers and surface treatments has been shown to reduce charge recombination at interfaces, leading to more stable and efficient devices. Despite these advancements, challenges remain, particularly in achieving long-term stability. Ion migration, which contributes to hysteresis and instability in perovskite solar cells, further exacerbates recombination losses over time [22]. Researchers continue to explore new materials and encapsulation techniques to address these issues, moving closer to the goal of commercializing perovskite solar technology.

By overcoming recombination challenges through defect passivation, interface engineering, and advanced fabrication techniques, HPSCs are steadily advancing toward their full potential. With continued research and innovation, these materials could soon rival and even surpass conventional silicon photovoltaics, paving the way for a future of high-efficiency, low-cost, and scalable solar energy solutions.

8.5 Ion migration in hybrid perovskite solar cells: a key challenge for stability

In methylammonium lead iodide, not only electronic charge carriers but also charged ionic species exhibit mobility, contributing to the overall transport properties of the material. Although ionic conductivity in halide perovskites was recognized several decades ago [23, 24], it is only in recent years that ion migration in hybrid perovskites gained significant interest – particularly following the unexpected observation of hysteresis in current–voltage (J–V) measurements [25]. This hysteresis, identified by the mismatch in J–V characteristics during forward and reverse voltage sweeps and at varying scan rates, raised questions about underlying mechanisms. Snaith et al. [25] proposed that the observed J–V hysteresis could stem from a combination of factors including (1) ferroelectric polarization, (2) charge trapping phenomena, and (3) migration of ions within the perovskite layer. Further systematic investigations by Unger and Tress demonstrated that the scan rate dependence of the J–V curves suggests that slow-moving ions are a primary cause of this behavior [17, 25].

Ferroelectric effects, which may originate from spontaneous polarization due to the rotational motion of MA ($CH_3NH_3^+$) ions within the PbI_6 octahedra, occur on ultrafast timescales (picoseconds) and are thus unlikely to significantly influence the slower electrical measurements [26]. Similarly, charge trapping and release happen on the order of nanoseconds and are considered too rapid to account for the hysteresis observed. A later study by Eames et al. [22] indicated that ionic species – most probably iodide-related complexes – respond to electric fields by migrating toward the interfaces between the perovskite and charge transport layers. This accumulation can screen the internal electric field, ultimately influencing charge collection and resulting in the characteristic hysteresis in the $J–V$ curves.

As a result, hybrid perovskites are now recognized for their dual nature, functioning both as semiconductors and as ionic conductors. Experimental and theoretical studies on activation energies for ion migration have pointed to iodide vacancies and interstitials as the dominant mobile defects contributing to this phenomenon [22].

8.6 Influence of microstructure on hybrid perovskite solar cell performance

The microstructure of HPSCs determines their efficiency, stability, and performance. In contrast to conventional silicon-based photovoltaics, perovskites have a unique polycrystalline structure. Factors like the spatial distribution of grains, crystallinity, density of defects, and quality of interfaces all critically affect the amount of recombination and charge transport processes [27]. Improvements in the microstructure can reduce charge losses, increase stability, and increase reliability of the perovskite solar cells over time.

The perovskite films' morphology also has microstructures, such as grain size, one of the most impactful on the performance of the film. As it was observed that larger grains are preferred as they lower the grain-boundary density which is responsible for charge recombination and loss in efficiency. Sherkar et al. [28] proved the existence of perovskite films which showed marked improvements in carrier lifetimes and charge transport due to large-grain structures. On the contrary, small grain sizes significantly increase the concentration of grain boundaries resulting in increased nonradiative recombination and reduced efficiency.

Optical and electronic functionalities of the perovskites are impacted by another critical factor – crystallinity. Films with high crystallinity tend to have higher performance due to better carrier mobility. On the contrary, low-crystallinity films have a performance deficiency due to a high density of defects, which act as charge traps [29]. Several methods have been developed to enhance crystallinity and create high-quality perovskite films including thermal annealing and solvent engineering [30].

The optoelectronic properties of HPSCs are highly affected by the microstructure of defects in perovskite, especially vacancy defects, interstitial defects, and substitutional defects. A vacancy defect's charge imbalance may lead to recombination losses due to the absence of halide or cation sites. Deep-level trap states that impede charge transports are caused by interstitial defects, which incorporate excess ions into the lattice. Changing the band structure or electronic properties of perovskite components by foreign atoms results in substitutional defects [27, 31]. To reduce the consequences of these changes, researchers have utilized additives such as small organic molecules, alkali metals, or 2D perovskite layers for adding defect passivation.

The interfaces within the perovskite absorber layer boundary and the transport charge layer have a profound impact on the performance metrics of the device. Geometrical imperfections result in a rise of interfacial charge resulting in accumulation of charges and recombination which ultimately diminishes efficiency. Iohexol and polymer-encapsulated SAMs as well as fullerene derivatives passivation strategies were used to mitigate interface optimization issues which help in improvement of charge extraction [32]. Moreover, the orientation of the perovskite grains at the boundaries plays a crucial role in determining the dynamics of charge carriers. Proper alignment between the charge carriers and the crystallographic framework reduces the energy barrier for carrier transport, thereby enhancing the power conversion efficiency (PCE). Specific advanced fabrication processes such as antisolvent engineering, vacuum deposition, and compositional tuning need to be done meticulously in order to achieve high levels of performance and stability in the perovskite solar cells [33]. These techniques result in producing defect-free, smoother interfacial, larger grains, and perovskite films which improve the efficiency as well as the long-term durability of the device.

Thus, optimizing the microstructure of perovskite solar cells is crucial in enhancing the efficiency and stability of the cells. The increase of grain size, enhancement of crystallinity, betterment of defect density, and the development of superior interfaces all contribute to the improvement of the reliability of perovskite photovoltaics. Further advances in engineering at the microstructural level will be pivotal in solving the issues at hand, thereby creating pathways toward innovative and highly efficient solar energy technologies.

8.7 Conclusion and future outlook

The development of HPSCs as a new and disruptive technology within photovoltaics is due to its unique combination of high PCE, low-cost fabrication, and tunable optoelectronic properties. In the last decade, efficiency improvements have been remarkable with laboratory devices exceeding 25% efficiency and competing against traditional silicon solar cells [34]. However, stability, scalability, and environmental concerns continue to be the primary hurdles in commercialization. HPSC's achievements highlight that perovskite devices still have poor stability compared to silicon solar cells, which can last for decades. Unlike silicon panels which can last for decades, HPSCs are more fragile due to their susceptibility to moisture, heat, light, and oxygen. The instability stems from ion migration, phase segregation, charge recombination, and interface degradation which all diminish the efficiency of the device over time. Recent decades have seen researchers make strides in improving this stability through composition engineering, defect passivation, and a combination of encapsulation techniques [27]. The introduction of triple cation perovskites (FA-Cs-MA), low-dimensional 2D/3D hybrid structures, and surface treatments has led to stronger perovskite films and better operational lifetimes.

Unlike other issues, "scalability" seems to be more of an economic problem than a technological one. Even though laboratory-scale perovskite solar cells have some of the highest efficiencies, they are not able to transcend the large-area, industrial-scale production barrier. As with many other techniques, spin-coating is not a viable option for mass production. More pragmatic approaches [31] like slot-die-coating, blade-coating, and vapor deposition need to be developed to fill that gap. Moreover, adding perovskites to the existing tandem solar cells is particularly advantageous for building integrated efficiency silicons because it enhances the overall efficiency while capitalizing on the existing silicon industry manufacturing base (Albrecht 2016). The use of lead-based perovskites also introduces issues regarding commercialization due to environmental and toxicity considerations. The use of lead in high-efficiency perovskite materials is problematic due to its poisonous nature and the severe risks it poses concerning the contamination of the environment if solar panels break down or are thrown away carelessly (Jiang et al., 2019). Strides are being made to create lead-free perovskites with other materials like tin (Sn), bismuth (Bi), and antimony (Sb); but at this point they have much greater inefficiency and instability (Ke et al., 2018). Focusing on developing strategies for lead-recycling, advancing encapsulation procedures, and inventing frameworks for green synthesis will be more valuable in addressing sustainability concerns.

8.7.1 Future outlook

The future of hybrid perosvkite solar cells lies in innovation integrating materials science, device engineering, and scalable hybrid perovskite manufacturing technologies. The field is expected to grow in the following key areas over the coming years:

1. **Enhancing stability:** Achieving lifetime operational stability on par with commercial-grade silicon solar cells by developing more stable perovskite compositions with advanced passivation and protective encapsulation techniques.
2. **Scalable fabrication:** Advancing from the current small-scale laboratory spin-coated devices to mass industrial production utilizing roll-to-roll and vapor deposition techniques.
3. **Tandem integration:** Optimization of perovskite-silicon tandem cells and perovskite-perovskite tandem cells to surpass the 30% efficiency barrier, allowing for cost-efficient high-performance solar modules.
4. **Sustainability and toxicity reduction:** Employing environmentally friendly fabrication and efficient recycling processes along with the development of lead-free perovskites to reduce the environmental impact.

5. **Commercialization and market adoption:** Faster transition of perovskite solar cells from research labs to real-world applications driven by collaboration across academic, industrial, or policy-oriented entities.

Addressing these issues is key to leveraging the potential of perovskite solar cells in transforming the renewable energy industry providing an efficient, affordable, and sustainable solution for global energy demands. Next-generation perovskite photovoltaic technologies are expected to be a major contributor to solar energy in the future and may even surpass or supplement traditional silicon technologies if they continue to achieve success.

References

[1]   U.S.E.I. Administration, Analysis of the Impacts of the Clean Power Plan, (2015).
[2]   Chini MK, Synthesis and properties of halide perovskites nanocrystals toward solar cells application, Energy Technology. 12 (2024) 2301444. https://doi.org/10.1002/ente.202301444.
[3]   Shivani K, Shambhavi B, Sadanand R, Lohia P, Investigating the performance of mixed cation mixed halide – based perovskite solar cells using various hole – transport materials by numerical simulation, Optical and Quantum Electronics. 53 (2021) 1–16. https://doi.org/10.1007/s11082-021-03262-7.
[4]   Hima A, Lakhdar N, Enhancement of efficiency and stability of CH3NH3GeI3 solar cells with CuSbS2, Optical Materials. 99 (2020) 109607. https://doi.org/10.1016/j.optmat.2019.109607.
[5]   NREL, Best-Research-Cell-Efficiencies-Rev210726.Pdf, (n.d.).
[6]   Akhiro K, Kenjiro T, Yasuo S, Tsutomu M, Organometal halide perovskites as visible-light sensitizers for photovoltaic cells, Journal of the American Chemical Society. 131 (2009) 6050–6051.
[7]   S.D. Stranks, G.E. Eperon, G. Grancini, C. Menelaou, M.J.P. Alcocer, T. Leijtens, L.M. Herz, A. Petrozza, H.J. Snaith, "Electron-hole diffusion lengths exceeding 1 micrometer in an organometal trihalide perovskite absorber", Science, 342 (2013), 341–344. https://doi.org/10.1126/science.1243982.
[8]   W. Yin, T. Shi, Y. Yan, "Unique properties of halide perovskites as possible origins of the superior solar cell performance", Advanced Materials, 26 (2014), 4653 –4658. https://doi.org/10.1002/adma.201306281.
[9]   F. Hao, C.C. Stoumpos, D.H. Cao, R.P.H. Chang, M.G. Kanatzidis, "Lead-free solid-state organic -inorganic halide perovskite solar cells", J. Mater. Chem. A, 2 (2014), 19186 –19190. https://doi.org/10.1039/C4TA04994B.
[10] S. Zhang, G. Xie, C. You, P. Li, "Organic -Inorganic Hybrid Devices –Perovskite-Based Devices", in: Introduction to Organic Electronic Devices, Springer, Singapore, 2022, pp. 283–307. https://doi.org/10.1007/978-981-19-3576-0_6.
[11] M. Saliba, T. Matsui, J.Y. Seo, K. Domanski, J.P. Correa-Baena, M.K. Nazeeruddin, S.M. Zakeeruddin, W. Tress, A. Abate, A. Hagfeldt, M. Grätzel, "Cesium-containing triple cation perovskite solar cells: improved stability, reproducibility and high efficiency", Energy Environ. Sci., 9 (2016), 1989 –1997. https://doi.org/10.1039/C5EE03874J.
[12] J.H. Noh, S.H. Im, J.H. Heo, T.N. Mandal, S.I. Seok, "Chemical management for colorful, efficient, and stable inorganic -organic hybrid nanostructured solar cells", Nano Lett. 13 (2013), 1764 –1769. https://doi.org/10.1021/nl400349b.

[13] Hao F, Stoumpos CC, Cao DH, Chang RPH, Kanatzidis MG, Lead-free solid-state organic-inorganic halide perovskite solar cells, Nature Photonics. 8 (2014) 489–494. https://doi.org/10.1038/nphoton. 2014.82.

[14] Geisz JF, France RM, Schulte KL, Steiner MA, Norman AG, Guthrey HL, Young MR, Song T, Moriarty T, Efficiency under 143 Suns concentration, Nature Energy. (n.d.). https://doi.org/10.1038/s41560-020-0598-5.

[15] S.N. Habisreutinger, T. Leijtens, G.E. Eperon, S.D. Stranks, R.J. Nicholas, H.J. Snaith, "Carbon nanotube/polymer composites as a highly stable hole collection layer in perovskite solar cells", Nanoscale 6 (2014), 8804–8810. https://doi.org/10.1039/C4NR03435J.

[16] Y. Hu,"Getting solid", Nat. Energy 1 (2016), Article 16042. https://doi.org/10.1038/nenergy.2016.42.

[17] Shi Z, Jayatissa AH, Perovskites-based solar cells: A review of recent progress, materials and processing methods, Materials (Basel). 11 (2018). https://doi.org/10.3390/ma11050729.

[18] A. Umar, P.K. Singh, D.K. Dwivedi, H. Algadi, "High power-conversion efficiency of lead-free perovskite solar cells: A theoretical investigation", J. Mater. Sci. Mater. Electron. 34 (2022), 334. https://doi.org/10.1007/s10854-022-08703-8.

[19] Shum K, Chen Z, Qureshi J, Yu C, Wang JJ, Pfenninger W, Vockic N, Midgley J, Kenney JT, Synthesis and characterization of CsSnI3 thin films, Applied Physics Letters. 96 (2010) 3–5. https://doi.org/10. 1063/1.3442511.

[20] Karthick S, Velumani S, Bouclé J, Experimental and SCAPS simulated formamidinium perovskite solar cells: A comparison of device performance, Solar Energy. 205 (2020) 349–357. https://doi.org/10. 1016/j.solener.2020.05.041.

[21] Jeon NJ, Noh JH, Yang WS, Kim YC, Ryu S, Seo J, Seok SI, Compositional engineering of perovskite materials for high-performance solar cells, Nature. 517 (2015) 476–480. https://doi.org/10.1038/ nature14133.

[22] C. Eames, J.M. Frost, P.R.F. Barnes, B.C. O'Regan, A. Walsh, M.S. Islam, "Ionic transport in hybrid lead iodide perovskite solar cells", Nat. Commun. 6 (2015), Article 7497. https://doi.org/10.1038/ ncomms8497.

[23] Halides THEP, Ionic conduction of, 11 (1983) 203–211.

[24] Shaikh SF, Kwon HC, Yang W, Mane RS, Moon J, Performance enhancement of mesoporous TiO2-based perovskite solar cells by ZnS ultrathin-interfacial modification layer, Journal of Alloys and Compounds. 738 (2018) 405–414. https://doi.org/10.1016/j.jallcom.2017.12.199.

[25] Snaith HJ, Abate A, Ball JM, Eperon GE, Leijtens T, Noel NK, Stranks SD, Wang JT, Wojciechowski K, Zhang W, Anomalous hysteresis in perovskite solar cells, (2014). https://doi.org/10.1021/jz500113x.

[26] Bakulin AA, Selig O, Bakker HJ, Rezus YLA, Mu C, Glaser T, Lovrincic R, Sun Z, Chen Z, Walsh A, Frost JM, Jansen TLC, Real-Time observation of organic cation reorientation in methylammonium lead iodide perovskites, (2015). https://doi.org/10.1021/acs.jpclett.5b01555.

[27] M. Abbas, X. Xu, M. Rauf, A. Ko, A.K.K. Kyaw, "A comprehensive review on defects-induced voltage losses and strategies toward highly efficient and stable perovskite solar cells", Adv. Energy Mater. 14 (2024), 2303431. https://doi.org/10.1002/aenm.2023034

[28] T.S. Sherkar, C. Momblona, L. Gil-Escrig, M. Sessolo, H.J. Bolink, L.J.A. Koster, "Recombination in perovskite solar cells: Significance of grain boundaries, interface traps, and defect states", ACS Energy Lett. 2 (2017), 1214 –1222. https://doi.org/10.1021/acsenergylett.7b00236.

[29] Baig F, Khattak YH, Marí B, Beg S, Ahmed A, Khan K, Efficiency enhancement of CH3NH3SnI3 solar cells by device modeling, Journal of Electronic Materials. 47 (2018) 5275–5282. https://doi.org/10. 1007/s11664-018-6406-3.

[30] Liu D, Traverse CJ, Chen P, Elinski M, Yang C, Wang L, Young M, Lunt RR, Aqueous-containing precursor solutions for efficient perovskite solar cells, Advanced Science. 5 (2018). https://doi.org/ 10.1002/advs.201700484.

[31] H. Choi, J. Jeong, H.-B. Kim, S. Kim, J.Y. Kim, "Methods of stability control of perovskite solar cells for high efficiency", Mater. Today Energy 20 (2021), 100686. https://doi.org/10.1016/j.mtener.2020. 100686.

[32] Jiang, Q., Zhao, Y., Zhang, X., Yang, X., Chen, Y., Chu, Z., Ye, Q., Li, X., Yin, Z., & You, J. (2019). Surface passivation of perovskite film for efficient solar cells. Nature Photonics, 13(7), 460–466. https://doi. org/10.1038/s41566-019-0398-2

[33] Saliba, M., Matsui, T., Seo, J.-Y., Domanski, K., Correa-Baena, J.-P., Nazeeruddin, M. K., Zakeeruddin, S. M., Tress, W., Abate, A., Hagfeldt, A., & Grätzel, M. (2016). Cesium-containing triple cation perovskite solar cells: improved stability, reproducibility and high efficiency. Energy & Environmental Science, 9(6), 1989–1997. https://doi.org/10.1039/C5EE03874J

[34] D. Feldman, K. Dummit, J. Zuboy, and R. Margolis, Summer 2023 Solar Industry Update, National Renewable Energy Laboratory (NREL), Aug. 3, 2023. Available: https://www.nrel.gov

Mrinmoy Kumar Chini* and Sadanand Maurya

Chapter 9
Challenges and future directions for perovskite-based tandem systems

Abstract: We are well aware of the exciting potential of perovskite materials espe-cially halide perovskite (HP)-based perovskites solar cells (PSCs). To commercialize and introduce highly efficient and stable PSCs, deep understanding of the mechanisms and fundamentals of energy conversion is most vital. The understanding of rationally behind the proposed perovskites materials with suitable optoelectronic properties in order to introduce high-efficiency and stable commercially viable devices is neces-sary. However, despite the potential aspects, few fundamental challenges and ob-stacles are yet to be addressed for the successful commercialization. Stability under ambient condition and upscaling has been one of the significant critical issues for HP-based PSCs. Therefore, it is important to discuss and understand the critical issues for the promising HP-based PSCs. In this chapter, we have discussed associated chal-lenges, degradation factors in perovskites, lead toxicity, and highlighted the various aspects and perspectives of current status of industrialization, device configurations, stability issues, upscaling aspects, reliability concerns for perovskite solar modules and the futuristic strategies with the future directions in the low cost and renewable PSC field.

Keywords: PSCs, tandem, halide perovskite, stability, efficiency

9.1 Introduction

The increased demand for renewable and sustainable energy for daily consumption in modern day's lifestyle signifies the importance of efficient energy conversion. Solar power generation is undoubtedly an important resource of a clean energy. The photo-conversion efficiency (PCE) of perovskite solar cells (PSCs) has been reached to over 26.1% within a short span of research [1]. This perovskite-based solar technology is the third-highest single-junction photovoltaic (PV) technology. It only lags behind crys-talline silicon and GaAs solar cells. The PSCs are attractive due to their unique intrin-

*Corresponding author: Mrinmoy Kumar Chini, Department of Applied Sciences, Galgotias College of Engineering and Technology, Greater Noida 201310, Uttar Pradesh, India,
e-mails: mrinmoychini@gmail.com; mrinmoychini@galgotiacollege.edu
Sadanand Maurya, Department of Applied Sciences, Galgotias College of Engineering and Technology, Greater Noida 201310, Uttar Pradesh, India

https://doi.org/10.1515/9783111726847-009

sic properties such as favorable bandgaps, excellent absorption coefficients, long carrier diffusion lengths, and simple device architectures [2]. Fundamental understanding of bulk HPs properties and various technological advancements plays a key role for the achieved high PCE. Along with significant PV performance, other key parameters such as solution processibility, low cost, and flexibility make PSCs with massive commercial potential for the grid connected system as well as in off-grid scenarios. The off-grid cases such as indoor PV and wearable electronics have the future for PSCs. Thus, the merits of both low cost and remarkable PCE makes halide perovskites (HPs) a revolutionary material for PV technology. Now, with the progress achieved with PSCs at lab scale, the next important steps are the effective industrialization and their commercialization. To achieve the successful commercialization, few key challenges such as up-scaling, remarkable environment stability and toxicity need to be addressed carefully. On further note, though small-area substrates have shown best device performances [3–5], upscaling to achieve high efficiency along with low batch-to-batch variation is the main challenge for the perovskite solar modules (PSMs). Usually, the PCE drops for large-area modules and the scalable methods for fabrication of all functional layers are limited. The second challenge is the intrinsic instability of perovskite material and environmental degradation of PSCs. Moreover, the toxicity of lead remains a major concern. Potential lead leakage poses risks to human health and water resources, which could block successful commercialization. While encapsulation helps limit leakage, exploring Pb-free or Pb-reduced materials serves as an excellent alternative supporting the sustainable development of PSCs. Recently, the achieved PCE for perovskite-Si tandem solar cells is 33.9% [1]. However, towards commercialization, the challenges such as the scale-up, stability, and cost control needs to be taken care for the commercialization of perovskite-Si tandem solar cells. In addition to that, for PSCs to be commercially viable, their levelized cost of energy (LCOE) must fall below that of traditional silicon-based solar cells.

One promising approach is to position PSCs in niche markets where their distinct advantages stand out. Their lightweight design, low-temperature processing, and mechanical flexibility make them ideal for use in building-integrated PV, portable electronics, wearable devices, and near-space technologies (Figure 9.1) [6]. Furthermore, their compatibility with roll-to-roll (R2R) manufacturing enables scalable, cost-efficient production, boosting their market appeal. While bulk HPs have driven major advances in PSCs, their phase instability under ambient conditions remains a key limitation. This has sparked growing interest in developing stable halide perovskite (HP) nanocrystals (NCs) as promising alternatives for durable, high-performance PSCs [7]. In this context, the challenges in developing HP NCs as stable, low-cost, and efficient photoactive materials offering a potential replacement for toxic, unstable Pb-based perovskites in future PSC integration are needed to be taken into consideration.

This book chapter provides a comprehensive overview of the commercialization pathway for PSCs. It begins with a brief summary of the current commercialization status, followed by a detailed discussion of three major challenges still hindering progress: upscaling, stability, and toxicity. The chapter then introduces effective encapsulation strategies and materials aimed at addressing these issues. Subsequently, it explores promising commercialization directions that distinguish PSCs from conventional silicon solar cells, such as flexible devices and indoor PV. The chapter concludes with a forward-looking perspective on the future of PSC commercialization.

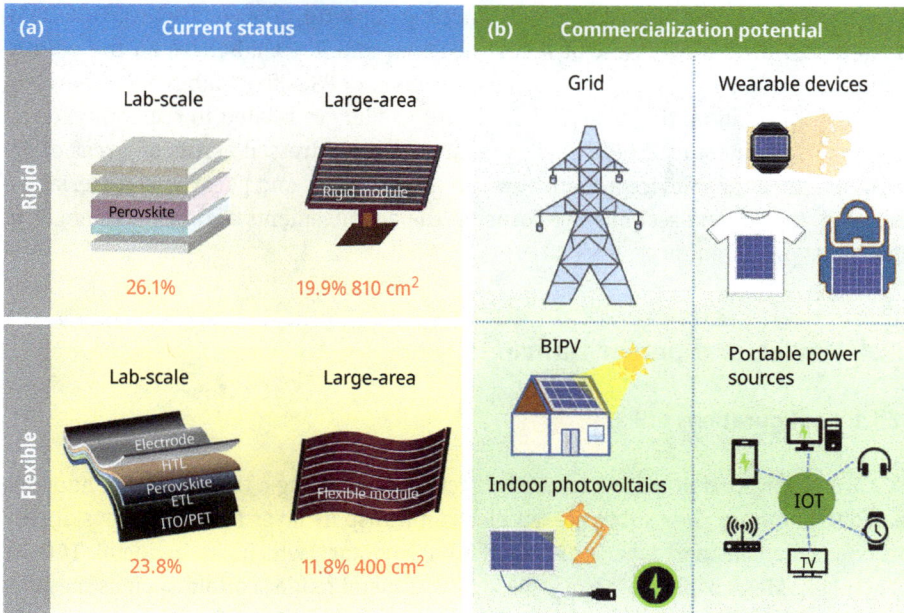

Figure 9.1: The commercialization path in terms of (a) current status and (b) promising applications of PSCs. Reproduced with permission from the reference [6]. Copyright © 2024 IOP Publishing.

9.2 Current status of industrialization

With the global push toward "zero carbon" and "carbon neutrality," the development of renewable energy particularly PV has become a shared priority. Among emerging technologies, PSCs have stood out, consistently breaking efficiency records and capturing widespread attention in the PV community. In industry, PSC startups have rapidly emerged, attracting significant funding and becoming a new focus for capital investment. Driven by advances in processing technologies and strong industrial backing, production lines are expanding quickly. Collaboration between academia and industry has led to notable improvements in both module efficiency and long-term stability positioning PSCs at the heart of solar energy's industrial transition [8, 9].

To design efficient and stable PSCs, a deep understanding of the fundamental principles behind energy conversion is essential. The targeted design and engineering of novel optoelectronic materials tailored to optimize structure–performance relationships play a key role in enabling high-performance devices.

Although the low-cost potential of PSCs makes them attractive for commercialization, challenges remain. A complex balance must be struck between achieving high efficiency, ensuring long-term stability, and scaling up manufacturing. This book chapter provides an overview of recent progress and key challenges on the path to commercializing PSCs. To support the advancement of PSCs from lab-scale research to industrial application, the prioritize areas are challenges related to scalability, durability, and sustainability. These challenges should be addressed across all areas of development including device architecture, interface design, and perovskite material engineering. In the next sections, we focus on the advancements and key challenges in the context of current progress.

9.2.1 Device and performance

9.2.1.1 Configurations of PSCs

PSCs are generally built with a layered structure comprising a transparent conductive oxide (TCO)-coated glass substrate, an electron transport layer (ETL), a perovskite absorber, a hole transport layer (HTL), and a back contact, which can be metal, TCO, or carbon [10]. Mesoscopic PSCs use a mesoporous metal oxide scaffold such as mesoporous TiO_2 on which the perovskite layer is deposited [11, 12]. Whereas, planar PSCs feature a flat perovskite layer placed between planar ETLs (e.g., compact TiO_2 and SnO_2) and HTLs (e.g., spiro-OMeTAD and PTAA) [13, 14]. Based on the deposition sequence, PSCs are classified as conventional (*n-i-p*) or inverted (*p-i-n*) architectures [15, 16]. Triple mesoscopic PSCs use a more complex scaffold with three layers: mesoporous TiO_2, a ZrO_2 spacer, and a carbon electrode [17]. These devices eliminate the need for HTLs by replacing metal contacts with carbon. PSCs are also ideal candidates for tandem configurations with mature PV technologies such as silico [18], CIGS [19], and low-bandgap perovskites [20]. Tandem architectures surpass the Shockley-Queisser efficiency limit for single-junction devices [21] and can tap into existing manufacturing infrastructure [22].

It is to be noted that small-area PSCs (<1 cm^2) are commonly fabricated via spin coating, but upscaling leads to notable drops in PCE across all device structures [12, 13, 23, 24]. To address this, alternative scalable techniques such as screen printing, slot-die coating, and spray coating are crucial for mid-sized modules fabrication [25]. Large-area PSC modules face challenges like ohmic loss due to the resistance of front TCO electrodes. While increasing the thickness of FTO layers can improve conductivity, it reduces optical transparency. Conductive grids or interconnected subcells help reduce resistance but can lower the module's active area. However, the geometrical fill factor (GFF), the ratio of active to total area, can improve with larger modules, as edge-related dead zones for encapsulation and framing become less significant in proportion [26].

9.2.1.2 Module design

Individual solar modules usually consist of series-connected cells and face specific design and fabrication challenges. To achieve high efficiency, it is essential to form uniform, pinhole-free perovskite films over large areas, use low-loss interconnects, and minimize inactive areas between cells to maximize the GFF. Scalable processing methods and post-patterning techniques such as laser scribing are promising approaches to enhance module performance [27]. In operation, partial shading can lead to mismatch losses and place stressed cells under reverse bias, which in PSCs can cause breakdown via tunneling triggered by interfacial ion accumulation. Therefore, careful design at both the device level to optimize interfacial properties and the module level is essential to prevent performance degradation under real-world conditions [28].

9.2.2 Stability challenges and mitigation strategies

One of the key challenges in evaluating the stability of PSCs is the lack of standardized testing protocols [29]. Some devices are encapsulated and others are not. The environment factors like humidity, temperature, and atmosphere are often inconsistent [30]. Standard protocols are also missing in outdoor measurements, thermal stability assessments, and cyclic experiments [31, 32]. However, inconsistency in testing and reporting makes proper comparison difficult. Studies sometimes provide only normalized performance data without reporting the original PV parameters. This can undermine the real efficiency of the PV devices. To ensure reliability and reproducibility, all the PV data and normalization details should be disclosed. The PSC community should adopt and follow standardized universal protocols, modeled after established benchmarks like IEC61215 or 61646 used in thin film and crystalline Si PVs to enable predictive lifetime modeling based on degradation kinetic studies [29]. Again, at the device level, instability often originates from reactive materials and interfacial degradation. The commonly used TiO_2, ETL, causes photocatalytic degradation under UV light [33]. Replacing TiO_2 with UV-stable alternatives such as $BaSnO_3$ [33] or introducing interfacial modifiers like CsBr, Sb_2S_3, or Cl atoms [34] can improve long-term stability. Organic HTLs like spiro-OMeTAD and PTAA are prone to moisture and thermal degradation. Stability can be improved by replacing these with inorganic HTMs, adding buffer layers, or using proper encapsulation techniques [30]. Metal electrodes such as Ag, Al, and Au also contribute to degradation. Ag and Al corrode due to ion migration, leading to visible color changes and loss in performance [35]. Au can diffuse into the HTL layer at high temperature and cause irreversible damages [36]. In this regard, Cr or reduced graphene oxide can be utilized to prevent metal diffusion and improve device lifetime [37]. In line with that indium tin oxide electrodes offer a more stable alternative by forming barrier against moisture and retaining volatile perovskite components [32]. On the other hand, recent studies have shown that triple mesoscopic PSCs using TiO_2/ZrO_2/carbon scaffolds and mixed cation perovskites can induce strong stability under harsh environmental conditions [38]. These devices have demonstrated promising performance with stability even after 1,000 h of illumination with an unfiltered Xe lamp. They also showed thermal stability at 85–100 °C for 2,160 h. These results are attributed to the elimination of unstable organic HTMs and metal electrodes as well as bulky organic ligands to form mixed cation. On further note, the stability of low dimensional perovskites is explored to enhance the stability. Here large organic ligands replace the cations and offer enhanced resistance to moisture and light degradation compared to bulk perovskites [39]. However, these bulky ligands hinder the charge transport and efficiency [40]. Shorter chain or conjugated cations offer enhanced stability and efficiency [41]. Additives such as thiophene, pyridine, and PCBM are known to enhance stability by forming protective surface layers [42].

Further improvements can originate from the tuning of perovskite composition. Replacing methylammonium (MA^+) with more stable cations formamidinium (FA^+) or cesium (Cs^+), and substituting iodide (I^-) with more stable halides like Cl^-, Br^-, or SCN^-, can adjust tolerance factors, strengthen Pb-I bonds, and increase the intrinsic stability of the perovskite crystal structure [43, 44].

On further note, various environment factors influence the stability of PSCs. The sunlight impacts the operational stability in terms of performance of PSCs. This can be attributed to light-induced lattice expansion and light soaking mechanism. Temperature effects the lattice structure as well as phase transition of perovskites. Oxygen (O_2) and moisture (H_2O) are known to induce decomposition of perovskites. The hydration of perovskites decomposes $MAPbI_3$ to MA, HI, and PbI_2. Along with terrestrial applications, solar cells are the important power source for spacecraft, satellites and space stations. In such condition, PSCs will undergo series of space stresses such as extreme temperature, ultra-high vacuum, and cosmic radiations [45, 46]. The stresses due to temperature and vacuum could be alleviated by proper encapsulation and device engineering [47]. But, cosmic radiations will considerably reduce the device lifetime and the long-term operational stability.

9.2.3 Upscaling and future applications

Several PV companies worldwide are actively scaling up the production of large-area PSC modules. They are producing large-scale modules with reasonable PCE and upscaling by supplying PSC specific materials and equipment. Despite this progress, large scale PSC production faces critical challenges. Film uniformity and morphology control remain difficult with both solution and vacuum-based deposition. Variations in solvent evaporation and precursor concentration can lead to inhomogeneous crystallization, ion migration, and film defects, like pin holes, can cause direct contact between ETLs and HTLs, leading to current leakage and voltage losses. Solvent toxicity is a major issue. Safer, low toxicity solvents and solvent-free method are under development [48]. Industrial-scale production requires closed environments, recycling systems, and proper waste gas treatment. On further note, life cycle assessments highlight the need to address the environmental impact of PSC manufacturing and disposal. Lead toxicity remains a concern, even though the total lead content in PSCs [49]. Lead in PSCs is found to be water soluble making it more environmentally mobile. Developing Pb-free perovskites with competitive efficiency remains a priority. Therefore, developing Pb-free alternatives with desirable properties such as strong optoelectronic characteristics, high stability, and tunable bandgap is high on priority. Recent research has focused on Bi- and Sn-based compounds, as well as elpasolite-structured double perovskites with varied anion in compositions, as potential alternatives of Pb-based materials [50, 51]. However, further investigations are needed into the synthesis of these Pb-free HPs, particularly those with mixed cations, and their integration into PSCs to evaluate their performance. Alternatively, additives that convert water soluble Pb into insoluble compounds could mitigate risk associated with Pb compounds in module damage. Other materials, such as gold and spiro-OMeTAD (HTM) also raise environmental concerns due to cost and solvent use [52]. This has led to interest in HTMs free and noble metal-free PSC architectures that maintain high PCE. PSC fabrication commonly involves toxic solvents like N, N-dimethylformamide, N-methyl-2-pyrrolidone, and chlorobenzene, as well as skin-penetrating solvents such as dimethyl sulfoxide. These pose environmental risks and limit the feasibility of ambient processing. Developing nonhazardous solvent systems [53] and greener processing methods using safer solution chemistry [48] offers significant potential and is a critical step toward more sustainable and scalable PSC manufacturing. Thus, recycling PSCs is essential to reduce environmental adverse impacts. In this regard, CdTe PV recycling models offer useful insights for Pb management. PSCs contain components such as frames, glass, encapsulants, HTMs, and electrodes. All of them need to be recycled. Methos such as solvent soaking, tape peeling, and recrystallization can recover PbI_2 and other materials [54]. Printable triple mesoscopic scaffolds can be re-employed by removing and reloading HPs [55]. In situ recycling employing methyl amine gas treatment has also been reported [56].

Effective end-of-life strategies are essential for managing the lead content in perovskite modules and minimizing environmental impact. Recycling methods such as dissolution and electrodeposition enable the recovery of lead [57], while incineration is a more sustainable disposal option than landfilling [58]. Future module designs should be developed alongside comprehensive cradle-to-grave life cycle assessments to fully evaluate their sustainability. Despite current challenges, the strong technological promise of PSCs and ongoing efforts toward commercialization are encouraging. By comparison, Si solar panels can achieve up to 85% recycling efficiency. More than 95% of the glass and frames can be reemployed. Remaining materials are extracted and remelted to obtain Si [59]. This is crucial, as an estimated 78 million tons of Si PV waste is expected by 2050.

9.2.4 Scale-up synthesis

One key challenge for industrial application of HPs is developing low-cost, scalable synthesis methods that produce NCs with uniform size, shape, and phase at high throughput. Common techniques like ligand-assisted reprecipitation and hot-injection rely on rapid precursor injection and fast mass transport to drive quick nucleation and growth. However, these methods are costly and less practical for large-scale production. To address this, ongoing research should focus on designing novel reactor systems and exploring low-cost, inert precursors. Scalable approaches like noninjection heating-up methods offer a promising route to producing high-quality, monodisperse NCs more efficiently.

9.2.5 Reliability concerns for PSMs

For commercial Si and thin film solar cells, international standards such as IEC61646 define endurance tests that module must pass before used in the practical field. These tests involve thermal, electrical, and mechanical reliability. Technologies like Si PVs usually offer warranties of 25 years and maintain roughly 80% of their original power output over that time. These lifespans are confirmed through long-term real world performance data. These test specifications for PV devices are largely based on standards developed for inorganic technologies. However, dedicated testing protocols for PSCs will be required based on their unique properties [60]. PSCs, however, still face critical challenges like stability issues limiting their durability. Few PSCs have passed tests like 1,000 h of damp heat and 10,000 h of light exposure with UV filters. But no PSC module has yet met the full IEC standards required for commercial application [60, 61]. Glass-glass encapsulation using polymeric layers and edge sealents is the industry-standard method for protecting PV modules from environmental stress.

Replacing reactive metal electrodes with TCOs allows PSCs to endure the encapsulation process, with some packaged cells successfully passing the IEC damp heat test [18]. To further improve mechanical durability especially for temperature cycling and flexible devices, development of flexible electrodes and flexible encapsulants is important for scaling up production and enabling flexible PSCs [62, 63].

Other stresses like mechanical strain from temperature cycling and high voltage conditions also affect PSCs. The perovskite materials are more fragile and expand more under heat, risking delamination. Tests show that using flexible encapsulant like EVA can help. High voltage can also cause performance loss through potential-induced degradation; a known issue in other PV technologies for PSCs because of ion migration and charge build up. To improve mechanical durability, several strategies such as cross-linking techniques to enhance cohesion between layers such as the perovskite absorber and charge transport layers, interfacial compatibilizers to strengthen adhesion across interfaces and raise the intrinsic fracture energy, and scaffold-based designs that offer extrinsic reinforcement and shielding to further improve mechanical robustness [64].

Another concern is partial shading, which creates reverse bias in shaded cells in PSCs, with breakdown occurring between −1 and −4 V. Local heating from shading can degrade the perovskite materials. So, further research is required to improve their resistance to handle such conditions. Potential solutions include bypass diodes within modules.

9.2.6 Economics

Maximizing the low LCOE and short energy payback time (EPBT) of PSCs depends on careful selection of materials and manufacturing processes. In high-throughput, low-capex R2R production, raw materials contribute over 75% of the total module cost [65]. High-temperature steps and complex, multi-stage syntheses further increase energy use and production costs. To retain PSCs' economic advantage, it is essential to develop low-temperature, simplified processes and materials guided by life cycle assessment studies [66] from the earliest stages of technology development.

9.3 Perspective and conclusion

PSCs can be produced using low-cost materials and simple fabrication methods. They have shown impressive efficiency in lab-scale devices and are steadily improving as researchers work to enhance their stability. PSCs offer both low material costs and a short EPBT. However, achieving long-term stability at the module level and ensuring reliable outdoor performance remain major challenges. The perovskite community needs to agree on standardized stability testing protocols that can accurately predict device lifetimes under real operating conditions. While encapsulation can reduce degradation from moisture and oxygen, performance loss from light and heat exposure must be addressed through better materials and interface design. Advances in compositional engineering, supported by machine learning and automated synthesis, will help accelerate progress. Scaling up PSCs from lab cells to larger modules currently leads to efficiency losses, but ongoing research and industrial development is expected to close this gap. Since cost per kilowatt-hour depends heavily on both efficiency and lifetime, the low raw material costs and simple manufacturing processes of PSCs suggest a clear path to lower production costs. In the long term, this could make PSCs a commercially competitive and sustainable solar technology. While crystalline silicon solar cells still dominate the PV market, their potential for further cost reduction and efficiency gains is increasingly limited. In contrast, PSCs offer significant room for improvement and hold strong promise as next-generation alternatives. Despite the challenges, continued research and investment could enable PSCs to transform the solar industry and support the transition to a more sustainable, low-carbon future. PSCs are well-positioned to play a major role in the global shift toward clean energy, with the potential to help displace fossil fuels as a next-generation solar technology.

References

[1] National renewable energy laboratory best research-cell efficiencies (available at: www.nrel.gov/pv/cell-efficiency.html) (Accessed 8 2023)

[2] Xing G, et al. Long-range balanced electron- and hole-transport lengths in organic-inorganic $CH_3NH_3PbI_3$, Science. 2013; 342: 344.

[3] Jeong M, et al. Stable perovskite solar cells with efficiency exceeding 24.8% and 0.3-V voltage loss, Science. 2020; 369: 1615.

[4] Jeong J, et al. Pseudo-halide anion engineering for α-FAPbI$_3$ perovskite solar cells, Nature. 2021; 592: 381.

[5] Min H, et al. Perovskite solar cells with atomically coherent interlayers on SnO_2 electrodes, Nature. 2021; 598: 444.

[6] Zhang L, et al. The issues on the commercialization of perovskite solar cells, Materials Futures. 2024; 3: 022101. 35.

[7] Chini MK. Synthesis and properties of halide perovskites nanocrystals toward solar cells application, Energy Technology. 2024; 12: 2301444.

[8] Xia J, et al. Tailoring electric dipole of hole-transporting material p-dopants for perovskite solar cells, Joule. 2022; 6: 1689.

[9] Nejand BA, et al. Scalable two-terminal all-perovskite tandem solar modules with a 19.1% efficiency, Nature Energy. 2022; 7: 620.

[10] Rong Y, et al. Challenges for commercializing perovskite solar cells, Science. 2018; 361: 1214.

[11] Kim H-S, et al. Lead iodide perovskite sensitized all-solid state submicron thin film mesoscopic solar cell with efficiency exceeding 9%, Scientific Reports. 2012; 2: 591.

[12] Yao K, et al. A copper-doped nickel oxide bilayer for enhancing efficiency and stability of hysteresis-free inverted mesoporous perovskite solar cells, Nano Energy. 2017; 40: 155.

[13] Tan H, et al. Efficient and stable solution-processed planar perovskite solar cells via contact passivation, Science. 2017; 355: 722.

[14] Bai Y, et al. Enhancing stability and efficiency of perovskite solar cells with cross linkable silane-functionalized and doped fullerene, Nature Communications. 2016; 7: 12806.

[15]   Liu T, et al. Inverted perovskite solar cells: Progresses and perspectives, Advanced Energy Materials. 2016; 6: 1600457.

[16]   Luo D, et al. Enhanced photovoltage for inverted planar heterojunction perovskite solar cells, Science. 2018; 360: 1442.

[17]   Mei A, et al. A hole-conductor-free, fully printable mesoscopic perovskite solar cell with high stability, Science. 2014; 345: 295.

[18]   Bush KA, et al. 23.6%-efficient monolithic perovskite/ silicon tandem solar cells with improved stability, Nature Energy. 2017; 2: 17009.

[19]   Bailie CD, et al. Semi-transparent perovskite solar cells for tandems with silicon and CIGS, Energy & Environmental Science. 2015; 8: 956.

[20]   Eperon GE, et al. Perovskite-perovskite tandem photovoltaics with optimized band gaps, Science. 2016; 354: 861.

[21]   Shockley W, et al. Detailed balance limit of efficiency of p-n junction solar cells, Journal of Applied Physics. 1961; 32: 510.

[22]   Green MA, et al. Commercial progress and challenges for photovoltaics, Nature Energy. 2016; 1: 15015.

[23]   Giacomo FD, et al. Up-scalable sheet-to-sheet production of high efficiency perovskite module and solar cells on 6-in. substrate using slot die coating, Solar Energy Materials and Solar Cells. 2017; 181: 53.

[24]   Chang C-Y, et al. An integrated approach towards the fabrication of highly efficient and long-term stable perovskite nanowire solar cells, Journal of Materials Chemistry A. 2017; 5: 22824.

[25]   Rong Y, et al. Toward industrial-scale production of perovskite solar cells: Screen printing, slot-die coating, and emerging techniques, Journal of Physical Chemistry Letters. 2018; 9: 2707.

[26]   Hinsch A, et al. Worldwide first fully up-scaled fabrication of $60 \times 100\ cm^2$ dye solar module prototypes, Progress in Photovoltaics: Research and Applications. 2012; 20: 698.

[27]   Taheri B, et al. Laser-scribing optimization for sprayed SnO_2-based perovskite solar modules on flexible plastic substrates, ACS Applied Energy Materials. 2021; 4: 4507.

[28]   Bowring A, et al. Reverse bias behavior of halide perovskite solar cells, Advanced Energy Materials. 2018; 8: 1702365.

[29]   Domanski K, et al. Systematic investigation of the impact of operation conditions on the degradation behaviour of perovskite solar cells, Nature Energy. 2018; 3: 61.

[30]   You J, et al. Improved air stability of perovskite solar cells via solution-processed metal oxide transport layers, Nature Nanotechnology. 2016; 11: 75.

[31]   Lin Y, et al. Enhanced thermal stability in perovskite solar cells by assembling 2D/3D stacking structures, Journal of Physical Chemistry Letters. 2018; 9: 654.

[32]   Cheacharoen R, et al. Design and understanding of encapsulated perovskite solar cells to withstand temperature cycling, Energy & Environmental Science. 2018; 11: 144.

[33]   Shin SS, et al. Colloidally prepared La-doped $BaSnO_3$ electrodes for efficient, photostable perovskite solar cells, Science. 2017; 356: 167.

[34]   Li W, et al. Enhanced UV-light stability of planar heterojunction perovskite solar cells with caesium bromide interface modification, Energy & Environmental Science. 2016; 9: 490.

[35]   Back H, et al. Achieving long-term stable perovskite solar cells via ion neutralization, Energy & Environmental Science. 2016; 9: 1258.

[36]   Domanski K, et al. Not all that glitters is gold: Metal-migration-induced degradation in perovskite solar cells, American Chemical Society Nano. 2016; 10: 6306.

[37]   Arora N, et al. Perovskite solar cells with CuSCN hole extraction layers yield stabilized efficiencies greater than 20%, Science. 2017; 358: 768.

[38]   Baranwal AK, et al. 100 °C Thermal stability of printable perovskite solar cells using porous carbon counter electrodes, ChemSusChem. 2016; 9: 2604.

[39] Tsai H, et al. High-efficiency two-dimensional Ruddlesden-Popper perovskite solar cells, Nature. 2016; 536: 312.

[40] Quan LN, et al. Ligand-stabilized reduced-dimensionality perovskites, Journal of the American Chemical Society. 2016; 138: 2649.

[41] Chen Y, et al. Tailoring organic cation of 2D air-stable organometal halide perovskites for highly efficient planar solar cells, Advanced Energy Materials. 2017; 7: 1700162.

[42] Noel NK, et al. Enhanced photoluminescence and solar cell performance via Lewis base passivation of organic-inorganic lead halide perovskites, American Chemical Society Nano. 2014; 8: 9815.

[43] Yang M, et al. Facile fabrication of large-grain CH3NH3PbI3-xBrx films for high-efficiency solar cells via CH3NH3Br-selective Ostwald ripening, Nature Communications. 2016; 7: 12305.

[44] Tai Q, et al. Efficient and stable perovskite solar cells prepared in ambient air irrespective of the humidity, Nature Communications. 2016; 7: 11105.

[45] Yang J, et al. Potential applications for perovskite solar cells in space, Nano Energy. 2020; 76: 105019.

[46] Tu Y, et al. Perovskite solar cells for space applications: Progress and challenges, Advanced Materials. 2021; 33: e2006545.

[47] Jiang Y, et al. Mitigation of vacuum and illumination-induced degradation in perovskite solar cells by structure engineering, Joule. 2020; 4: 1087.

[48] Noel NK, et al. A low viscosity, low boiling point, clean solvent system for the rapid crystallisation of highly specular perovskite films, Energy & Environmental Science. 2017; 10: 145.

[49] Park N-G, et al. Towards stable and commercially available perovskite solar cells, Nature Energy. 2016; 1: 16152.

[50] Abate A. Perovskite solar cells go lead free, Joule. 2017; 1: 659.

[51] Chini MK, et al. Lead-free, stable mixed halide double perovskites $Cs_2AgBiBr_6$ and $Cs_2AgBiBr_{6-x}Cl_x$ – A detailed theoretical and experimental study, Chemical Physics. 2020; 529: 110547.

[52] Lunardi MM, et al. A life cycle assessment of perovskite/silicon tandem solar cells, Progress in Photovoltaics: Research and Applications. 2017; 25: 679.

[53] Bu T, et al. Synergic interface optimization with green solvent engineering in mixed perovskite polar cells, Advanced Energy Materials. 2017; 7: 1700576.

[54] Binek A, et al. Recycling perovskite solar cells to avoid lead waste, ACS Applied Materials & Interfaces. 2016; 8: 12881.

[55] Ku Z, et al. A mesoporous nickel counter electrode for printable and reusable perovskite solar cells, Nanoscale. 2015; 7: 13363.

[56] Yang WS, et al. Iodide management in formamidinium-lead halide-based perovskite layers for efficient solar cells, Science. 2017; 356: 1376.

[57] Poll CG, et al. Electrochemical recycling of lead from hybrid organic–inorganic perovskites using deep eutectic solvents, Green Chemistry. 2016; 18: 2946.

[58] Serrano-Lujan L, et al. Tin- and lead-based perovskite solar cells under scrutiny: An environmental perspective, Advanced Energy Materials. 2015; 5: 1501119.

[59] Weckend S, et al., "End-of-life management: Solar photovoltaic panels" (IRENA and IEA-PVPS, 2016); http://iea-pvps.org/fileadmin/dam/public/report/technical/IRENA_IEAPVPS_End-of-Life_Solar_PV_Panels_2016.pdf.

[60] Jordan DC, et al. Photovoltaic failure and degradation modes, Progress in Photovoltaics: Research and Applications. 2017; 25: 318.

[61] Jordan DC, et al. Photovoltaic degradation rates-an analytical review, Progress in Photovoltaics: Research and Applications. 2013; 21: 12.

[62] Cheacharoen R, et al. Design and understanding of encapsulated perovskite solar cells to withstand temperature cycling, Energy & Environmental Science. 2018; 11: 144.

[63] Park M-H, et al. Flexible lamination encapsulation, Advanced Materials. 2015; 27: 4308.

[64] Watson BL, et al. Scaffold-reinforced perovskite compound solar cells, Energy & Environmental Science. 2017; 10: 2500.

[65] Song Z, et al. A technoeconomic analysis of perovskite solar module manufacturing with low-cost materials and techniques, Energy & Environmental Science. 2017; 10: 1297.

[66] Chang NL, et al. A manufacturing cost estimation method with uncertainty analysis and its application to perovskite on glass photovoltaic modules, Progress in Photovoltaics: Research and Applications. 2017; 25: 390.

Suchi Priyadarshani and Roshan Raghavendra Rao

Chapter 10
Perovskite photovoltaic panels in buildings

Abstract: The photovoltaic (PV) industry is experiencing a significant transition from silicon (Si)-based panels to perovskite panels. This shift is driven by the superior efficiency and potential cost-effectiveness of perovskite materials. Building-integrated photovoltaics (BIPVs) are at the forefront of this transition, playing a pivotal role in promoting sustainable building practices. By integrating renewable energy sources directly into the building structure, BIPVs help reduce reliance on fossil fuels for energy and conventional building materials, thereby contributing to environmental sustainability. Given the advancements in PV technology, BIPVs are expected to adopt perovskite panels soon. Consequently, it is imperative to examine the challenges associated with this emerging technology when utilized in BIPVs.

This chapter provides a comprehensive analysis of the current knowledge regarding perovskites and their potential impact on the performance of BIPVs from environmental, economic, and social perspectives. It also investigates the anticipated variations in the functional performance of buildings employing perovskite panels, emphasizing their effects on real-time electricity generation, health, and comfort aspects within such buildings. Additionally, the risks and threats to ecological diversity posed by the replacement of Si-based PV panels with perovskite panels have also been discussed. This chapter provides the current state of knowledge and highlights the critical areas that need further research and attention for the successful integration of perovskite panels into building structures.

Keywords: Perovskite, BIPVs, electricity generation, crystalline, structural stability

10.1 Introduction

Here, we provide a detailed view of the application of perovskite photovoltaic (PV) panels in building-integrated photovoltaic (BIPV) systems, comparing them with traditional crystalline silicon (c-Si) panels in terms of construction, performance, durability, and environmental impact. Durability is a critical factor for the adoption of perovskite panels in buildings. Building-applied photovoltaics (BAPV) and BIPV represent two distinct approaches. BAPV refers to PV modules that are added onto an existing building envelope such as rooftop or ground-mounted panels attached post-construction. In contrast,

Suchi Priyadarshani, Indian Institute of Science, Bengaluru, India, e-mail: suchip@iisc.ac.in
Roshan Raghavendra Rao, Indian Institute of Science, Bengaluru, India, e-mail: roshanrao@iisc.ac.in

https://doi.org/10.1515/9783111726847-010

BIPV modules replace or become part of the building materials such as solar façades, skylights, or glazing. By definition, a BIPV product is a PV module that also serves as a construction element; if removed, it would need to be replaced by an equivalent building component [1, 2]. In practice, BIPV systems aim to provide both the aesthetic and functional roles of a building material while generating electricity (e.g., solar roof tiles, semitransparent window modules, or curtain walls). Conversely, BAPV installations are retrofits, such as rigid silicon panels mounted on rails above an existing roof. The unique properties of perovskite PV – ultrathin, lightweight, flexible, and tunable in appearance – make it suitable for both modes. For instance, perovskite films can be laminated into double-glazed façades or applied as cladding overlays (BIPV) but can also be adhered to existing exterior surfaces (BAPV).

In each scenario, they must meet the relevant building and electrical standards, but the defining difference lies in integration versus attachment [1, 2]. Hence, an emphasis has been made here to compare different aspects of PV panels with the c-Si-based PV panels currently used in buildings as the benchmark. Various strategies for improving the durability of perovskite solar cells (PSCs), including the use of robust encapsulation techniques to prevent moisture ingress and enhance stability, are also highlighted. PSCs have garnered significant attention due to their potential for high efficiency and lower production costs compared to conventional silicon (Si)-based solar cells. Also an examination of perovskite panels, focusing on their construction, performance under various environmental conditions, and their implications for BIPV systems is made here.

c-Si PV panels utilize pure silicon wafers with a thickness of approximately 150–200 µm, while PV panels employ organometal halide perovskite compounds with a thickness ranging from 300 to 900 nm. The manufacturing process for c-Si panels is energy-intensive, involving silicon refining and wafer slicing, whereas perovskite panels are fabricated using solution-based thin-film techniques at lower temperatures.

The perovskite PV panels are typically lighter and can be manufactured in flexible formats, making them suitable for various applications, particularly in BIPV settings. The performance of perovskite panels is influenced by environmental factors such as temperature, humidity, and UV exposure. Higher temperatures can increase the bandgap of perovskites, slightly reducing power output.

A brief review of the historical cost trends of silicon and perovskite solar panels, noting significant reductions in silicon costs over the decades, is presented. It projects that perovskite panels could achieve competitive pricing, potentially below \$0.20/W by 2030 if technical challenges are overcome.

In summary, while PV panels offer promising advantages in terms of cost, efficiency, and design flexibility for building applications, their long-term durability and environmental impacts require further investigation and innovation. It is emphasized that a need for robust encapsulation and recycling strategies to mitigate risks associated with lead and other materials used in PSCs is required.

10.2 Comparison of perovskite and c-Si photovoltaic panel construction

PV panel design is key for their use in BIPV and BAPV. Comparing c-Si and perovskite PV technologies helps stakeholders understand their structural, functional, and aesthetic integration within buildings, aiding decisions on adaptability, performance, and long-term use. PV panels based on c-Si and those utilizing perovskite thin-film cells exhibit substantial differences in their construction and materials. Table 10.1 summarizes the key differences between c-Si and perovskite solar panels regarding their construction and suitability.

Table 10.1: Key differences between crystalline silicon and perovskite solar panels regarding their construction and suitability.

Aspect	Crystalline silicon PV panels	Perovskite PV panels
Active material	A single-element crystalline silicon wafer with a semiconductor p–n junction has a silicon active layer thickness of approximately 150–200 µm. It requires very pure silicon, with a purity of 99.9999% [3].	Organometal halide perovskite compounds, such as lead halide perovskite, serve as light absorbers. They typically have a film thickness of 300–900 nm (ultrathin coating) [4].
Auxiliary layers	The structure includes an antireflective coating on the cell, an EVA polymer encapsulant, and cells mounted on a back sheet and front glass. Silicon cells are connected with metallic ribbons.	The structure comprises multiple functional layers including an electron transport layer and a hole transport layer around the perovskite absorber, which serve to extract charges. A transparent conductive oxide front contact is applied on glass or plastic substrates, while metallic or carbon back contacts are used. Various designs, such as n–i–p and p–i–n configurations, have been developed to optimize performance [4].
Electrode materials	Front grid electrodes are typically made of silver, while rear electrodes are generally aluminum. Busbars and ribbons, often copper or coated materials, connect cells within a module. These metals contribute to the cost, particularly silver, but their use is well-established and widely accepted in the industry.	The front electrode is usually ITO/FTO, and the back electrode is often a thin metal like gold (used in high-end lab cells) or silver/carbon (for cost-effective scaling). Gold increases efficiency but is costly. Research is focused on developing affordable conductive inks or TCO-based electrodes for perovskite cells [4].

Table 10.1 (continued)

Aspect	Crystalline silicon PV panels	Perovskite PV panels
Encapsulation	Cells are enclosed in EVA (or similar), sandwiched between a glass front and a protective back sheet, and sealed to prevent moisture ingress. This durable packaging is industry standard and ensures modules last for decades.	Encapsulation is crucial because of moisture and oxygen sensitivity. Common methods include glass or multilayer polymer barriers with edge sealing. Extra layers and desiccants may be required for longevity, increasing weight and cost. Improved encapsulants are being developed [5].
Manufacturing process	Semiconductor fabrication involves multiple steps and high temperatures. The process includes silicon refining at temperatures exceeding 1,000 °C, crystal growth, wafer slicing, diffusion or implantation of dopants, and high-temperature metallization firing. It requires cleanrooms and expensive equipment. The process is highly automated for mass production but is energy-intensive [3].	Solution-based thin-film fabrication uses wet chemical deposition at or near room temperature, avoiding crystal ingots or wafer sawing. It can be performed in ambient or glovebox conditions without the high temperatures or energy input required for silicon processing. However, it still requires uniform large-area coating control for scaling [3].
Process complexity	The process is intricate yet well-established, involving multiple steps and materials within a stable supply chain. Silicon production lines offer substantial throughput, necessitating significant initial capital investment. While economies of scale have successfully reduced costs, the manufacturing process remains relatively complex.	The production process is simpler and potentially significantly lower in cost per area, due to fewer high-precision steps and suitability for continuous manufacturing methods such as roll-to-roll. While it is theoretically easier to initiate production on smaller scales, achieving high yield and consistent quality at scale remains an active challenge. Additionally, the manufacturing process requires lower energy and material inputs per cell [6].
Module structure	Rigid and heavy. Cells are brittle and need a sturdy glass frame for support. Typical panel thickness is around 3–4 mm (including glass, cells, and encapsulant) and weighs about 10–15 kg/m^2. They lack flexibility and will crack if bent.	Thin and lightweight. When constructed on a flexible substrate, modules can be exceptionally thin (submillimeter) and lightweight (a few hundred grams per square meter for the active layers; total weight depends on encapsulation). Flexible configurations are possible, with the ability to bend to a few centimeters in radius or be rollable. If laminated in glass for protection, the weight becomes comparable to that of silicon panels. However, glass is not strictly required for functionality [7].

Table 10.1 (continued)

Aspect	Crystalline silicon PV panels	Perovskite PV panels
Durability and lifespan	Silicon cells exhibit long-term stability, typically maintaining their performance with less than a 20% reduction over 25–30 years. The cells themselves remain stable, with most degradation resulting from encapsulant aging or microcracks. These cells are proven to perform reliably in various climates once sealed, being robust under UV exposure, heat, cold, and moisture [8].	Perovskite panels are still limited in their durability. Laboratory cells can function for a few years with specialized encapsulation; however, a 20+ year lifespan for mass-market applications has not yet been demonstrated. These panels are highly sensitive to moisture, oxygen, and thermal stress without protection. Ongoing improvements in materials and sealing techniques are addressing these issues, but currently, perovskite panels have a shorter useful life compared to silicon panels [8].
Ideal applications	Conventional installations include rooftop solar, ground-mounted solar farms, and any application requiring reliable long-term output. Due to their rigidity and weight, they are best suited for installations where mounting structures can support them and no flexibility is needed. They are appropriate for open areas, residential and commercial roofs, among other settings. These installations are also found in most consumer solar products today.	Emerging uses include portable and wearable solar applications, building-integrated PV, and aerospace needs where light weight and flexibility are vital. It can be used on curved surfaces, vehicles, or tents where silicon is impractical. It can also enhance traditional panels by adding a layer to boost performance. Broad adoption in mainstream installations awaits improved stability [9].

10.3 Influence of environmental factors on performance of perovskite-based BIPV and BAPV

10.3.1 Temperature fluctuations

Raising cell temperature increases the perovskite bandgap (E_g) slightly, reducing J_{sc} and influencing V_{oc} and fill factor. E_g rises by tens of meV over tens of °C [10]. However, PSCs have low temperature coefficients (–0.08 to –0.17%/°C) [10], so their power output falls less with heat compared to silicon cells ($\approx$–0.3%/°C). Triple-cation PSCs showed TPCE $\approx$ –0.08%/°C, outperforming Si, CIGS, and GaAs. Carbon-based perovskite cells maintained ≥70–100% efficiency over ~4–6 months in temperate climates, indicating gradual loss under real temperature cycling [11]. Despite benign TPCE, organic-inorganic perovskites decompose at high heat. Pure $MAPbI_3$ or $FAPbI_3$ films start losing organic components above ~85–100 °C [12], producing volatile products like methylammonium iodide and HI. Extreme summer rooftop temperatures can irreversibly degrade BIPV modules if not managed. Thermal cycling can cause strain, delamination, or cracks due to mismatched expansion between layers. Rigorous IEC thermal cycle tests (200 cycles −40/+85 °C) are needed for durability, and system design should ensure good heat dissipation. Integrated BIPV may trap more heat than ventilated BAPV panels on sloped roofs. Passive cooling or heatsinking can reduce peak cell temperatures.

10.3.2 Humidity and rainfall

Perovskites exhibit hygroscopic behavior. Water molecules infiltrate and chemically decompose the absorber; for example, H_2O forms strong hydrogen bonds with organic cations, disrupting Pb–I bonds and producing PbI_2, HI, CH_3NH_2, and so on [12]. Even slightly acidic rainwater or elevated humidity levels can initiate these reactions. In pure water, $MAPbI_3$ dissolves entirely, indicating complete degradation [13]. As a result, any prolonged exposure to moisture (rain, dew, condensation) rapidly deteriorates unprotected films. Effective moisture barriers are essential for protecting perovskite materials. Industrial encapsulation techniques (such as glass-glass lamination with EVA or POE plus edge seal) can withstand IEC damp-heat tests (85 °C/85% RH) and extended outdoor exposure [14]. For instance, a properly laminated perovskite cell maintained ≥95% of its initial efficiency after 1,566 h at 85/85 RH and showed virtually no loss over 10 months outdoors. In contrast, simple "glue + glass" covers (lab encapsulation) failed the damp-heat test due to moisture ingress. BIPV installations often feature building-grade outer layers (e.g., tempered glass facades or roofing membranes) that can act as one barrier, but the cell must be sealed on all sides. Important design measures include low-PWV polymer edge seals (e.g., butyl) and high-adhesion interlayers. Some designs also incorporate desiccants or gas-filled insulating layers. Rain can clean off dust but also repeatedly wet module edges. For BIPV facades, water runoff management is necessary to prevent pooling at joints; sealed perimeters and drainage paths are required. In BAPV (such as roof panels), rain typically self-cleans, and moisture evaporates quickly. Nevertheless, driving rain and freeze/thaw cycles can force moisture into microcracks or interfaces, making damp-heat and humidity-freeze testing (part of IEC 61215) crucial. PSCs that pass such tests (e.g., 1,000+ h at 85/85) generally use edge-sealed glass stacks [12]. Trials in humid climates (e.g., tropical outdoor tests) demonstrate that unprotected PSCs fail quickly, whereas well-encapsulated modules (often featuring carbon back-electrodes with no metal corrosion pathways) can last several months. The most stable trials, which adhere to strict glass lamination protocols, maintain power >90% for over 6 months [14].

In conclusion, moisture represents the primary degradation factor for PSCs, and blocking it is critical for ensuring their longevity.

10.3.3 UV exposure

Intense ultraviolet radiation (below approximately 400 nm) can cause damage to PSCs, particularly those incorporating titanium dioxide (TiO_2) electron-transport layers. Ultraviolet light can generate metastable trap states and initiate photocatalytic reactions; in a study, exposure of methylammonium lead iodide ($MAPbI_3$)/TiO_2 cells exclusively to UV led to the formation of trap charges that resulted in the conversion of some perovskite to lead iodide (PbI_2) [15]. This predominantly affected the fill factor and short-circuit current. Notably, a significant portion of this performance loss (>60%) was recoverable under full-spectrum (visible) illumination, as the additional carriers neutralized the traps. However, real-world deployments expose PSCs to full sunlight and moisture along with UV, leading to more severe long-term effects. The repeated formation of UV-induced traps, especially in the presence of oxidative species (O_2), contributes to irreversible aging. Formal testing protocols, such as IEC UV preconditioning, recognize UV as a stress factor. Many stability protocols include pre-irradiation of devices with UV (e.g., UV lamps) prior to damp-heat testing to simulate worst-case scenarios [12]. In BIPV) and BAPV modules, front-cover glass often absorbs some UV radiation, particularly if it is filtered or low-iron, thereby reducing the UV dose received by the cells. Additional mitigation strategies involve using UV-stable transport layers; for instance, replacing TiO_2 with tin oxide (SnO_2) or carbon can significantly enhance UV tolerance. Furthermore, incorporating UV-absorbing polymers or coatings can protect underlying layers. Building integration designs can take advantage of this approach, such as utilizing UV-blocking glazing or tinted encapsulants in facade modules, where visible transparency is not critical. Ultraviolet radiation introduces considerable photochemical stress. Although not as immediately destructive as moisture or heat, its cumulative impact and synergy with other factors (such as oxygen) necessitate shielding PSCs from deep-UV through appropriate material selection or filtering glasses [15].

10.3.4 Soiling and surface contamination

Dust, pollen, soot, and other particulates on the surface of PV cells reduce light absorption and consequently power output. This is a well-documented issue for all PV systems. A thin layer of dirt can decrease daily yield by 5–20%, depending on environmental conditions. Furthermore, dust can cause local shading that results in hotspots or localized heating in modules. There is limited literature on the chemical effects of soiling on PSCs, but one potential risk is the deposition of hygroscopic salts (e.g., NaCl and sulfates) under moist conditions, which may attract moisture and accelerate degradation beneath the dust layer. Acidic deposits from air pollution might also chemically attack unprotected perovskite materials. BIPV facades are generally near-vertical, thus accumulating less dust and benefiting from better rain wash-off compared to horizontal panels. However, vertical installations may not dry as quickly and can trap pollutants in grooves or at joints. BAPV roof panels tend to be tilted (often ~20°–45°), allowing rain to effectively wash off dust. Some BIPV modules, especially those on tall walls, can be more challenging to clean or inspect, prolonging the presence of soiling. Conversely, translucent or semitransparent BIPV elements, such as window-integrated PSCs, may visually reveal soiling, prompting cleaning. Regular cleaning schedules are recommended in dusty environments for both BIPV and BAPV systems. The development of anti-soiling or hydrophobic coatings for PSCs shows promise, with certain nanotextured coatings demonstrating efficacy in laboratory tests. Designing systems for easy access or incorporating automated cleaning solutions (e.g., by rainwater diversion) can enhance long-term energy yield.

10.3.5 Durability considerations

Protecting the perovskite stack is of utmost importance. Industry practices, as evidenced by durability studies, involve glass-glass lamination with edge sealants [12]. This method effectively seals out moisture and oxygen while holding the layers together under thermal cycling conditions. Encapsulant materials such as EVA and POE must be optically clear in the required spectrum and exhibit low permeability. It is crucial that all four edges of the cell are sealed (commonly with butyl rubber) not just the front and back. For BIPV, integration with building materials can permit the use of thicker or multilayer barriers (e.g., a 3D-printed PV roof shingle could incorporate foam or vapor barriers). To enhance stability, inorganic or fully inorganic perovskites should be used when possible. For instance, FA-Cs mixed perovskites have better moisture resistance compared to MA-based variants. Using noncorrosive electrodes (such as carbon or nickel instead of silver or gold) prevents metal migration under heat and humidity. Employing UV-stable and nonreactive transport layers (ETL/HTL) like SnO_2 or dense oxide films can extend the lifespan of the system. Internal additives or cross-linkers can bind the film to suppress water-induced swelling. Many stable PSCs utilize polymeric hole conductors with very low permeability. To mitigate thermal stress, it is essential to match coefficients of thermal expansion. Using a glass substrate or steel, instead of polymers where feasible, or employing low-temperature lamination can prevent initial strain [12]. In large modules, incorporating split-junctions or expansion joints is advisable. For very large installations, active cooling systems (liquid or air channels behind BIPV panels) might be considered. For BIPV facades, incorporating shade controls or selective surfaces (such as fritting patterns) can limit midday overheating or glare, thereby reducing cooling loads and UV intensity. Building-integrated modules can leverage angle-optimized azimuth (vertical versus sloped) to balance power generation and comfort. While tilt in BAPV is usually optimized for energy yield, BIPV windows might prioritize transparency or shading over efficiency. Designing dust-shedding surfaces, such as those with hydrophobic or photocatalytic coatings (as seen in self-cleaning glass), can be beneficial. Panels should be accessible for cleaning crews or robotic washers. In areas prone to frequent sandstorms or bird droppings, planning for anti-soiling protection is necessary. BIPV/PV products must comply with both building and PV standards. This entails passing IEC tests (damp heat, thermal cycling, UV exposure, and mechanical load) and adhering to building codes (fire safety, wind load, and impact resistance). Designers must test full modules in climatic chambers and in situ. Field monitoring is critical; recent long-term outdoor data (6–12 months in variable climates) indicate that well-encapsulated PSCs can maintain stability [11]. With careful design, durability comparable to conventional PV can be achieved. PSCs that pass IEC 85/85 with glass encapsulation have operated for over 10 months outdoors with minimal degradation. Advances such as low-thermal-expansion substrates, multibarrier coatings, and 2D/3D hybrid structures (enhancing phase stability) are being implemented.

10.4 Functional performance of perovskite BIPV and BAPV

10.4.1 Current projects

PV technology is currently undergoing field testing in various demonstration projects globally. In Europe, Skanska Poland has collaborated with Saule Technologies to coat the Spark office tower in Warsaw with printed perovskite films – semiconductor ink-jet-printed modules laminated into the double-glazed façade – beginning in 2018. This initiative was reported as the first commercial-scale BIPV deployment of perovskite PV [16, 17]. In Japan, Sekisui Chemical and NTT DATA commenced trials of film-type PSCs on vertical walls in 2023. The first phase, initiated in April 2023, involved the installation of test panels on a Sekisui Research and Development building to verify installation methods and structural safety under wind loads [18, 19]. A second phase is planned for NTT DATA's Shinagawa TWINS data-center tower. Additionally, Sekisui demonstrated perovskite BIPV at its Osaka headquarters by installing a large panel on the 12th-floor roof in October 2023, which is expected to be operational by July 2024 and engineered to withstand high wind loads for at least 20 years [20]. In Japan, Panasonic Housing & Life has developed a prototype of "power-generating glass" by directly coating glass panels with perovskite. They commenced a demonstration lasting over 1 year in 2023 at a model house in Fujisawa SST, a smart-town development, to test the long-term durability and graded-transparency designs [21]. Meanwhile, in Taiwan, Taiwan Perovskite Solar Corp (TPSC) introduced a perovskite solar window named "Windows ZERO" at the Tokyo Smart Energy Week in February 2025. This demonstration product is a light-transmitting window capable of noise reduction, rain-repulsion, and thermal insulation while generating electricity, specifically designed for high-rise and dense urban applications [22]. In China, several pilot projects are underway. In early 2025, Microquanta semiconductor connected a 1.0-MW perovskite rooftop system at Qinghai University, using their flat-glass modules. This is among the first megawatt-scale perovskite PV installations, directly supplying solar power to the campus [23]. Chinese developer Utmo Light showcased its first commercial perovskite module (UL-M12-G1) at the SNEC PV trade show in June 2024. These modules can be tinted any color and assembled into large glass sheets for BIPV. They passed full IEC durability testing and are already on sale, with installations in China and Japan. Utmo Light plans to scale up to multi-MW by year-end and is building a 1 GW/year production line for full-size BIPV panels [24]. These examples demonstrate that perovskite PV is now transitioning from lab testing to real-building applications on roofs, facades, and windows.

Perovskite devices can be modified for specific transparency and color. Researchers have adjusted the halide composition of perovskites to produce different visible colors, enabling colored modules. Additionally, patterning techniques [25] (micropatterning or micro-meshing) can create partly transparent cells for windows or sunshades. This means perovskite PV can be customized to match a building's design: for example, a façade could be covered in tinted PV panels that generate power. Panes of "solar glass" can incorporate graded translucency or coatings while hosting the PV layer. Perovskite modules are thin and flexible, enabling unique shapes and mounting options. Panasonic utilizes inkjet printing and laser patterning on glass to customize size, transparency, and design [21]. Similarly, Saule uses roll-to-roll printing on plastic for lightweight films laminated into double-glazed windows or façades [17]. These laminated BIPVs (PVC between glass) fit standard construction methods and meet building codes due to their minimal weight, as demonstrated in the Saule/Spark installation. Apart from their aesthetic appeal, perovskite BIPV systems can fulfil additional architectural functions. For instance, the TPSC "Windows ZERO" product is an energy-generating window that also offers insulation and noise reduction [22]. Generally, a semitransparent or thin-film PV layer will decrease solar heat gain and can be combined with thermal breaks or coatings. Essentially, architects can utilize perovskite PV as they would any advanced façade element: it serves as multifunctional cladding that complies with weatherproofing and fire codes while generating solar power. According to one industry survey, advancements in cell interconnection techniques (such as shingling) and new materials (such as perovskites for customizable shapes and colors) are enhancing design flexibility in BIPV applications [2]. In summary, contemporary design approaches for perovskite BIPV focus on modular "solar glass" panels of varying transparencies and colors, flexible film laminates suitable for curved or unconventional surfaces, and system integration into standard building assemblies [17, 21].

10.4.2 Structural stability and strength

Buildings are designed to keep occupants safe and ensure a sturdy living space, generally defined by walls and roof. Most of the modern buildings as on date are framed structures, where only the columns, beams, roofs, and so on are responsible for taking the load of the building. Other elements of the building like walls and fenestrations are for enclosure of space and not for major load-bearing purposes. This arrangement in buildings allows for use of light-weight thin materials as walls. Many modern buildings consist of full glass surfaces, which provide light as well as aesthetically pleasing. PV panels can easily be used in place of the glass as wall elements, fenestration, and so on. Also, in roofs, PV panels can be integrated appropriately in building, without considerable loading on them. For the purpose of use in huge farms, exposed to extremities of weather conditions, wind, snow, and so on, PV panels are tested for many aspects of structure and stability before deployment on field. These tests are governed by industry standard compliance codes (IEC 61215/61730 for silicon PV) [26]. Perovskite modules as on date also must comply with the same qualification standards as silicon PV (IEC 61215/61730). As per this standard, accelerated tests are conducted to ensure structural stability and strength under extreme damp-heat exposures (85 °C/85% RH, 1,000+ h), thermal cycling (−40 °C to +85 °C, 200+ cycles), humidity-freeze, UV preconditioning, and mechanical load. The PV panels are also tested simultaneously for their electrical efficiency under these circumstances, where the pass criteria are less than 5% power loss [27]. Recent advancements in perovskite technology combining robust encapsulation with intrinsic stabilizers have yielded many prototype devices that survive thermal cycling and damp heat with minimal loss, suggesting enhanced durability and environmental safety [28]. Given the construction difference in Si-based PV and perovskites, their strength criterions are different. As far as Si-based PV panels are concerned, they are often laminated with toughened glass, which acts as a protective layer for the outdoor surface in buildings that prevent the penetration of moisture/rain indoors. Also, given the stability of glass at extreme temperature and humidity variations, the modules are intact and do not affect the functional performance of the building, while the energy generation can be varying. Perovskite films need substrates for strength due to their brittleness [29]. These substrates can be glass as well as polymers. As on dates, technological advancements suggest that flexible perovskite cells on polymer substrates retain ~81% efficiency after 20,000 bending cycles and ~94% after 3,000 cycles, with degradation from cracks [30]. Challenges like preventing microcracks and delamination persist. Current research focuses on tougher compositions, fracture-resistant layers, and robust lamination. As per industry standards, Si-based PV panels should be safe for hailstones of approximately 25 mm in diameter striking the glass at high speeds. Standard modules are expected to endure multiple impacts, typically around 9–10 hits. Perovskite cells exhibit similar characteristics to silicon cells in this regard, where severe impacts can cause the glass to fracture. Properly encapsulated perovskite modules have demonstrated good resistance to such conditions. In a study published by *Nature Energy*, various encapsulations were tested, revealing that a rigid epoxy-laminated panel retained significantly less lead runoff (375 times less) after breakage compared to a conventional glass/resin panel [31]. Perovskite modules need lamination for impact absorption. Tests show polymer-encapsulated panels leak significantly less lead after hail compared to glass/resin laminates [31]. This reduces cracking and contains fragments. However, without strong protection, hail can still crack the perovskite film and allow moisture in. While specific standards on perovskite modules are not available, there is a need for development of technical benchmarks to ensure that the panels can withstand loads without causing the glass to break or the panel to bend. The stresses from wind suction and snow weight can flex the module; thus, the glass and encapsulant layers must be robust enough to resist cracking under such pressures. The criterion for acceptability limits for electrical power degradation, and strength are still being evaluated or perovskites, and separate standards are being developed [28, 31]. In the past BIPVs majorly consisted of PV roof, ensuring unhindered sunlight exposure and design for maximum electricity generation. Recent practices in construction have also been promoting the integration of PV as wall facades with high solar exposure and practical potential for electricity generation. To this end, tinted PV are being used to provide additional benefit with regards to building aesthetics. Also, conventionally, it was extremely difficult to integrate the Si-based PV modules into complex geometries, given its size and shape constraints. Given the flexible nature of the potential polymer substrate in perovskite modules and its lightweight nature, this can potentially be integrated into complex geometries and open up many design opportunities for architects and designers.

10.4.3 Weathering aspects

Weathering refers to the deterioration of building elements. This phenomenon is commonly observed in the exterior surfaces of buildings such as paint and claddings. Weathering of walls or façades often results in structural elements being exposed to extreme heat and moisture, leading to structural defects and cracks. For example, paints can react with moisture to form flaky and brittle substances through a process known as efflorescence. This not only creates visible patches but also exposes the wall to further deterioration from heat and moisture. Although weathering might appear superficial, it has the potential to impact other building elements significantly. In the case of BIPVs, where PV panels are used on walls or roofs, deterioration can occur at two levels. The first level involves weathering or damage within the PV panel system, such as issues with frames, leakages between panels, and the stability of adhesives. These problems are similar to those in other panel systems integrated into glass façades or curtain walls, and solutions for remediation are readily available. The second level concerns the degradation of the PV modules themselves. PV panels degrade over time, with experimental studies highlighting typical degradation modes resulting from cycles of extreme temperature and humidity exposure, module failure, and so on. This degradation can manifest physically, biologically, or chemically. For example, discoloration and hotspots in silicon-based PV panels are often observed alongside moisture ingress [27]. Perovskite PV degradation involves environmental stress (moisture, O_2, UV, and heat), material instabilities (mobile ions, phase changes, and chemistry), and packaging performance. Rainwater, salt spray, and pollutants can attack cell edges or cause potential-induced degradation. Protocols for industry standard tests like ammonia and salt-fog tests (IEC 62716 and TS 63209) [32, 33] are in place to test Si-based PV panels for weathering aspects before deployment on field. While specific test protocols for perovskites are not yet in place, current research envisaged for its development soon. Extreme operating temperatures (≥ 85 °C, < -40 °C) cause irreversible degradation over time [28]. Common failure points include the module perimeter, microcracks in the encapsulant, and delamination between layers. Thermal cycling (daily heating/cooling) adds mechanical stress due to different expansion coefficients of the perovskite layer and surrounding films, leading to microcracks or delamination. This mechanical fatigue can create paths for moisture ingress [34]. Practically, in all major deterioration and weathering pathways, effective sealing of PV panel modules is critical. Overall, field durability requires inert encapsulants with effective sealing against both mechanical and chemical ingress. Outdoor trials show that perovskite stability is improving but still lagging silicon regarding electrical efficiency and stability under dynamic temperature and humidity extremes, which raises concerns about their suitability for applications like BIPV. Current degradation studies on perovskite modules are limited, and extensive examination is necessary before they can be used in applications where there is a significant potential threat to human safety.

10.4.4 Indoor environmental quality

Indoor environmental quality (IEQ) is an emerging area of research that has gained a lot of relevance in the last decade due to its potential impacts on occupants' comfort, health, and overall productivity. IEQ has multiple facets and determine many realms of comfort that can directly or indirectly impact an experience of a space [35]. The COVID-19 pandemic highlighted the importance of IEQ pointing out the relevance of ventilation, monitoring of pathogens, and so on, which otherwise was not a usual activity with regards to air conditioning systems in public spaces and offices. Materials used for construction are in constant contact with the air surrounding it, that is, indoor air enclosed in the room. These are sometime physical interactions, like impact of temperature and humidity causing change in form of material, melting, and distortion [36]. Often chemical interactions between the air and building materials can lead to particle disintegration and pollution of indoor air and so on.

10.4.4.1 Impact on visual and acoustic comfort

Damages in Si-based PV panels are generally not visible through naked eye. In case of physical damages, for instance, black-green patches due to excessive moisture ingress related damage on the back sheet can lead to an unpleasant sight. Generally, Si-based PV panels generally come with limited colors; however, given that the perovskite modules can be manufactured using a variety of substrates, they provide key benefits with regard to the pigments and tints [37] giving the architects and designers a lot of design flexibility and choice. Also, in case of Si-based PV panels, the arrangement of the silicon wafers on the laminate in an orthogonal grid fashion leaves thin slits for the ingress of light making it a transmitter of light [38]. In case of perovskite modules, because there are no practically opaque elements being used, there is a potential to control the light ingress through the panels during manufacturing, which could also be beneficial in harnessing uniform daylight into any enclosed space.

Typically, windows target moderate visible transmission (around 20–30%) to provide ample daylight without causing glare [13]. Optimal transmission for vertical BIPV windows is around 50–70%, while near-clear glass can lead to glare issues. Perovskite BIPV glazing performs like solar-control glass, reducing sun glare and spreading light evenly, thus enhancing comfort and color rendition, supporting productivity by balancing illumination [39]. Perovskite cells are also effective at responding to diffuse and angled light due to their multilayer thin-film structure, which can trap light through internal reflections and efficiently convert a wide range of wavelengths without relying on direct beam incidence [13]. Two main approaches identified are internal modifications (changing the perovskite absorber or stacking interlayers) and external modifications (applying coatings or films). Perovskite layers may have a slight tinted hue due to their bandgap. By adjusting the perovskite composition (such as mixing iodine/bromine or incorporating tin to reduce the bandgap), the absorption spectrum can be tailored to best match the specific application [40]. A "wide-gap" perovskite (~1.8–2.0 eV) will absorb more blue light and exhibit a lighter color, whereas a "narrow-gap" formulation (~1.2–1.3 eV) could capture light within the near-infrared range [41]. This tunability allows engineers to select a perovskite formula that either maximizes energy capture by utilizing a broader range of the solar spectrum or prioritizes transparency and color as needed. Overall, the optical profile of perovskite BIPVs is highly customizable, making them ideal for integration into building exteriors without compromising design or daylighting requirements. The acoustic properties of building materials are dictated by their ability to absorb, transmit, and reflect sound waves. Key factors influencing these properties include material density, porosity, thickness, and the presence of air gaps [42]. Materials that are porous and dense tend to absorb sound more effectively, while those that are nonporous and dense are better at transmitting sound. To this end, PV panels are generally thin and do not really have a great significance in regulating indoor noise; however, a few studies have highlighted the shape and reflective properties of PV panel and its use as an urban noise barrier [42]. Perovskite BIPV panels are typically laminated into insulated glazing units, so their mass and cavity thickness drive performance as a wall, window, or roof element [43]. For BIPV systems with ventilated cavities, such as solar air-heating facades, fan or vent noise can become audible inside the building. This issue stems from HVAC integration rather than PV glass itself. Silencers and strategic duct placement are necessary to mitigate fan noise in ventilated BIPV facades [42]. Research into BIPVs and the role of PV itself for acoustic design is limited.

10.4.4.2 Impact on thermal comfort

Thermal comfort is one of the key aspects of building design and has a huge impact on both occupant's comfort as well as building energy demands. PV panels are generally thin, lightweight with very low thermal mass, resulting in a very high thermal transmittance [38]. In building applications, Si-based PV panels offer high U-value combined with very low thermal damping and time lag. This makes it not suited for places where thermal insulation is necessary, making them a very poor choice in tropical areas [38]. However, with appropriate use of a combination of materials sandwiched with PV panels can help to optimize the required U-value of wall assemblies. This is a challenge due to fire safety issues and other construction challenges. With respect to perovskite-based PV panels, as of now, there have not been any studies practically measuring thermal properties of a full panel module.

Advanced coatings, like low-emissivity (low-E) coatings, can reduce U-values to ~0.3–1.3 W/m²K for double/triple units. Cutting-edge combinations, such as vacuum-insulated glazing with low-E and PV, can achieve $U \approx 0.14$ W/m²K, surpassing most conventional window standards [43]. Well-designed perovskite PV windows can match or exceed the insulation of high-performance insulating units. Given the importance of PV panels' thermal performance in BIPVS, it is critical to gain knowledge in this area before the use of perovskite-based PV panels in buildings. Si-based PV panels use the non-IR spectrum of light for electricity generation, while IR is responsible for heating up an indoor space, which often results in higher mean radiant temperatures indoors in BIPVs. The bandwidth of perovskite-based PV panels is larger than the presently used Si-based PV panels, which might result in a lesser IR light transmittance indoors, leading to benefits in this regard [13]. However, this needs more experimental data and tests at full scale to support the argument. In hot climates, tinted PV glazing blocks intense sunlight and reduces overheating, as studied in Riyadh [43]. During cooler seasons, designers may prioritize the power output from PV or add heating to make up for lower solar gains. Advanced designs may include dynamic shading or integrated PV-thermal elements to balance these effects.

10.4.4.3 Impact on indoor air quality (air pollutants and ventilation) and health

Perovskite BIPV materials raise safety concerns due to lead (Pb) and organic halides used as raw materials in these panels [13]. Proper sealing is essential to contain toxins. In cases of breakage or fire, polymers may release toxic fumes, so robust sealing and recycling are recommended. Even though these material panels are sealed between glass layers, lead release to about 0.05–1 mg/L is often observed in rain tests, leaching up to ~4 mg/L in case of improper encapsulation [44]. Lead halide perovskites are toxic; leached Pb can harm plants and humans. Standard encapsulants do not neutralize Pb, so careful disposal and encapsulation are necessary. Solid organic components in perovskite cells have low volatility under normal conditions; however, temperature, humidity, and radiation can accelerate off gassing, deteriorating the indoor air quality [44]. Decomposition from fire or extreme heat can produce toxic gases, and it is fatal in many cases. Fire safety and heat avoidance are important design considerations.

Conventional Si-based PV also contains hazardous elements; however, the presence of toxic metals in perovskite makes it further dangerous to be used. Safety standards with respect to PV panels refer to tests confirming safe levels of ions in leaching tests [44]; however, such tests for perovskite modules are yet to be designs. Indoor air quality has recently been highlighted as an important determinant of productivity and health. Building materials are one of the contributors toward indoor air pollutants. Given its specific importance toward health of inhabitants, it is essential that perovskite BIPV must ensure air around it to be safe and free from toxins. More scientific examinations are essential in this regard to develop protocols for ensuring safe indoor spaces with employment of perovskite as building elements.

10.4.4.4 Effects on planetary health and ecology

PV panels offer the potential for high-efficiency solar power with potentially lower manufacturing costs and energy inputs compared to conventional silicon panels. However, like any technology, they have environmental and ecological impacts throughout their life cycle. This assessment addresses key areas of concern: the manufacturing footprint, resource extraction and ecotoxicity, end-of-life (EoL) challenges, and broader ecological consequences [44]. Producing perovskite solar modules generally requires less energy and results in lower greenhouse gas emissions than making traditional c-Si panels. This is because perovskite cells can be fabricated via low-temperature solution processing or printing, avoiding the energy-intensive silicon wafer production steps. Life cycle assessments (LCAs) suggest that multi-c-Si modules have the highest environmental impact, primarily due to the electricity and heat required for silicon refining and wafering [45]. In contrast, perovskite manufacturing can utilize mild processing conditions and inexpensive equipment, reducing the energy payback time (the time for a panel to generate the energy used in its production) to as little as a few months in some scenarios. Nonetheless, perovskite PV production is not without environmental impact; careful attention must be given to the materials and processes used. Key environmental considerations in manufacturing perovskite modules include energy use and emissions, proper solvent and chemical use, material selection (availability, extraction, toxicity, recovery, and recycling), and appropriate handling of EoL. Addressing the EoL of perovskite PV panels is essential to prevent environmental contamination and to reclaim valuable materials in a sustainable manner. As an emerging technology, perovskite PV does not yet produce a large waste stream from spent modules, making it imperative to plan for their proper disposal or recycling now. The primary concerns at EoL include the potential release of toxic substances (such as lead) if modules are discarded improperly, and the loss of critical materials (such as indium or glass) if we fail to recover and reuse them. Perovskite PV, being a thin-film technology, may be easier to recycle than silicon PV. Researchers are developing recycling methods specifically for perovskite solar modules. A study [46] demonstrated an aqueous-based recycling process that recovers all major components of a perovskite device, including glass, electrode, metal contacts, and perovskite materials. This process involves soaking modules in solutions to separate layers without harsh chemicals. Applying circular economy principles at the design stage simplifies EoL management for perovskite PV. Using soluble interlayer adhesives and avoiding incompatible material mixtures can facilitate disassembly and recycling. Incorporating lead-absorbing layers ensures safer handling during recycling. Extending module lifespan to 20–25 years improves environmental amortization and allows more time to develop recycling systems. Considering a "service" model where companies lease and take back panels for refurbishing or recycling aligns with the idea of manufacturers retaining product lifecycle responsibility [47]. To this end, use of perovskite-based PV panels, as well as EOL perovskites in BIPV, can potentially unravel a framework for extended use of these panels, until the recycling technology is environmentally and economically viable [38, 47].

10.5 Sustainability aspects of perovskite photovoltaic panels

10.5.1 Cost trends in perovskite and c-Si photovoltaic panels

Solar PV technology has experienced significant cost reductions over the past five decades. Si-based PV panels have evolved from a niche, high-cost power source in the 1970s to one of the most economical forms of electricity available today. In recent years, PSCs have emerged as a promising alternative, offering the potential for even lower costs and diverse applications. This analysis reviews historical cost trends for silicon and perovskite PV panels from the 1970s through 2025 and projects future cost predictions beyond 2025.

Since the 1970s, the cost of silicon PV has decreased exponentially. Table 10.2 presents a summary of representative module prices per watt for silicon PV over time, alongside the timeline for the emergence of perovskite PV. In the mid-1970s, silicon module prices were approximately \$100/W. However, by 2010, mass production and technological advancements had reduced typical module prices to approximately \$2.15/W. This downward trend continued into the 2010s, and by 2021, silicon module prices averaged about \$0.27/W, representing a nearly 90% reduction from 2010 to 2021.

By the mid-2020s, silicon PV costs reached new lows. In early 2024, an oversupply in the industry reduced mainstream module prices to approximately \$0.15/W, with some bids reported below \$0.09/W [48]. Even outside of temporary surplus conditions, silicon modules in 2025 generally cost only a few tenths of a dollar per watt – considerably less than a decade earlier. This decline in unit price has made solar power one of the most cost-effective sources of electricity globally. The price reduction aligns with Swanson's law, which observes that solar module prices fall about 20% for every doubling of cumulative shipped volume [49]. This learning-curve dynamic, along with ongoing technology improvements and economies of scale, has driven the long-term cost trend for silicon PV.

Looking beyond 2025, the central question is how perovskite solar panels will compete with established silicon PV technology in terms of cost and performance.

If technical progress continues, perovskite PV modules are projected to achieve significantly lower costs. Some industry roadmaps predict module costs below \$0.20/W soon. A notable projection by a Japanese consortium (Sekisui Chemical) targets perovskite module costs around ¥10/W (approximately \$0.14/W) by 2040. These costs would position perovskites among the most economical energy technologies available. For comparison, silicon PV prices are already trending in the \$0.10–\$0.20/W range and may decline slightly further but could face diminishing returns after reaching massive scale. Silicon module prices might stabilize at a few cents per watt due to raw material and overhead costs. Perovskites have the potential to undercut even these prices through simpler manufacturing processes, provided they can be produced in large volumes

Table 10.2: Representative PV module prices per watt over time, comparing silicon and perovskite technologies.

Year	Silicon PV module price (USD/W)	Perovskite PV module price (USD/W)
1975	≈$115 (very high cost, niche use) [50]	N/A (technology not yet developed)
2010	≈$2.1 (mass production era begins) [50]	N/A (in research stage)
2021	≈$0.27 (mainstream adoption) [50]	Not commercial (lab prototypes only)
2024	~$0.15 (typical); lows ≈ $0.09 (oversupply) [48]	~$0.38–$0.50 (pilot production scale) [51]
2030 (projected)	~$0.10–$0.20 (further gradual decline)	<$0.20 [52]
2040 (projected)	Silicon possibly leveling off	≈$0.14 [53]

with high yield and longevity. Thus, perovskite PV has a possible pathway to becoming cost-competitive or cost-superior to silicon in the 2030s, assuming that the remaining challenges are addressed. As of 2025, perovskite PV is transitioning from laboratory research to market deployment. Early commercial products have begun shipping; for example, Oxford PV, a leading developer of perovskite-silicon tandem modules, delivered its first commercial panels (24.5% efficient tandem modules) to a utility-scale project in 2024. Several companies in China, such as UtmoLight, GCL, MicroQuanta, and Renshine Solar, are commissioning pilot production lines and planning multi-gigawatt factories for both single-junction perovskite and tandem cells. The late 2020s are likely to witness a scale-up if these pilots succeed, potentially leading to the first gigawatt-scale perovskite manufacturing plants and real-world deployments of hundreds of megawatts. By 2030, the industry aims for perovskite panels to be widely available for various applications, from BIPV to utility farms. However, this timeline assumes current hurdles, such as stability and yield, and are largely overcome within the next 5–10 years. Silicon PV will remain a competitive and continually improving technology during this period, so perovskite will need to demonstrate not only cost advantages but also reliability to gain significant market share. Present indications suggest that perovskites will complement silicon in the near term (e.g., tandem modules or portable and BIPV products where flexibility is valued), with the potential to gradually capture a larger market share if they prove economically superior in the long run.

10.5.2 Other economic factors

a) **Material costs and abundance**: Key PSC materials vary in cost and availability. Lead is inexpensive (~$2–$3/kg) and global reserves are large (~90 Mt) [54]. In contrast, indium (for TCO) costs ~$240/kg (2023) [55] and tellurium (CdTe) costs ~$80/kg [56], reflecting their rarity. Perovskite cells use minimal amounts of each material per area (<1 g/m^2 of Pb [57] and only nanometer-scale thickness of electrodes), so raw material cost per panel is low, but scaling to TW levels would increase demand. Precious metals like Au/Ag (used in lab devices) are very expensive ($60,000/kg Au and $700/kg Ag) but used in thin films; industry is actively replacing them with Ni, Cu, or carbon contacts [58]. Silicon PV requires multi-kilogram silicon wafers (with abundant silica, but costs for energy), silver gridlines, and aluminum frames, which have moderate costs but require significant material flows. CdTe PV processes are somewhat cheaper and have fewer material inputs per watt but rely on Te (scarce) and Cd (toxic) [56, 57].

b) **Supply chain reliability**: Many perovskite components come from byproduct streams in established industries. For example, lead is mostly recycled from auto batteries, and indium is recovered from zinc smelters [55]. However, reliance on a few suppliers presents a risk: for instance, China produces approximately 66% of refined tellurium and a major share of indium. If PSC deployment increased rapidly, the demand for ITO or tellurium could exceed current production levels. Developers address this by avoiding scarce materials (e.g., using fluorine-doped tin oxide (FTO) instead of ITO [57] or new polymers and graphene electrodes) and by planning recycling processes. In contrast, silicon PV uses very common materials (Si, Al, and Cu) with well-established global supply chains, and CdTe PV achieved TRL by integrating Cd recycling into their economics [57]. Any new PSC supply chain will require similar robustness; fortunately, the demonstrated recycling of battery lead into PbI$_2$ indicates a circular route for PSC materials [59].

c) **Large-scale deployment:** At the utility scale, PV technology aims for an exceptionally low levelized cost of electricity (<$30/MWh projected [57]) due to its solution processing and thin layers. The absorber and other layers in perovskite PV are extremely thin, resulting in significantly lower raw material usage per watt compared to silicon. For instance, a 1-μm perovskite film contains approximately 1.4 g/m^2 of lead [60], whereas a 200-μm silicon wafer contains about 4,000 g/m^2 of silicon. This thinness implies that producing a terawatt of PSCs would require only thousands of tons of lead, which is minimal compared to annual lead production. However, scaling up will necessitate vast quantities of back-end materials, such as glass, polymers, and framing, as well as any rare inputs like indium or tellurium. PV manufacturers are already familiar with these constraints, as cadmium telluride (CdTe) scale is largely limited by tellurium supply. Analysts caution that, without recycling, large PV expansion could strain all critical metals [57]. PSCs could mitigate some of these issues through mass production and reuse, as theoretical models indicate that recycling-enabled PSCs would have significantly shorter payback times than Si-based systems [44].

10.5.3 Environmental impact

a) **Raw material extraction:** Perovskite cells utilize lead-halide compounds (e.g., CH$_3$NH$_3$PbI$_3$) as the absorber, typically deposited on glass with a transparent conducting oxide (TCO, e.g., indium tin oxide, ITO). Lead is abundant, mined globally at approximately 5 million tons per year [54], and relatively cost-effective; however, lead mining and smelting are energy-intensive and environmentally detrimental processes. Indium (for ITO) and precious metals such as gold (Au) and silver (Ag) are much scarcer and have high environmental costs [57, 61]. Life cycle studies indicate that the TCO-coated glass substrate, along with its high-temperature processing, significantly contributes to the embodied energy and emissions associated with PSC manufacturing [44]. In contrast, silicon PVs employ extremely abundant quartz (SiO$_2$), requiring highly energy-intensive refining, while CdTe PV involves the use of cadmium (which is toxic) and tellurium (a rare byproduct of copper) [57].

b) **Toxicity:** The lead in PSCs can be harmful if released. A typical perovskite layer contains approximately 0.5–1.5 g of Pb/m^2 [57, 60] (e.g., a 500-nm film has less than 1 g/m^2), and lead halides are water-soluble. If modules break due to hail, floods, or fire, lead can leach into soil or water [57]. Workers exposed during manufacture or repair must also be protected. Researchers indicate that Pb iodide (a degradation product) is relatively insoluble on its own, but mixed perovskite salts dissolve more readily, posing a risk. Tin-based (Pb-free) perovskites avoid lead but still have toxicity concerns (e.g., Sn^{2+} oxidizes to toxic Sn^{4+}) [44]. For comparison, CdTe panels contain cadmium (a known carcinogen), but the PV industry manages cadmium by recycling modules and noting that fixed cadmium in stable CdTe is less prone to leaching than perovskite lead. Silicon PV does not contain heavy metal absorbers (mostly glass, silicon, aluminum, and copper); its wastes are not classified as hazardous.

c) **Life cycle footprint:** PV systems can achieve a very low carbon and energy footprint if optimized effectively. With aggressive recycling, LCA modelling indicates that PSCs can reach energy payback times as low as approximately 0.1 year and around 13 gCO$_2$-eq/kWh [44], which is significantly lower than typical silicon PV systems (1–2.4 years or approximately 22–38 gCO$_2$-eq/kWh). However, in practice, PSCs often involve energy-intensive steps such as spin coating and annealing, as well as the use of toxic solvents like DMF and DMSO, which increase their footprint unless recycling is implemented. Published LCAs of commercial modules indicate that silicon PV produces tens of gCO$_2$-eq/kWh, while CdTe modules can achieve down to approximately 10 gCO$_2$-eq/kWh in sunny climates [60] due to substantial initial processing requirements. Therefore, perovskites have the potential to outperform other types of PV systems if material and process efficiency are maximized and the devices maintain functionality for extended periods.

d) **Recyclability and disposal:** Currently, perovskite modules are not produced with recycling considerations in mind. However, recent research and development indicate that nearly complete recovery is achievable. A 2025 study demonstrated an aqueous process that reclaims perovskite, transport layers, glass, and metals, enabling the reuse of more than 96% of materials [46]. This recycling approach reduces resource use by approximately 96.6% and toxicity impacts by about 68.8% compared to landfilling [46]. In practice, broken or decommissioned PSCs would likely be treated as hazardous waste, like lead batteries, due to the inability of encapsulation to completely prevent lead solubility [57]. In comparison, cadmium telluride (CdTe) panels already have a take-back recycling program (First Solar) that recovers cadmium and tellurium and prevents toxic emissions [57]. Conversely, silicon-panel recycling primarily recovers bulk glass and aluminum, with most silicon being lost. Effective recycling of PSCs or the implementation of on-device lead traps will be necessary to mitigate disposal risks.

10.5.4 Social aspects

a) **Regulations and commercialization:** Perovskite BIPV systems must comply with both PV and building standards. In many jurisdictions, PV modules are required to meet the IEC 61215 performance tests and the IEC 61730 safety tests. Additionally, BIPV products may require CE marking or UL certification and adherence to building product codes including fire safety and wind load regulations. It is noteworthy that Chinese industry has initiated formal certification processes for perovskite modules. In 2023, Hangzhou Microquanta announced that its mass-production perovskite module successfully passed all IEC 61215/61730 stability tests conducted by VDE, reportedly making it the first in the world to achieve this milestone [2]. In 2024, Utmo Light reported that its 0.72-m^2 module successfully cleared IEC testing and obtained TÜV certification [2]. Furthermore, Utmo Light subjected its panels to 2,300 h of intense UV and heat exposure, equivalent to approximately 12 years of real-life conditions, without degradation [24]. These results demonstrate that perovskite modules can meet traditional PV durability standards. Regulatory frameworks are evolving, for instance, China's "Action Plan for Smart PV" (2021–2025) explicitly promotes long-life, high-safety BIPV, and under this initiative, glass-glass perovskite modules have already been installed as demonstration projects. In Europe and the USA, no perovskite BIPV products have yet achieved full certification, but companies are aiming to meet established standards such as EN 50583 for BIPV. Commercial readiness is progressing. Utmo Light began selling its perovskite modules (priced around $0.19/W) in mid-2024, offering them as complete building glass panels with warranties (10-year product and 25-year performance) [24]. Saule Technologies is seeking CE and IEC approvals in Poland/Europe, and Oxford PV (UK) is shipping tandem modules for utility use, though not specifically for BIPV. Several Chinese firms, including Micro-quanta and Utmo Light, are scaling to GW/year production by the mid-2020s. In summary, perovskite BIPV is transitioning from lab research to market availability. Some companies now offer certified modules and warranty-backed products. Regulatory compliance, including IEC/UL testing, is being demonstrated in pilot lines, and building codes can be met due to the modules' light weight and encapsulation. With global policy support, such as Japan's 20 GW perovskite target by 2040, and ongoing major pilot installations, perovskite BIPV is approaching commercial adoption and is likely to enter niche markets first before expanding into mainstream construction.

10.6 Conclusions

A comprehensive view on the environmental, economic, and social implications of utilizing perovskite panels in BIPVs is provided. Despite the promising electrical efficiency of perovskites, their adoption in real-time scenarios under varying environmental stressors remains challenging. Additionally, when integrated into buildings, it is imperative to address toxicity risks appropriately to ensure occupant health, comfort, and safety.

With the transition from Si-based to perovskite-based PV panels, the development of performance standards for perovskites is essential. These performance indicators will differ from those of Si-based panels due to the distinct nature of materials used in perovskites and the associated manufacturing technology. Beyond cost benefits, a preparedness plan for the EoL management of perovskite-based panels is necessary to mitigate unforeseen impacts of toxic elements used in these panels.

While perovskite-based BIPVs hold promise for more effectively harnessing renewable energy sources, their practical applications require holistic and careful scrutiny. This view reveals the potential benefits and risks including the impact on ecological diversity, cost-effectiveness, and health and comfort aspects within buildings.

References

[1] Building Integrated Photovoltaics (BIPV) | WBDG – Whole Building Design Guide [Internet]. [cited 2025 May 23]. Available from: https://www.wbdg.org/resources/building-integrated-photovoltaics-bipv

[2] Technology Collaboration Programme by International Energy Agency Photovoltaic Power Systems Programme PVPS Task 1 Strategic PV Analysis and Outreach. 2024 [cited 2025 May 23]; Available from: www.iea-pvps.org

[3] How are Solar Cells Made? Silicon vs. Perovskite Production – Solaires Entreprises Inc [Internet]. [cited 2025 May 23]. Available from: https://www.solaires.net/blog/blog_post/how_are_solar_cells_made/

[4] Perovskite Solar | Perovskite-Info [Internet]. [cited 2025 May 23]. Available from: https://www.perovskite-info.com/perovskite-solar

[5] Afre RA, Pugliese D. Perovskite solar cells: A review of the latest advances in materials, fabrication techniques, and stability enhancement strategies, Micromachines 2024, Vol 15, Page 192 [Internet]. 2024 Jan 27 [cited 2025 May 23]; 15(2): 192. Available from: https://www.mdpi.com/2072-666X/15/2/192/htm.

[6] McGovern L, Garnett EC, Veenstra S, Van der Zwaan B. A techno-economic perspective on rigid and flexible perovskite solar modules, Sustain Energy Fuels [Internet]. 2023 Oct 24 [cited 2025 May 23]; 7(21): 5259–5270. Available from: https://pubs.rsc.org/en/content/articlehtml/2023/se/d3se00828b.

[7] Perovskite Solar Cells vs Silicon Solar Cells | Ossila [Internet]. [cited 2025 May 23]. Available from: https://www.ossila.com/pages/perovskite-solar-cells-vs-silicon-solar-cells

[8] What are the advantages of perovskite solar cells over silicon solar cells – BLOG – Tongwei Co., Ltd., [Internet]. [cited 2025 May 23]. Available from: https://en.tongwei.cn/blog/93.html

[9] Perovskite solar panels: An expert guide [2025] [Internet]. [cited 2025 May 23]. Available from: https://www.sunsave.energy/solar-panels-advice/solar-technology/perovskite

[10] Moot T, Patel JB, Mcandrews G, Wolf EJ, Morales D, Gould IE, et al. Temperature coefficients of perovskite photovoltaics for energy yield calculations. 2021 [cited 2025 May 23];6:51. Available from: http://pubs.acs.org/journal/aelccp

[11] Perovskite solar cells stable for 6 months under real outdoor operating conditions – IRAMIS [Internet]. [cited 2025 May 23]. Available from: https://iramis.cea.fr/en/nimbe/2024/09/perovskite-solar-cells-stable-for-6-months-under-real-outdoor-operating-conditions/

[12] Zhang D, Li D, Hu Y, Mei A, Han H. Degradation pathways in perovskite solar cells and how to meet international standards, Communications Materials 2022 3:1 [Internet]. 2022 Aug 29 [cited 2025 May 23]; 3(1): 1–14. Available from: https://www.nature.com/articles/s43246-022-00281-z.

[13] Roy A, Ghosh A, Bhandari S, Sundaram S, Mallick TK. Perovskite Solar Cells for BIPV application: A review, Buildings 2020, Vol 10, Page 129 [Internet]. 2020 Jul 13 [cited 2025 May 23]; 10(7): 129. Available from: https://www.mdpi.com/2075-5309/10/7/129/htm.

[14] Emery Q, Remec M, Paramasivam G, Janke S, Dagar J, Ulbrich C, et al. Encapsulation and outdoor testing of perovskite solar cells: Comparing industrially relevant process with a simplified lab procedure, ACS Appl Mater Interfaces [Internet]. 2022 Feb 2 [cited 2025 May 23]; 14(4): 5159–5167. Available from: https://pubs.acs.org/doi/pdf/10.1021/acsami.1c14720.

[15] Lee SW, Kim S, Bae S, Cho K, Chung T, Mundt LE, et al. UV degradation and recovery of perovskite solar cells, Scientific Reports 2016 6:1 [Internet]. 2016 Dec 2 [cited 2025 May 23]; 6(1): 1–10. Available from: https://www.nature.com/articles/srep38150.

[16] Skanska launches "first" perovskite solar cell film on buildings [Internet]. [cited 2025 May 23]. Available from: https://optics.org/news/9/1/20

[17] Saule technologies and skanska change construction industry – saule technologies [Internet]. [cited 2025 May 23]. Available from: https://sauletech.com/saule-technologies-and-skanska-change-construction-industry/

[18] Commencement of Japan's first demonstration test of perovskite solar cells installed on exterior walls of buildings | SEKISUI CHEMICAL CO.,LTD [Internet]. [cited 2025 May 23]. Available from: https://www.sekisuichemical.com/news/2023/1384316_40406.html

[19] NTT DATA to Conduct Demonstration Testing for Film-Type Perovskite Solar Cells on Exterior Building Walls | NTT DATA Group [Internet]. [cited 2025 May 23]. Available from: https://www.nttdata.com/global/en/news/press-release/2023/february/ntt-data-to-conduct-demonstration-testing-for-film-type-perovskite-solar-cells-on-exterior-building

[20] Zhao J China Advances to Gw-Scale Mass Production of Perovskite Solar Cells-Aims to Exceed 30% Conversion Efficiency with Tandem Cells.

[21] Panasonic Holdings Corporation to Start the World's First* Long-term Implementation Demonstration Project for the Building Integrated Perovskite Photovoltaics Glass in the Fujisawa Sustainable Smart Town | Innovations/Technologies | Company | Press Release | Panasonic Newsroom Global [Internet]. [cited 2025 May 23]. Available from: https://news.panasonic.com/global/press/en230831-2

[22] Net zero city, smart city expos demonstrate Taiwan's green transformation strengths | NEWS | reccessary [Internet]. [cited 2025 May 23]. Available from: https://www.reccessary.com/en/news/net-zero-city-expo-showcases-taiwan-green-transition-strength

[23] China solar pv news snippets – January 30, 2025 [Internet]. [cited 2025 May 23]. Available from: https://taiyangnews.info/markets/china-solar-pv-news-snippets-high-tech-enterprise-tag-for-tongwei-subsidiary-more

[24] Commercial perovskite solar modules at SNEC 2024 trade show – pv magazine International [Internet]. [cited 2025 May 23]. Available from: https://www.pv-magazine.com/2024/06/13/commercial-perovskite-solar-modules-at-snec-2024-trade-show/

[25] Bati ASR, Zhong YL, Burn PL, Nazeeruddin MK, Shaw PE, Batmunkh M. Next-generation applications for integrated perovskite solar cells, Communications Materials 2023 4:1 [Internet]. 2023 Jan 5 [cited 2025 May 23]; 4(1): 1–24. Available from: https://www.nature.com/articles/s43246-022-00325-4.

[26] IEC. International Standard Norme Internationale Photovoltaic (PV) module safety qualification-Part 2: Requirements for testing Qualification pour la sûreté de fonctionnement des modules photovoltaïques (PV)-Partie 2: Exigences pour les essais International Electrotechnical Commission Commission Electrotechnique Internationale. 2023.

[27] Azmi R, Ugur E, Seitkhan A, Aljamaan F, Subbiah AS, Liu J, et al. Damp heat-stable perovskite solar cells with tailored-dimensionality 2D/3D heterojunctions, Science (1979) [Internet]. 2022 Apr 1 [cited 2025 May 23]; 376(6588): 73–77. Available from: /doi/pdf/10.1126/science.abm5784.

[28] Mariani P, Molina-García MÁ, Barichello J, Zappia MI, Magliano E, Castriotta LA, et al. Low-temperature strain-free encapsulation for perovskite solar cells and modules passing multifaceted accelerated ageing tests, Nature Communications 2024 15:1 [Internet]. 2024 May 29 [cited 2025 May 23]; 15(1): 1–15. Available from: https://www.nature.com/articles/s41467-024-48877-y.

[29] Emery Q, Dagault L, Khenkin M, Kyranaki N, De Araújo WMB, Erdil U, et al. Tips and tricks for a good encapsulation for perovskite-based solar cells, Progress in Photovoltaics: Research and Applications [Internet]. 2025 Apr 1 [cited 2025 May 23]; 33(4): 551–559. Available from: /doi/pdf/10.1002/pip.3888.

[30] Zhang M, Li Z, Gong Z, Li Z, Zhang C. Perspectives on the mechanical robustness of flexible perovskite solar cells, Energy Advances [Internet]. 2023 Mar 16 [cited 2025 May 23]; 2(3): 355–364. Available from: https://pubs.rsc.org/en/content/articlehtml/2023/ya/d2ya00303a.

[31] Jiang Y, Qiu L, Juarez-Perez EJ, Ono LK, Hu Z, Liu Z, et al. Reduction of lead leakage from damaged lead halide perovskite solar modules using self-healing polymer-based encapsulation, Nature Energy [Internet]. 2019 Jul 1 [cited 2025 May 23]; 4(7): 585–593. Available from: https://www.nature.com/articles/s41560-019-0406-2.

[32] IEC 62716:2013 | IEC [Internet]. [cited 2025 May 23]. Available from: https://webstore.iec.ch/en/publication/7392

[33] IEC TS 63209-2:2022 | IEC [Internet]. [cited 2025 May 23]. Available from: https://webstore.iec.ch/en/publication/65283

[34] Degradation and Failure Modes in New Photovol-taic Cell and Module Technologies 2025 PVPS Task 13 Reliability and Performance of Photovoltaic Systems. [cited 2025 May 23]; Available from: www.iea-pvps.org

[35] Priyadarshani S, Mani M, Maskell D. Influence of building typology on Indoor humidity regulation, REHVA European HVAC [Internet]. 2021 Dec 21 [cited 2025 May 23]; 2021(6): 48–52. Available from: https://researchportal.bath.ac.uk/en/publications/influence-of-building-typology-on-indoor-humidity-regulation.

[36] Miah MH, Rahman MB, Nur-E-Alam M, Islam MA, Shahinuzzaman M, Rahman MR, et al. Key degradation mechanisms of perovskite solar cells and strategies for enhanced stability: Issues and prospects, RSC Adv [Internet]. 2025 Jan 7 [cited 2025 May 23]; 15(1): 628–654. Available from: https://pubs.rsc.org/en/content/articlehtml/2025/ra/d4ra07942f.

[37] Ghosh A, Mesloub A, Touahmia M, Ajmi M. Visual comfort analysis of semi-transparent perovskite based building integrated photovoltaic window for hot desert climate (Riyadh, Saudi Arabia), Energies 2021, Vol 14, Page 1043 [Internet]. 2021 Feb 17 [cited 2025 May 23]; 14(4): 1043. Available from: https://www.mdpi.com/1996-1073/14/4/1043/htm.

[38] Rao RR Exploring End-of-Life Photovoltaic (PV) Panel as a Building Material: A Case of Crystalline Silicon PV. 2023 [cited 2025 May 23]; Available from: https://etd.iisc.ac.in/handle/2005/6279

[39] Cannavale A, Ayr U, Martellotta F. Energetic and visual comfort implications of using perovskite-based building-integrated photovoltaic glazings, Energy Procedia [Internet]. 2017 Sep 1 [cited 2025 May 23]; 126: 636–643. Available from: https://www.sciencedirect.com/science/article/pii/S1876610217337621.

[40] Wang H, Li J, Dewi HA, Mathews N, Mhaisalkar S, Bruno A. Colorful perovskite solar cells: Progress, strategies, and potentials, Journal of Physical Chemistry Letters [Internet]. 2021 Feb 4 [cited 2025 May 23]; 12(4): 1321–1329. Available from: /doi/pdf/10.1021/acs.jpclett.0c03445.

[41] Zhang W, Anaya M, Lozano G, Calvo ME, Johnston MB, Míguez H, et al. Highly efficient perovskite solar cells with tunable structural color, Nano Lett [Internet]. 2015 Mar 11 [cited 2025 May 23]; 15(3): 1698. Available from: https://pmc.ncbi.nlm.nih.gov/articles/PMC4386463/.

[42] Di Bella A, De Carli M, Elarga H, Granzotto N. Thermal and acoustic analysis of innovative integration of pv modules in façade envelopes. 2026.

[43] Mert Cüce A, Attia M, Cüce E. Perovskite photovoltaic glazing systems: A pathway to low/zero carbon buildings. Sustainable and Clean Buildings [Internet]. 2025 [cited 2025 May 23]; 1(1): Available from: https://avesis.erdogan.edu.tr/yayin/3729405f-d849-49a3-a6d4-54fc2d413538/perovskite-photovoltaic-glazing-systems-a-pathway-to-low-zero-carbon-buildings.

[44] Torrence CE, Libby CS, Nie W, Stein JS. Environmental and health risks of perovskite solar modules: Case for better test standards and risk mitigation solutions, iScience [Internet]. 2023 Jan 20 [cited 2025 May 23]; 26(1): 105807. Available from: https://www.sciencedirect.com/science/article/pii/S2589004222020806?via%3Dihub.

[45] Rossi F, Rotondi L, Stefanelli M, Sinicropi A, Vesce L, Parisi ML. Comparative life cycle assessment of different fabrication processes for perovskite solar mini-modules, EPJ Photovoltaics [Internet]. 2024 [cited 2025 May 23]; 15: 20. Available from: https://www.epj-pv.org/articles/epjpv/full_html/2024/01/pv230066/pv230066.html.

[46] Xiao X, Xu N, Tian X, Zhang T, Wang B, Wang X, et al. Aqueous-based recycling of perovskite photovoltaics, Nature 2025 638:8051 [Internet]. 2025 Feb 12 [cited 2025 May 23]; 638(8051): 670–675. Available from: https://www.nature.com/articles/s41586-024-08408-7.

[47] Rao RR, Priyadarshani S, Mani M. Examining the use of End-of-Life (EoL) PV panels in housing and sustainability, Solar Energy [Internet]. 2023 Jun 1 [cited 2025 May 23]; 257: 210–220. Available from: https://www.sciencedirect.com/science/article/pii/S0038092X23002633?via%3Dihub.

[48] A tumultuous year for solar and batteries – Pv magazine International [Internet]. [cited 2025 May 23]. Available from: https://www.pv-magazine.com/2024/12/30/a-tumultuous-year-for-solar-and-batteries/

[49] How Solar Panel Efficiency and Cost Changed Over Time – Renogy United States [Internet]. [cited 2025 May 23]. Available from: https://www.renogy.com/blog/solar-panel-efficiency-and-cost-over-time/?srsltid=AfmBOooiDjshEYBRthhTlsxYKAQsM3erCP8bEGnOA5HoE58QkKkj3Ps4

[50] History of solar energy prices | AVENSTON [Internet]. [cited 2025 May 23]. Available from: https://avenston.com/en/articles/pv-cost-history/

[51] Reality Check: Can Perovskite Really Outshine Silicon PV? – CleanTechnica [Internet]. [cited 2025 May 23]. Available from: https://cleantechnica.com/2025/04/26/reality-check-can-perovskite-really-outshine-silicon-pv/

[52] Perovskite Solar Cells {2025} | 8MSolar [Internet]. [cited 2025 May 23]. Available from: https://8msolar.com/perovskite-solar-cells/

[53] Perovskite solar cells: Progress continues in efficiency, durability, and commercialization – The American Ceramic Society [Internet]. [cited 2025 May 23]. Available from: https://ceramics.org/ceramic-tech-today/perovskite-solar-cells-progress-2025/

[54] International Lead And Zinc Study Group (ILZSG). The World Lead Factbook 2023 [Internet]. 2023 [cited 2025 May 23]. Available from: https://www.ilzsg.org/wp-content/uploads/SitePDFs/1_ILZSG%20World%20Lead%20Factbook%202023.pdf

[55] Stewart, A.A., 2024, Indium: U.S. Geological Survey Mineral Commodity Summaries 2024, p. 90–91.

[56] Flanagan, D.M., 2024, Tellurium: U.S. Geological Survey Mineral CommoditySummaries 2024, p. 178–179.

[57] Charles RG, Doolin A, García-Rodríguez R, Villalobos KV, Davies ML. Circular economy for perovskite solar cells – Drivers, progress and challenges, Energy Environ Sci [Internet]. 2023 Sep 13 [cited 2025 May 23]; 16(9): 3711–3733. Available from: https://pubs.rsc.org/en/content/articlehtml/2023/ee/d3ee00841j.

[58] Li M, Park SY, Wang J, Zheng D, Wostoupal OS, Xiao X, et al. Nickel-doped graphite and fusible alloy bilayer back electrode for vacuum-free perovskite solar cells, ACS Energy Lett [Internet]. 2023 Jul 14 [cited 2025 May 23]; 8(7): 2940–2945. Available from: /doi/pdf/10.1021/acsenergylett.3c00852.

[59] Suo J, Yang B, Prideaux S, Pettersson H, Kloo L. From lead–acid batteries to perovskite solar cells – Efficient recycling of Pb-containing materials, RSC Sustainability [Internet]. 2025 Feb 5 [cited 2025 May 23]; 3(2): 1003–1008. Available from: https://pubs.rsc.org/en/content/articlehtml/2025/su/d4su00470a.

[60] Benefits and risks of lead halide perovskite photovoltaics – pv magazine International [Internet]. [cited 2025 May 23]. Available from: https://www.pv-magazine.com/2019/11/14/benefits-and-risks-of-lead-halide-perovskite-photovoltaics/

[61] Substitute for Gold Layer in Perovskite Clears Way for Cheaper Commercialization | NREL [Internet]. [cited 2025 May 23]. Available from: https://www.nrel.gov/news/detail/program/2023/substitute-for-gold-layer-in-perovskite-clears-way-for-cheaper-commercialization

Neha Kumari, Vishal Sharma, Kamna, Sahil Kumar, Sapna Thakur,
Gun Anit Kaur, and Mamta Shandilya*

Chapter 11
Wearable and flexible perovskite solar cells

Abstract: Perovskite solar cells (PSCs) are emerging as one of the most promising next generation photovoltaic (PV) technologies with significant breakthroughs in efficiency, materials engineering, and structural design. Since its first publication in 2009, with power conversion efficiency (PCE) of 3.8%, PSCs have topped 26% in single junction devices and more than 34% in tandem arrangements, very close to the theoretical Shockley–Queisser limit. This chapter gives a thorough overview of PSC development, device design, fabrication process, and current improvement with a focus on major breakthroughs such as interface engineering, passivation tactics and tandem integration. When compared to other solar technologies such as silicon, CIGS, CdTe, and organic PVs, PSCs have benefits in adjustable bandgaps, flexibility, low-temperature manufacturing and high light absorption. Despite issues like stability and lead toxicity, developing solutions including 2D/3D hybrid structures, roll-to-roll printing and lead-free perovskites are opening the path for scalable and sustainable commercialization. The chapter finishes by examining PSC's future prospects and possibilities for revolutionizing the global renewable energy environment.

Keywords: Perovskite solar cells, wearable, flexible, interface engineering, lead free

11.1 Introduction

Global population growth is increasing energy consumption, necessitating the rapid adoption of renewable energy sources to meet global energy demands and mitigate negative environmental impacts [1]. The global demand for sustainable energy solutions has pushed solar energy to the forefront of renewable technology, as it is clean, abundant, and environmentally friendly, using various technologies [2]. Solar photovoltaic (PV) technology is a promising renewable energy source that captures sunlight and converts it into electrical power. It has improved efficiency and lower prices, making it a potential solution to mitigate fossil fuel usage and greenhouse gas emis-

*Corresponding author: Mamta Shandilya, School of Physics and Materials Science,
Shoolini University, Bajhol, Solan, 173229, Himachal Pradesh, India,
e-mails: mamtashandilya@shooliniuniversity.com; mamta2882@gmail.com
Neha Kumari, Vishal Sharma, Kamna, Sahil Kumar, Gun Anit Kaur, School of Physics and Materials
Science, Shoolini University, Bajhol, Solan 173229, Himachal Pradesh, India
Sapna Thakur, Department of Biotechnology, Eternal University, Baru Sahib 173101, India

https://doi.org/10.1515/9783111726847-011

sions [3, 4]. Solar cells convert light photons into voltage or energy through the PV effect [5]. In 1954, the crystalline silicon-based solar cell was developed, resulting in a 4.5% power conversion efficiency (PCE) . Researchers have since explored low-cost device structures and materials, leading to the development of second-generation solar cells, GaAs, CdTe, InP, and CIGs. The third generation with dye-sensitized structure emerged in the 1990s, and organic PV (OPV) were introduced in the 2000s [6]. Research in nanomaterials is focusing on finding cheap, low-cost solar devices. Currently, crystalline silicon solar cells dominate, but researchers are seeking high-efficiency, low-cost PV technology. Perovskite materials are gaining interest due to their excellent performance, low-cost raw materials, and easy processing [7]. The discovery of PV activity in hybrid organic-inorganic lead-halide materials has resulted in a significant new area of research: perovskite solar cells (PSC) [8, 9]. This adverb refers to solar cell absorber materials with perovskite crystal structures, which were originally based on $CaTiO_3$ [10]. Perovskite materials have economic benefits such as plentiful starting materials, easy processing processes, and inexpensive energy costs for device fabrication [11]. In the 1990s, Mitzi and colleagues studied the optical characteristics of organic-inorganic perovskites and reported that the material's high exciton properties imply its potential usage in light-emitting diodes (LEDs), transistors, and solar cells [12, 13]. The journey of employing perovskites in solar cells started in 2006 with the usage of $CH_3NH_3PbBr_3$ as the sensitizer in liquid DSSC, although the attained efficiency was just 2.2% [14]. In 2009, the same group first reported the use of perovskite material with enhanced efficiency of 3.8% and 3.1% when employing iodine-based liquid electrolytes with $CH_3NH_3PbI_3$ and $CH_3NH_3PbBr_3$ perovskites, respectively, as the sensitizer [8]. While, for the first time in 2012, Kim et al. reported all-solid-state PSCs which showed the PCE of 9.7% [15]. Recently, PSCs has gained significant interest of researchers owing to their low manufacturing expenses and significant efficiency. Progress has been made in scaling up PSC device production, with the solution processable nature of the starting ingredients allowing for the printing of the absorber layer [16, 17]. To create extremely stable and efficient PSCs, compositional engineering of perovskite materials is a useful strategy. Crystallographic stability and the development of the three-dimensional (3D) crystal structure are determined by the radii of each component in the perovskite material via the Goldschmidt tolerance factor (t). Thus, different sized cations and anion, such as Cs, methylammonium (MA), formamidinium (FA), I, Br, and Cl, may be used to create perovskite crystals, which will cause a change in the bandgap. Perovskites based on MA were initially employed as sensitizers in liquid-state solar cells in 2009; they produced a PCE of 3.81% with exceptionally low stability. PSCs have advanced significantly since they were first used in solar cells as visible light-harvesting materials in 2009 [8]. Over the previous decade, the PCE of solar devices based on perovskite has increased from 3.5% achieved by Kojima et al. [8] to 25.8% by [18, 19]. PSCs have advanced quickly because of their inexpensive manufacturing methods, which frequently use solution-based approaches that work well with flexible substrates. This adaptability not only lowers production costs, but also creates opportunities for PSC integration into a range of applications, such as portable electronics and building-integrated PVs [20, 21]. In current developments, PSCs are centered on using novel materials to increase sustainability, stability, and efficiency. Advanced interlayers (such as zwitterionic materials), 2D/3D hybrid structures, and lead-free perovskites like $FASnI_3$ are being investigated to lessen toxicity and increase lifetime. Commercialization is being advanced by flexible and tandem designs with materials discovery [22].

11.2 Early discoveries and origins

Perovskites are remarkable materials that have revolutionized PVs; they can be synthetic or natural, and the German mineralogist Gustrav Rose named the first natural perovskite after his Russian colleague Lev Perovsky in 1839 [23]. While there are many different types of perovskites, ranging from conductors to insulators, the ones used in solar cells and other optoelectrical applications are typically semiconductor hybrid inorganic–organic perovskites, which have a direct bandgap, strong light absorption characteristics, high charge-carrier mobility, and a small exciton binding energy [24]. In general, perovskite materials have the formula ABX_3, and they have the same crystal structure as $CaTiO_3$. X is an anion, whereas A and B are cations with distinct atomic radii (A is bigger than B). Their ideal cubic crystal structure consists of corner-sharing BX_6 octahedra, with the A-cation situated in the interstitial space formed by the 3D structure's eight neighboring octahedra [20]. The basic perovskite's typical unit cell structure is depicted in Figure 11.1(a). In solar cells, perovskite materials are a type of organic inorganic metal halide compound with a perovskite structure. The metal cation B (Pb^{2+}, Sn^{2+}, etc.) and halogen anion X (Cl^-, Br^-, or I^-, or a coexistence of several halogens) occupy the core and apex of the octahedra, respectively, while Group A (methylammonium, CH_3NH^{3+}, MA^+, or formamidinium, $CH(NH_2)^{2+}$, FA^+) is situated in the vertex of the face-centered cubic lattice. The metal-halogen octahedra are connected to create a stable network structure in 3D. Typical perovskite cubic lattice is shown in Figure 11.1(b) [25].

Depending on the temperature change, perovskite materials go through distinct phases. Upon cooling below 100 K, the perovskite exhibited a stable orthorhombic (γ) phase. The tetragonal (β) phase began to emerge and replace the initial orthorhombic (γ) phase when the temperature was raised to 160 K. When the temperature increased to around 330 K, a stable cubic (α) phase began to replace the tetragonal (β) phase [27]. The three crystal structures are all shown in Figure 11.2.

The entire performance of perovskite-based optoelectronic devices is determined by the efficiency of conversion between photons and charge carriers. As a result, for the past decade, research has concentrated on examining the electrical and optical characteristics of these devices [29]. The development of PSCs has seen remarkable progress over the years. The development of perovskite materials in optoelectronics started in the 1990s

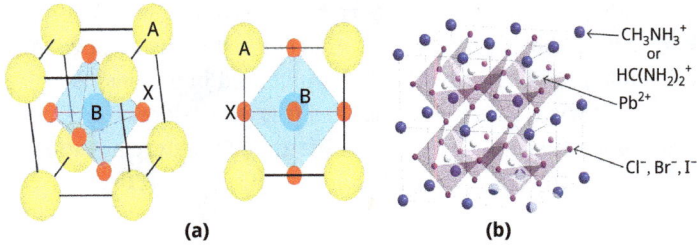

Figure 11.1: (a) Structure of basic perovskite's typical unit cell structure and (b) typical perovskite cubic lattice [26].

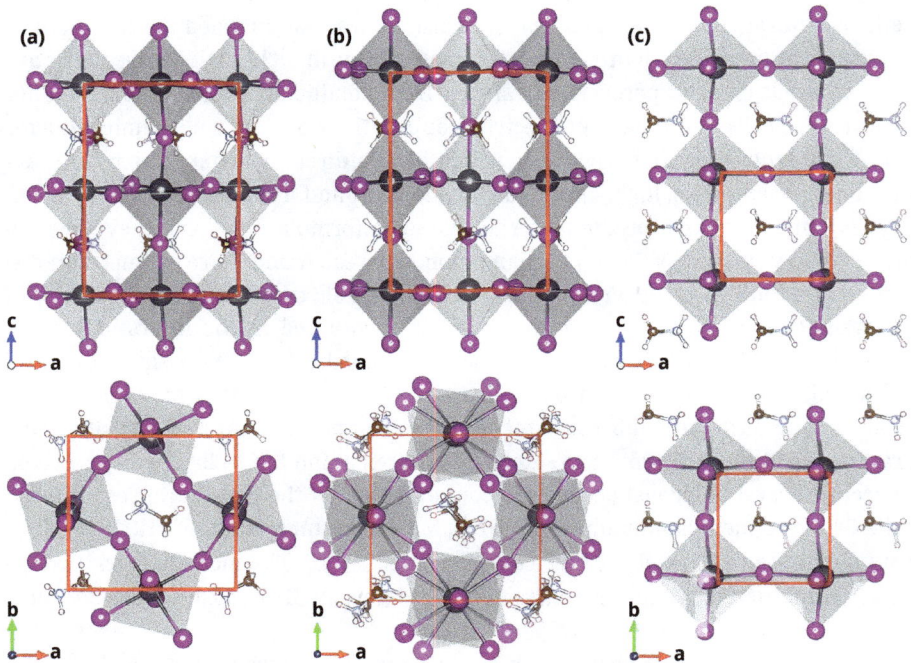

Figure 11.2: Comparison of the (a) orthorhombic, (b) tetragonal, and (c) cubic perovskite phases derived from structural optimization for MA PbI3. The top row displays the ac-plane, while the bottom row shows the ab-plane [28].

with David Mitzi's groundbreaking research on organic–inorganic hybrid perovskites, which showed promise in field-effect transistors, LEDs, and solar cells because of their advantageous optoelectronic characteristics [12, 30]. In 2009, Kojima et al. added methylammonium lead halide perovskites ($CH_3NH_3PbI_3$ and $CH_3NH_3PbBr_3$) as sensitizers to dyesensitized solar cells (DSSCs), obtaining PCEs of 3.8% and 3.1%, respectively. However, because perovskites are soluble in liquid electrolytes, the devices degraded quickly [8].

In 2011, Im et al. used perovskite quantum dots in DSSCs to improve the PCE to 6.5%, however the stability problems remained [31]. In 2012, Kim et al. made a major advancement by substituting a solid-state hole transport material (HTM, Spiro-OMeTAD) for the liquid electrolyte. This led to enhanced device stability and a PCE of 9.7% [15]. Furthermore, a two-step sequential deposition technique was created by Burschka et al. in 2013, which improved film morphology and produced a 15% PCE [32]. Additional progress was made in 2014 by Im et al., who achieved a PCE of 17.01% by optimizing the size of MAPbI$_3$ crystals [33]. Thus, significant progress has been made so far in PSCs traced back to the 2009 and till now. In recent years, significant advancements have been made in the discipline. Researchers added phenylethylammonium in 2023 to improve stability and performance; after 200 h of continuous operation, the device maintained 89% efficiency and achieved a PCE of 22.02% [34]. In a recent work published in 2025, Bai et al. demonstrated a high PCE of 24.72% by introducing a new polyionic liquid (PTA) as a buried interface modifier in inverted PSCs. After more than 1,000 h of exposure to heat and light, the PTA-treated devices showed exceptional thermal and operational durability by cooperatively passivating both shallow and deep-level faults, retaining more than 88% of their initial efficiency [35]. These advancements highlight the revolutionary potential of perovskites in upcoming optoelectronic applications and demonstrate the impressive advancements in perovskite research [36].

11.3 The advent of perovskite solar cells

From 2009 to 2012, the field of PV research experienced a pivotal shift with the advent of PSCs, which emerged as a strong contender to the established silicon-based technologies. This transformative period was characterized by groundbreaking developments, initiated by Tsutomu Miyasaka's innovative research and culminating in the creation of solid-state designs that effectively tackled early stability problems [36, 37]. In the early stages of development, PSCs faced considerable challenges due to the instability caused by liquid electrolytes. Initially, when perovskite materials were used as sensitizers in dye-sensitized solar cells with liquid electrolytes, the devices showed instability as the perovskite materials degraded in the liquid environment [38]. This issue prompted researchers to develop all-solid-state PSCs, which eliminated the need for liquid electrolytes and significantly improved stability [39]. Using liquid electrolytes in the initial PSCs led to their swift degradation, limiting their operation to mere seconds (Kojima et al., 2009). The devices' performance dropped by 80% as the perovskite quantum dots dissolved in the liquid electrolyte after 10 min of continuous exposure to light. These advancements helped boost the PCE from 3.8% to over 10% in the early development phase [31]. In 2012, Park and his team improved the stability of the devices to 500 h under ambient conditions by replacing the liquid hole transport layer (HTL) with a solid-state layer made from 2,2',7,7'-tetrakis[N,N-di(4-methoxyphenyl)amino]-9,9'-spirobifluorene (spiro-MeOTAD). As a result, they increased the efficiency to 9.7% by fine-tuning the thickness of the titanium oxide electron transport layer (ETL) [15].

11.4 Rapid progress and efficiency improvements

PSCs have advanced significantly in terms of architecture and composition, resulting in increased efficiency and ease of production. Early PSC designs were influenced by dye-sensitized solar cells (DSSCs), which used perovskites as both light absorbers and charge carriers [40, 41]. One of the first important designs was the mesoporous structure, when perovskite infiltrated a mesoporous metal oxide scaffold such as TiO_2 or AI_2O_3. This arrangement enhances film homogeneity, promotes electron transport, and lowers resistance. The mesoporous structure has a compact TiO_2 layer that blocks holes, followed by a perovskite capping layer and HTM [42]. Planar PSCs eliminate the scaffold and use a flat perovskite layer sandwiched between compact TiO_2 and HTM, simplifying manufacturing and lowering processing temperatures [43]. Another structure, the inverted planar design, lays perovskite on hole-conducting layers NiO, collecting holes via the bottom electrode [44]. High-performance PSCs commonly use mesoporous TiO_2 scaffolds to enhance light absorption and device performance. Perovskite compositions evolved from methylammonium lead iodide ($MAPbX_3$), a highly efficient light absorber with limited stability and tunability [14, 45]. Researches created perovskite with mixed cation and halide ions [46, 47], including formamidinium (FA^+), caesium (Cs^+), and bromide (Br^-) or chloride (Cl^-). These adjustments improved thermal and environmental stability, optimized bandgaps and increased crystallinity. As a result, PCE topped 20% and devices demonstrated lower hysteresis and better long-term operating stability [48]. Fabrication processes like as spin-coating and vapor deposition are critical for producing efficient PSCs. Spin-coating is commonly utilized for creating compact TiO_2 blocking layers, mesoporous TiO_2 scaffolds, perovskite absorbers, and HTMs [49]. It is a low-cost and simple approach for precisely controlling thickness and homogeneity by altering factors such as spin speed and solution concentration. This method guarantees full pore filling in mesoporous materials, reducing leakage and improving charge transfer [50]. In contrast, vapor deposition processes like vacuum thermal evaporation provide fine control over film thickness and homogeneity, resulting in high-purity layers with clean interfaces. These strategies assist to decrease flaws, improve crystallinity, and increase device stability [51]. Procedures like as annealing and antisolvent treatments help to improve grain development and film structure.

The use of perovskite in solar cells began in 2006 using $CH_3NH_3PbBr_3$ as a sensitizer in liquid DSSCs. However, the devices demonstrated low efficiency and instability due to perovskite dissolution in liquid electrolytes [14]. Although efficiency increased marginally in 2009, volatility remained as issue [8]. In 2012, liquid electrolytes were substituted with solid-state HTMs considerably improving stability and performance [40]. The most extensively used HTM, Spiro-OMeTAD, initially had restricted charge mobility. Researchers boosted its characteristics by using cobalt-based dopants and ionic additions, which increased efficiency but occasionally reduced stability [52]. Since then, the emphasis has switched to identifying more efficient and stable HTMs. Inspired by organic PVs, novel *p*-type materials like as PTAA, FDT, and triazatruxene have demonstrated promising results, providing both high efficiency and increased stability [53], making solid HTMs a significant step in PSC construction.

11.5 Addressing challenges

PSCs have made impressive efficiency gains but stability, environmental safety and scalability remain problems. One of the key problems is PSC's long-term durability, as perovskite materials are susceptible to moisture, oxygen, heat, and UV radiation, resulting in active layer deterioration and lower device performance [54, 55]. Ion migration, phase segregation or disintegration of the perovskite structure are all possible modes of degradation. The usage of lead-based perovskites raises environmental and health problems. Lead-free alternatives, such as tin (Sn), bismuth (Bi), and antimony (Sb)-based perovskite are being developed but their efficiency and stability remain inferior to lead based equivalents [56]. Additionally, many PSCs display hysteresis in current-voltage characteristics. Interfacial flaws, ion migration, and charge buildup all contribute to device dependability via hysteresis [57]. The scalability and repeatability pose considerable challenges to commercialization. While laboratory scale systems work well, scaling up while keeping constant film quality and efficiency is difficult [58]. Variations in manufacturing techniques, ambient conditions and material purity all have an impact on repeatability. Addressing these difficulties is critical for converting PSCs from laboratory research to reliable, environmentally acceptable, and economically viable solar technology.

11.6 Recent developments

PSCs have gained a notable amount of attention over the past decade. The first PSC was reported in 2009 with a PCE of 3.8% [59]. Since then, until 2012, not many improvements were made for perovskite-sensitized solar cells. After that, continuous efforts to improve efficiency and stability by doping, optimizing material compositions, modifying structure, surface passivation, enhanced encapsulation, incorporating additives, using different fabrication techniques, optimizing metal electrodes, creating interfacial bilayers, and so on, are being considered [60]. The main components of PSCs are substrate, TCO, ETL, perovskite absorber layer, HTL, metal electrodes, and advanced layers like passivation and dipole layers. TCO is the front electrode that lets light in while swiftly collecting charges. Commonly used TCO are Indium tin oxide (ITO) and fluorine-doped tin oxide (FTO), In 2019, Zirconium-doped indium oxide electrodes were employed in four-terminal perovskite/silicon tandem cells, replacing tin-doped indium oxide, achieving a PCE of 26.2% [61]. In the year 2021, Jen et al. utilized a narrow bandgap organic heterojunction to enhance the light absorption range and to improve the short circuit current, elevating the PCE to 21.73% [62]. Fully printed flexible PSCs achieved 17.0% efficiency using a dipole interlayer to reduce recombination losses [63]. Figure 11.3 shows the evolution of different PSCs since 2021.

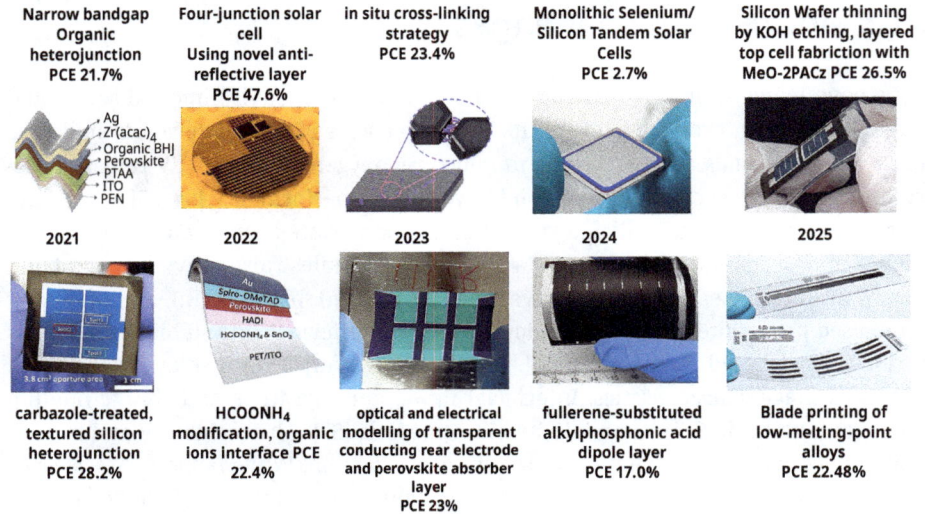

Figure 11.3: Evolution of different PSCs over the years.

In 2023, triple-halide perovskite silicon tandem with piperazinium iodide was reported to improve band alignment and charge extraction, reducing recombination, achieving 32.5% certified efficiency [64]. Table 11.1 depicts the comparison and key findings of different PSCs over the years.

In 2024, First monolithic selenium/silicon tandem solar cells were fabricated using ZnMgO, achieving 1.68 V open-circuit voltage but limited by electron transport barriers. Replacing ZnMgO with TiO_2 improved charge flow, increasing efficiency tenfold to 2.7% [70]. Grain boundary healing and cross-linkable layers: By passivating grain boundaries, recombination is minimized, and charge transport is improved, leading to higher efficiency in PSCs. Cross-linkable layers strengthen the stability and longevity of perovskite films, ensuring they perform reliably over time. An indoor perovskite solar cell efficiency of 42% under 3,000 K LED at 1,000 lux was achieved for a through improved defect passivation [73]. In 2022, a quadruple-junction solar cell

Table 11.1: Comparison and key findings of different PSCs over the years.

S. no.	Key findings	Year	PCE (%)	Working conditions	References
1.	Efficient, and mechanically durable PSCs by utilizing low-bandgap organic bulk-heterojunction	2021	21.7	1,000 h illumination (1-sun) inert atmosphere	[62]
2.	Excellent transmittance and conductivity by Cr seed incorporation	2021	28.3	Optical field simulation	[65]
3.	Improved phase stability of perovskite/silicon tandem solar cell	2021	28.2	Outdoor environment, 100% RH (250 h), 1-sun illumination, 85/85 damp-heat test for 500 h	[66]
4.	Reduced defects by guanidinium cation substitution	2022	17.23	–	[67]
5.	Improved stability and efficiency by interface engineering	2023	32.5	–	[64].
6.	Outperformed mono-facial counterparts, enhanced energy yield	2023	23	Bifacial illumination	[68]
	Suppressed hole accumulation and low charge recombination loss by sub-nanometer dipole interfaces in hybrid perovskite/organic solar cells	2024	24.0	–	[69]
7.	Ten-fold increase in efficiency for monolithic selenium/silicon tandem solar cells	2024	2.7	SCAPS-1D simulations	[70].
8.	Improved energy alignment by tin oxide surface modification	2024	17.0	1,200 h illumination (1-sun), nitrogen atmosphere	[63]
9.	Improved long-term stability by dual-interface passivation of inverted f-PSCs	2025	21.58	Indoor light, 3,000 h in N_2 atmosphere	[71]
10.	Fast charge extraction, higher stability	2025	22.48	Retention of 80% of the initial PCE after 1,500 h aging.	[72]

reached an efficiency of 47.6% using a novel anti-reflective layer under concentrated sunlight (665 suns) [74]. In 2021, Jiang et al. achieved 28.2% efficiency in four-terminal perovskite/silicon tandem cells. As of December, efficiency for perovskite/silicon tandems reached 29.8% [75]. Efficiencies have made significant leaps over the years, still, long-term stability remains a key issue as PSCs face serious degradation when exposed to external factors like moisture, air, water, light, heat, mechanical stress, chemical reactions and electric field. Several strategies like optimizing metal electrodes, controlling internal stress, incorporating additives, creating interfacial bilayers, integrating 2D perovskite layers, and compositional engineering have yielded encouraging outcomes in enhancing stability and performance [76]. Bilayers between charge transport and perovskite layers further enhance stability. For instance, 1,2-EDT (ethanedithiol) coatings improve moisture resistance, with minimal degradation under humid conditions [77].With the increasing demand for sustainable and adaptable energy solutions, all-perovskite tandem cells with customized bandgaps and flexible PSCs (F-PSCs) are leading the way in innovation, promising not only to enhance the efficiency of solar energy conversion but also to open new avenues for integration into lightweight wearable devices, lightweight, printable alternatives, portable electronics, building-integrated PVs, healthcare offering noninvasive, real-time data collection features, automotive applications, and even in space related applications, where weight and efficiency are crucial [78, 79].

11.6.1 Tandem solar cells (perovskite–silicon and perovskite–perovskite)

Perovskite-based tandem solar cells, particularly those combining perovskites with crystalline silicon (c-Si), have advanced well beyond the laboratory, surpassing 29% efficiency [80]. Tandem solar cells work by stacking multiple PV layers, with the top cell absorbing high-energy photons and allowing lower-energy ones to reach a bottom cell optimized for infrared absorption as could be seen in Figure 11.4. The two main configurations, 2 T and 4 T, offer different benefits: 2 T integrates well into monolithic structures, while 4 T provides more flexibility for current matching.

Several innovations have driven efficiency improvements, such as precise tuning of the perovskite bandgap, reducing interfacial recombination, and enhancing charge transport layers. Perovskite materials like $Cs_{1-x}FA_xPb(I_{1-x}Br_x)_3$, which offer tunable bandgaps, have shown strong performance and enhanced durability [82]. Light absorption has been significantly enhanced by using anti-reflective coatings (ARCs), photonic crystal structures, and plasmonic nanoparticles. MgF_2 were widely used as ARCs and were often modeled using SCAPS-1D, enabling efficiencies up to 21.62% [83]. Han et al. achieved a 22.43% PCE in perovskite/CIGS tandem cells by using a transparent 100 nm ITO top contact and a CMP-treated ITO interlayer to avoid shorting. They also introduced a TPFB-doped PTAA HTL, which enhanced the open-circuit voltage (V_{OC}), fill factor (FF), and short-circuit current (J_{SC}). The device maintained 88% of its efficiency after 500 h of aging, with partial recovery after rest [84]. Tang et al. developed a mixed-solvent passivation strategy for 4 T perovskite/CIGS tandem cell using a 1:1 ratio IPA:toluene solution to apply GABr, effectively reducing surface recombination without damaging the perovskite layer resulting in a high efficiency of 25.5%. In 2024, a novel sub-nanometer dipole interface layer (B3PyMPM) was applied to the perovskite surface, which alleviated the energy barrier between the perovskite and the

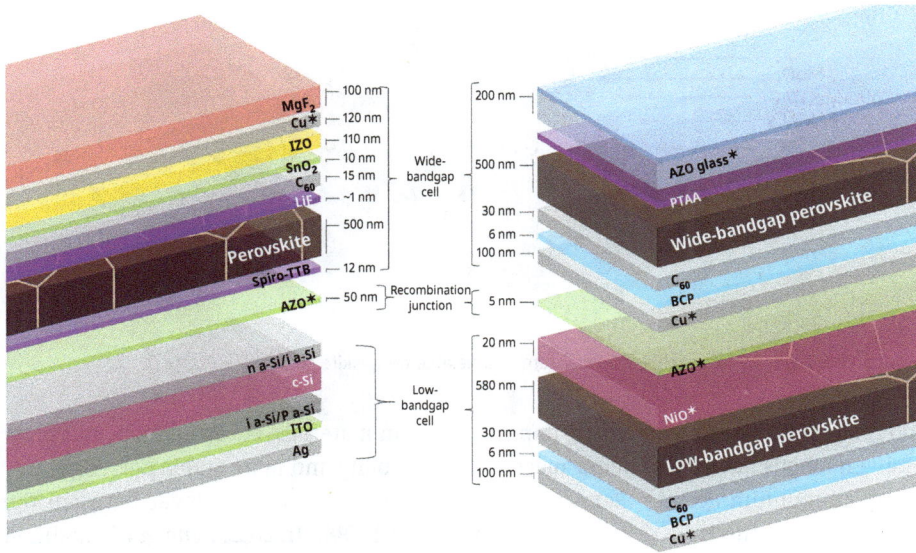

Figure 11.4: Schematic of perovskite–silicon tandem solar cell (left), perovskite–perovskite tandem [81].

bulk heterojunction and suppressed charge accumulation, and enhanced near-infrared absorption, achieving 24% efficiency. By precisely determining the filtered spectrum and current matching points a three-terminal tandem device reached an efficiency of 26.24% [85]. Silicon-perovskite tandem solar cells, with a theoretical efficiency of 43%, exceed the 33.7% Shockley–Queisser limit [86]. Plasmonic nanoparticles such as silver and gold nanospheres have shown promise in enhancing localized surface plasmon resonance (LSPR), which improves light trapping while minimizing recombination losses. Among the various strategies explored, hybrid light-trapping structures that combine ARCs with photonic crystals have demonstrated the most effective approach for optimizing efficiency. These advancements reinforce the viability of PSCs, positioning them as key contenders in the evolution of PV technology.

11.6.2 Flexible and printable perovskite solar cells

Flexible and printable PSCs are redefining how and where solar power can be used, thanks to their lightweight nature and compatibility with bendable substrates. Layer-by-layer schematic diagram of a flexible PSC F-PSC is shown in Figure 11.5. Unlike traditional rigid panels, these devices can be produced using low-temperature, roll-to-roll techniques, making them ideal for wearable tech, portable power, and building-

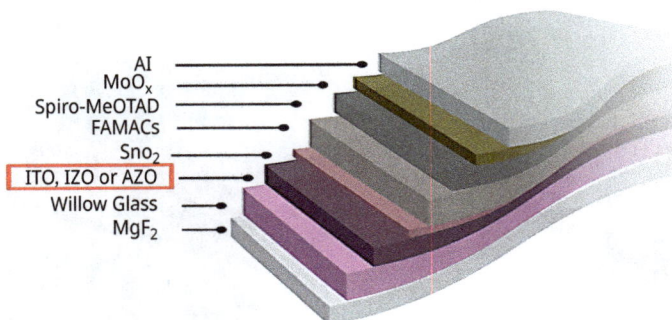

Figure 11.5: Layer-by-layer schematic diagram of a flexible perovskite solar cell F-PSC [87].

integrated PVs. Over recent years, improvements in materials, encapsulation, and processing methods have led to better mechanical durability and more stable power output. Pentylammonium acetate (PenAAc) was employed as a modified layer between the perovskite and the HTL, resulting in a PCE of 23.35% [88]. In 2022, *Zeng et al.* modified the SnO$_2$ ETL with a multifunctional modification material called ammonium citrate (TAC). The device based on TAC-doped ETL attained a higher power conversion efficiency (PCE) of 21.58 from 19.75% [89]. In 2024, Ren *et al.* improved the PCE of f-PSC to 25.09% (24.90% certified) by adding controllable growth of a SnO$_2$ electron transport layer through constant pH chemical bath deposition (CBD), which helped control the growth of SnO$_2$, maintaining 90% of the initial PCE after 10,000 bending cycles [90]. Recently, Shishido et al. [91] have achieved the highest efficiency ever recorded for a flexible perovskite-silicon tandem solar cell that is 26.5% by thinning a silicon wafer via potassium hydroxide etching, followed by silicon nitride's (SiN$_x$) protection film through chemical vapor deposition and wafer dicing. A variety of printing methods, such as blade coating, slot-die coating, spray coating, flexographic printing, gravure printing, screen printing, and inkjet printing, have demonstrated significant effectiveness in controlling film formation and enhancing PCE to exceed 21%. Chen et al. [92] reported high-performance printable CsPbI$_3$ solar cells made by blade-coating. SnO$_2$, CsPbI$_3$, and Spiro-OMeTAD layers in ambient conditions, with an efficiency of 19%. By applying slot–die coating as an active-layer deposition technique for MAPbI$_3$, 20.83% PCE is achieved [93]. Screen-printed flexible perovskite devices show a maximum of 15.89% efficiency [94]. With the help of the inkjet printing deposition approach, flexible perovskite showed the highest PCE of 21.65% for small-area cells. Using the flexographic printing method to fabricate flexible solar cells, a maximum PCE of 15.3% is achieved [95]. Flexible perovskite made by blade-coating technique has shown the highest efficiency of 21.09% [96]. The future of printable PSCs looks promising, with ongoing efforts to boost their efficiency, stability, and scalability for large-scale production. Their flexibility and lightweight design make them perfect for integration into everything from wearable tech to building materials.

11.6.3 Interface engineering and defect passivation strategies:

Enhancing the stability of PSCs requires a multifaceted approach targeting interfacial engineering, strain relief, additive incorporation, and molecular-level passivation strategies. Interfacial bilayers play a dual role by aiding defective passivation and serving as a barrier from outside influences, thereby significantly improving long-term stability. 2D perovskite layers aids in slowing the recombination of charges and making it easier to extract charges effectively [97]. Oh et al. [67] reported 5% guanidinium (GAþ)-substituted $GA_xMA_{1-x}PbI3$ solar cells exhibited a peak efficiency of 17.23%, outperforming both undoped and highly GA-doped counterparts. Incorporation of poly(4-vinylpyridine) into $CH_3NH_3PbI_3$ PSC, to reduce trap states at grain boundaries, along with a 30% enhancement in PCE. Self-assembled monolayers are gaining attention for surface properties, their ability to reduce interfacial recombination and improve efficient charge transfer. Zhang et al. [71] demonstrated that dual-interface passivation using in-situ 2D/3D/2D heterojunctions can maintain 86% of the initial PCE after 10,000 bending cycles, achieving 21.58% efficiency in flexible PSCs and 24.08% for rigid PSCs. With continued material and structural improvements, the gap between lab-scale performance and commercial viability of PSCs is steadily reducing.

11.7 Comparative analysis with other PV technologies

In this section, we compare PSC with other PV technologies to assess their efficiencies, environmental impact and commercial potential. PSCs have demonstrated impressive efficiencies especially in laboratory conditions, thus evaluating them alongside other PV technologies helps us identify their strengths and limitations. While silicon PV is well-established with mature infrastructure, PSCs offer high efficiency with potential for further improvements, low manufacturing costs through solution-based processes and scalable techniques like roll-to-roll printing, and flexibility, making them suitable for applications such as portable electronics and building-integrated PVs, potentially enabling broader application scenarios. PSCs have seen rapid progress, with a record 26.5% efficiency for a flexible perovskite-silicon tandem device [91], which are comparable to silicon solar cells (26.81%) with the added benefit of fine-tuning of properties for different operating conditions and applications. CIGS and CdTe trail slightly, achieving a PCE of 23.4% and 21.0%, respectively [98]. Due to thin, solution-processed layers, PSCs, CdTe, and CIGS require fewer raw materials than silicon solar cells, reducing resource strain and production energy. PSCs offer a high EROI potential, unlike OPVs, with limited energy returns, due to their lower performance. From a manufacturing standpoint, PSCs have low-temperature processing (below 150 °C), low energy demands, and compatibility with scalable techniques like roll-to-roll printing.

In contrast, silicon requires high-temperature processing (above 1,000 °C), and CIGS and CdTe rely on vacuum-based or intermediate-temperature methods. PSCs remain vulnerable to moisture, oxygen, and UV radiation, which accelerate degradation, and the presence of lead complicates disposal and recycling [99]. On the other hand, silicon shows high environmental resilience, with well-established existing recycling infrastructure. CIGS and CdTe involve hazardous elements like cadmium and selenium that require strict handling. PSCs also exhibit excellent light absorption with thin layers and a tunable bandgap between 1.2 and 2.3 eV, enabling tailored applications such as tandems. Silicon's fixed 1.1 eV bandgap limits its flexibility, while CIGS and CdTe offer narrower or fixed ranges, making them somewhat flexible [100].

11.7.1 Potential for commercial deployment

Over the last few decades, different PV technologies have transitioned from laboratory to commercial applications. PSCs have a relatively brief history compared to other established PV technologies; however, they are developing at a much faster pace [101]. Silicon solar cells have long dominated the market, but their limited flexibility, brittle nature, and poor resistance to stress restrict their use in many futuristic applications, one such is full-scale flexible PV modules. CdTe and CIGS exhibit flexibility, but suffer from low PCE, toxicity, and high cost [100]. PSCs, on the other hand, are one of the most promising technologies due to their characteristics, such as high PCE, under the same bandgap settings, low-temperature fabrication, flexibility, adaptability, and cost-effectiveness. While PSCs have made impressive strides in small-scale devices, replicating that success in a large-scale application remains a tough hurdle. Stability, scalability, sustainability, and storage are must to keep in mind for their successful commercialization. Enhancing the durability of PSCs can be effectively achieved using 2D/3D and double perovskite structures, mesoporous carbon-based architectures, polymer interfacial layers, and organic composite materials [102]. Considering tandem solar cells like perovskite-silicon can further boost efficiencies and offer a competitive edge over conventional PVs. Addressing lead toxicity and environmental concerns is essential for wider acceptance of PSCs. To meet cost and scalability demands, lab-scale fabrication should be performed repeatedly under real-world conditions. In 2024, Weerasinghe *et al.* reported flexible perovskite modules prepared by a full roll-to-roll process at ambient temperature with an efficiency of 11%, exhibiting great potential in large-scale f-PSCs module production [103]. With continued progress in improving stability, scalability, and environmental safety, their successful integration into the PV industry is increasingly attainable.

11.8 Future outlook

The future of PSCs appears highly promising, as researchers continue to advance their efficiency and scalability [104]. Single-junction PSCs in laboratory settings have already achieved efficiencies exceeding 26%, while tandem configurations, especially perovskite-silicon tandems, have surpassed 34%, approaching the theoretical Shockley–Queisser limit. These impressive efficiencies, combined with the potential for cost-effective, low-temperature solution processing, make PSCs strong contenders for commercial PV applications [105]. However, their commercialization depends on overcoming ongoing challenges, particularly in terms of long-term operational stability and the use of toxic lead. Recent efforts to address these issues include the creation of durable encapsulation methods, innovative interface engineering techniques, and the exploration of alternative materials like tin-based perovskites to reduce toxicity [106, 107]. Moreover, new designs such as multi-layer ETLs and the incorporation of ionic salts have shown significant improvements in device stability and efficiency, with some inverted PSCs achieving 26% efficiency while retaining over 98% of their performance after 2,100 h at elevated temperatures [108]. On a global scale, PSCs are anticipated to play a crucial role in meeting renewable energy targets due to their lightweight, flexible nature and suitability for building-integrated PVs. Countries like Japan and Germany are already investing in scalable perovskite technologies to enhance national energy resilience and meet climate objectives [109]. With continued interdisciplinary efforts, PSCs are expected to complement and, in some instances, surpass traditional silicon technologies, making a substantial contribution to the global transition to clean energy [110].

11.9 Conclusion

PSCs have advanced rapidly, rising from 3.8% efficiency in 2009 to more than 26% today and even higher in tandem designs. Their lightweight nature, flexible structure, and low cost, low-temperature production distinguish them from existing solar technologies. PSCs are highly interesting for emerging applications like as wearable electronics, building-integrated systems, and perhaps space exploration. Researches addressed significant difficulties like as instability and lead content by improving materials, interface designs and encapsulation approaches. Flexible, roll-to-roll manufacturing has also moved them closer to mass production. While concerns persist, particularly long-term stability and environmental safety, continued research and worldwide investment are driving consistent progress. PSCs are well on their way to becoming a significant element of the world's journey towards clean and sustainable energy, given their combination of high efficiency, adaptability, and the possibility for cost-effective manufacture.

References

[1]   Pandey SV, et al. The circuitry landscape of perovskite solar cells: An in-depth analysis. Journal of Energy Chemistry, 2024; 94: 393–413.

[2]   Kahandal, S.S., et al., *Perovskite solar cells: fundamental aspects, stability challenges, and future prospects*. Progress in Solid State Chemistry, 2024. 74: p. 100463.

[3]   Bhat I, Prakash R. LCA of renewable energy for electricity generation systems—a review, Renewable and Sustainable Energy Reviews. 2009; 13(5): 1067–1073.

[4]   Kumar S, et al. Applications of solid waste-derived carbon nanomaterials in solar cell, In: Waste Derived Carbon Nanomaterials. 2025; ACS Publications, 2: 123–144.

[5]   Panagoda L, et al. Advancements in photovoltaic (Pv) technology for solar energy generation, Journal of Research Technology & Engineering. 2023; 4(30): 30–72.

[6]   Soonmin H, et al. Overview on different types of solar cells: An update, Applied Sciences. 2023; 13(4): 2051.

[7]   Afroz M, et al. Perovskite Solar Cells: Progress, Challenges, and Future Avenues to Clean Energy. 2025; Solar Energy, vol. 287: 113205.

[8]   Kojima A, et al. Organometal halide perovskites as visible-light sensitizers for photovoltaic cells, Journal of the American Chemical Society. 2009; 131(17): 6050–6051.

[9]   Kumar S, et al. Surface modification of carbon nanofiber with C20H38O11 polymer by spun calcination method, Journal of Inorganic and Organometallic Polymers and Materials. 2024; 34(1): 336–345.

[10]   Gouitaa N, et al. Effect of Cl substitution on the microstructure, dielectric and optical properties of CaCu3Ti4O12 ceramics. Results in Physics, 2025; 68: 108072.

[11]   Ball JM, et al. Low-temperature processed meso-superstructured to thin-film perovskite solar cells, Energy & Environmental Science. 2013; 6(6): 1739–1743.

[12]   Mitzi DB, Chondroudis K, Kagan CR. Organic-inorganic electronics, IBM Journal of Research and Development. 2001; 45(1): 29–45.

[13]   Kumari, N., et al., *Environmentally sustainable techniques for rGO synthesis: focus on spun calcination and clean technology advances*. Journal of Inorganic and Organometallic Polymers and Materials, 2024. 35: p. 1–25.

[14]   Calió L, et al. Hole-transport materials for perovskite solar cells, Angewandte Chemie International Edition. 2016; 55(47): 14522–14545.

[15]   Kim H-S, et al. Lead iodide perovskite sensitized all-solid-state submicron thin film mesoscopic solar cell with efficiency exceeding 9%, Scientific Reports. 2012; 2(1): 591.

[16]   Heo JH, Im SH. Highly reproducible, efficient hysteresis-less CH 3 NH 3 PbI 3− x Cl x planar hybrid solar cells without requiring heat-treatment, Nanoscale. 2016; 8(5): 2554–2560.

[17]   Alberola-Borràs J-A, et al. Relative Impacts of Methylammonium Lead Triiodide Perovskite Solar Cells Based on Life Cycle Assessment. 2018; Solar Energy Materials and Solar Cells, vol. 179: 169–177.

[18]   Liu Z, et al. Efficient and stable FA-rich perovskite photovoltaics: From material properties to device optimization, Advanced Energy Materials. 2022; 12(18): 2200111.

[19]   Wang R, et al. Prospects for metal halide perovskite-based tandem solar cells, Nature Photonics. 2021; 15(6): 411–425.

[20]   Shah AUI, Meyer EL. Perovskite-based Solar Cells in Photovoltaics for Commercial Scalability: Current Progress, Challenges, Mitigations and Future Prospectus. 2025; Solar Energy, 286: 113172.

[21]   Kumar, S., et al., *Interface engineering of composite systems: Focusing on the compatibility of reduced graphene oxide and Bi0. 8La0. 1Ba0. 1Fe0. 9Ti0. 1O3 hybrid systems*. Materials Today Sustainability, 2024. 27: p. 100813.

[22]   Faini F, et al. Hybrid halide perovskites, a game changer for future solar energy?, MRS Bulletin. 2024; 49(10): 1059–1069.

[23] Le H. Elaboration of New Hole Transporting Materials for Hybrid Perovskite Solar Cells. 2018; Université de Cergy Pontoise.

[24] Brenner TM, et al. Hybrid organic—inorganic perovskites: Low-cost semiconductors with intriguing charge-transport properties, Nature Reviews Materials. 2016; 1(1): 1–16.

[25] Zhou, D., et al., *Perovskite-based solar cells: materials, methods, and future perspectives*. Journal of Nanomaterials, 2018(1): p. 8148072.1–15.

[26] Shi Z, Jayatissa AH. Perovskites-based solar cells: A review of recent progress, materials and processing methods, Materials. 2018; 11(5): 729.

[27] Whitfield P, et al. Structures, phase transitions and tricritical behavior of the hybrid perovskite methyl ammonium lead iodide, Scientific Reports. 2016; 6(1): 35685.

[28] Korshunova K, et al. Thermodynamic stability of mixed Pb: Sn methyl-ammonium halide perovskites, Physica Status Solidi (B). 2016; 253(10): 1907–1915.

[29] Wu S, et al. The evolution and future of metal halide perovskite-based optoelectronic devices, Matter. 2021; 4(12): 3814–3834.

[30] Mitzi DB. Introduction: Perovskites. 2019; ACS Publications, 3033–3035.

[31] Im J-H, et al. 6.5% efficient perovskite quantum-dot-sensitized solar cell, Nanoscale. 2011; 3(10): 4088–4093.

[32] Burschka J, et al. Sequential deposition as a route to high-performance perovskite-sensitized solar cells, Nature. 2013; 499(7458): 316–319.

[33] Im J-H, et al. Growth of CH3NH3PbI3 cuboids with controlled size for high-efficiency perovskite solar cells, Nature Nanotechnology. 2014; 9(11): 927–932.

[34] Cai Q, et al. Stable perovskite solar cells with 22% efficiency enabled by inhibiting migration/loss of iodide ions, Physical Chemistry Chemical Physics. 2023; 25(9): 6955–6962.

[35] Bai Y, et al. Enhancing efficiency and stability of inverted perovskite solar cells through synergistic suppression of multiple defects via poly (ionic liquid)-buried interface modification. Journal of Materials Science & Technology, 2025; 212: 281–288.

[36] Roy P, et al. A review on perovskite solar cells: Evolution of architecture, fabrication techniques, commercialization issues and status. Solar Energy, 2020; 198: 665–688.

[37] Ahmad, W., et al., *Revolutionizing Photovoltaics: From Back-Contact Silicon to Back-Contact Perovskite Solar Cells*. Materials Today Electronics, 2024. 9: p. 100106.

[38] Chowdhury TA, et al. Stability of perovskite solar cells: Issues and prospects, RSC Advances. 2023; 13(3): 1787–1810.

[39] Fu Q, et al. Recent progress on the long-term stability of perovskite solar cells, Advanced Science. 2018; 5(5): 1700387.

[40] Lee MM, et al. Efficient hybrid solar cells based on meso-superstructured organometal halide perovskites, Science. 2012; 338(6107): 643–647.

[41] Chen Q, et al. Planar heterojunction perovskite solar cells via vapor-assisted solution process, Journal of the American Chemical Society. 2014; 136(2): 622–625.

[42] Etgar L, et al. Mesoscopic CH3NH3PbI3/TiO2 heterojunction solar cells, Journal of the American Chemical Society. 2012; 134(42): 17396–17399.

[43] Jiang Q, et al. Planar-structure perovskite solar cells with efficiency beyond 21%, Advanced Materials. 2017; 29(46): 1703852.

[44] Luo D, et al. Enhanced photovoltage for inverted planar heterojunction perovskite solar cells, Science. 2018; 360(6396): 1442–1446.

[45] Li Z, et al. Scalable fabrication of perovskite solar cells, Nature Reviews Materials. 2018; 3(4): 1–20.

[46] Ono LK, Juarez-Perez EJ, Qi Y. Progress on perovskite materials and solar cells with mixed cations and halide anions, ACS Applied Materials & Interfaces. 2017; 9(36): 30197–30246.

[47] Xu F, et al. Mixed cation hybrid lead halide perovskites with enhanced performance and stability, Journal of Materials Chemistry A. 2017; 5(23): 11450–11461.

[48]   Alta F, Asu E. National Renewable Energy Labs (NREL) efficiency chart, 2019.

[49]   Heo JH, Song DH, Im SH. Planar CH3NH3PbBr3 hybrid solar cells with 10.4% power conversion efficiency, fabricated by controlled crystallization in the spin-coating process, Advanced Materials. 2014; 26(48): 8179–8183.

[50]   Liu X, et al. Boosting the efficiency of carbon-based planar CsPbBr3 perovskite solar cells by a modified multistep spin-coating technique and interface engineering. Nano Energy, 2019; 56: 184–195.

[51]   Leyden MR, et al. High performance perovskite solar cells by hybrid chemical vapor deposition, Journal of Materials Chemistry A. 2014; 2(44): 18742–18745.

[52]   Burschka J, et al. Tris (2-(1 H-pyrazol-1-yl) pyridine) cobalt (III) as p-type dopant for organic semiconductors and its application in highly efficient solid-state dye-sensitized solar cells, Journal of the American Chemical Society. 2011; 133(45): 18042–18045.

[53]   Li S, et al. A brief review of hole transporting materials commonly used in perovskite solar cells, Rare Metals. 2021; 40(10): 2712–2729.

[54]   Ball JM, et al. Optical properties and limiting photocurrent of thin-film perovskite solar cells, Energy & Environmental Science. 2015; 8(2): 602–609.

[55]   Sha, W.E., et al., The efficiency limit of CH3NH3PbI3 perovskite solar cells. Applied Physics Letters, 2015. 106(22) p. 221104. 1–14.

[56]   Ke W, Kanatzidis MG. Prospects for low-toxicity lead-free perovskite solar cells, Nature Communications. 2019; 10(1): 965.

[57]   Kang DH, Park NG. On the current–voltage hysteresis in perovskite solar cells: Dependence on perovskite composition and methods to remove hysteresis, Advanced Materials. 2019; 31(34): 1805214.

[58]   Correa-Baena J-P, et al. Promises and challenges of perovskite solar cells, Science. 2017; 358(6364): 739–744.

[59]   Zhang Q, et al. Advancing all-perovskite two-terminal tandem solar cells: optimization of wide-and narrow-bandgap perovskites and interconnecting layers. 2025; Energy & Environmental Science.

[60]   Afre RA, Pugliese D. Perovskite solar cells: A review of the latest advances in materials, fabrication techniques, and stability enhancement strategies, Micromachines. 2024; 15(2): 192.

[61]   Aydin E, et al. Zr-doped indium oxide (IZRO) transparent electrodes for perovskite-based tandem solar cells, Advanced Functional Materials. 2019; 29(25): 1901741.

[62]   Wu S, et al. Low-bandgap organic bulk-heterojunction enabled efficient and flexible perovskite solar cells, Advanced Materials. 2021; 33(51): 2105539.

[63]   Dong L, et al. Fully printed flexible perovskite solar modules with improved energy alignment by tin oxide surface modification, Energy & Environmental Science. 2024; 17(19): 7097–7106.

[64]   Mariotti S, et al. Interface engineering for high-performance, triple-halide perovskite–silicon tandem solar cells, Science. 2023; 381(6653): 63–69.

[65]   Yang D, et al. 28.3%-efficiency perovskite/silicon tandem solar cell by optimal transparent electrode for high efficient semitransparent top cell. Nano Energy, 2021; 84: 105934.

[66]   Liu J, et al. 28.2%-efficient, outdoor-stable perovskite/silicon tandem solar cell, Joule. 2021; 5(12): 3169–3186.

[67]   Oh J, et al. Guanidinium cation substitution effects on perovskite solar cells, Applied Science and Convergence Technology. 2022; 31(6): 161–163.

[68]   Jiang Q, et al. Highly efficient bifacial single-junction perovskite solar cells, Joule. 2023; 7(7): 1543–1555.

[69]   Lee MH, et al. Suppressing hole accumulation through sub-nanometer dipole interfaces in hybrid perovskite/organic solar cells for boosting near-infrared photon harvesting, Advanced Materials. 2024; 36(47): 2411015.

[70]   Nielsen R, et al. Monolithic selenium/silicon tandem solar cells, PRX Energy. 2024; 3(1): 013013.

[71]   Zhang H-W, et al. Dual-interface passivation to improve the efficiency and stability of inverted flexible perovskite solar cells by in-situ constructing 2D/3D/2D perovskite double heterojunctions. Journal of Power Sources, 2025; 629: 235973.

[72]   Jiang B, et al. Blade Printing of Low-melting-point Alloys as Back Electrodes for High-efficiency and Stable Inverted Perovskite Solar Cells. 2025; Energy & Environmental Science.

[73]   Shi ZE, et al. Enhanced indoor perovskite solar cells: Mitigating interface defects and charge transport losses with polyarene-based hole-selective layers, Advanced Energy Materials. 2025; 15(6): 2404234.

[74]   Corentin, J., et al. A novel parallel interconnection approach to reduce shading losses on submillimeter concentrated photovoltaic technologies. in 2024 IEEE 74th Electronic Components and Technology Conference (ECTC). 2024. IEEE. 2205–2210

[75]   Tockhorn P, et al. Nano-optical designs for high-efficiency monolithic perovskite–silicon tandem solar cells, Nature Nanotechnology. 2022; 17(11): 1214–1221.

[76]   Qin J, et al. Towards operation-stabilizing perovskite solar cells: Fundamental materials, device designs, and commercial applications, InfoMat. 2024; 6(4): e12522.

[77]   Kim, H.J., G.S. Han, and H.S. Jung, *Managing the lifecycle of perovskite solar cells: Addressing stability and environmental concerns from utilization to end-of-life.* eScience, 2024. 4(2): p. 100243.

[78]   Niranjan D, et al. Current status and applications of photovoltaic technology in wearable sensors: A review. Frontiers in Nanotechnology, 2023; 5: 1268931.

[79]   Guk K, et al. Evolution of wearable devices with real-time disease monitoring for personalized healthcare, Nanomaterials. 2019; 9(6): 813.

[80]   Khan F, et al. Perovskite-based tandem solar cells: Device architecture, stability, and economic perspectives. Renewable and Sustainable Energy Reviews, 2022; 165: 112553.

[81]   Tian X, Stranks SD, You F. Life cycle energy use and environmental implications of high-performance perovskite tandem solar cells, Science Advances. 2020; 6(31): eabb0055.

[82]   Liu X, et al. Over 28% efficiency perovskite/Cu (InGa) Se 2 tandem solar cells: Highly efficient sub-cells and their bandgap matching, Energy & Environmental Science. 2023; 16(11): 5029–5042.

[83]   Dipta, S.S., A. Uddin, and G. Conibeer, Enhanced light management and optimization of perovskite solar cells incorporating wavelength dependent reflectance modeling. Heliyon, 2022. 8(11). 1–12.

[84]   Han Q, et al. High-performance perovskite/Cu (In, Ga) Se2 monolithic tandem solar cells, Science. 2018; 361(6405): 904–908.

[85]   Shrivastav N, et al. Design and simulation of three-junction all perovskite tandem solar cells: A path to enhanced photovoltaic performance. Materials Letters, 2024; 362: 136169.

[86]   Jiang Y, et al. Infrared PbS quantum dot–lead halide perovskite combinations for breaking the Shockley–Queisser Limit, Solar RRL. 2025; 9(1): 2400743.

[87]   Jung HS, et al. Flexible perovskite solar cells, Joule. 2019; 3(8): 1850–1880.

[88]   Liu J, et al. Recoverable flexible perovskite solar cells for next-generation portable power sources, Angewandte Chemie. 2023; 135(40): e202307225.

[89]   Park SY, Zhu K. Advances in SnO2 for efficient and stable n–i–p perovskite solar cells, Advanced Materials. 2022; 34(27): 2110438.

[90]   Qiu J, Shi Y, Zhang F. The rise of flexible perovskite photovoltaics, Device. 2024; 2(5): 100371.

[91]   Shishido H, et al. High-Efficiency Perovskite/Silicon Tandem Solar Cells with Flexibility. 2025; Solar RRL, 202400899.

[92]   Chen C, et al. Perovskite solar cells based on screen-printed thin films, Nature. 2022; 612(7939): 266–271.

[93]   Li J, et al. 20.8% slot-die coated MAPbI3 perovskite solar cells by optimal DMSO-content and age of 2-ME based precursor inks, Advanced Energy Materials. 2021; 11(10): 2003460.

[94]   Chen C, et al. Screen-printing technology for scale manufacturing of perovskite solar cells, Advanced Science. 2023; 10(28): 2303992.

[95]   Huddy JE, Ye Y, Scheideler WJ. Eliminating the perovskite solar cell manufacturing bottleneck via high-speed flexography, Advanced Materials Technologies. 2022; 7(7): 2101282.

[96]   Lee K-M, et al. High-performance perovskite solar cells based on dopant-free hole-transporting material fabricated by a thermal-assisted blade-coating method with efficiency exceeding 21%. Chemical Engineering Journal, 2022; 427: 131609.

[97]   Wu G, et al. Surface passivation using 2D perovskites toward efficient and stable perovskite solar cells, Advanced Materials. 2022; 34(8): 2105635.

[98]   Teknetzi I. Environmentally friendly approaches for recycling of CIGS solar cells, 2025.

[99]   Dou J, Bai Y, Chen Q. Challenges of lead leakage in perovskite solar cells, Materials Chemistry Frontiers. 2022; 6(19): 2779–2789.

[100]   Matsui T, et al. Progress and Limitations of Thin-film Silicon Solar Cells. 2018; Solar Energy, vol. 170: 486–498.

[101]   Lee SW, et al. Historical analysis of high-efficiency, large-area solar cells: Toward upscaling of perovskite solar cells, Advanced Materials. 2020; 32(51): 2002202.

[102]   Hadadian M, Smått J-H, Correa-Baena J-P. The role of carbon-based materials in enhancing the stability of perovskite solar cells, Energy & Environmental Science. 2020; 13(5): 1377–1407.

[103]   Weerasinghe, H.C., et al., *The first demonstration of entirely roll-to-roll fabricated perovskite solar cell modules under ambient room conditions*. Nature Communications, 2024. 15(1): p. 165 1–12.

[104]   Park NG. Research direction toward scalable, stable, and high efficiency perovskite solar cells, Advanced Energy Materials. 2020; 10(13): 1903106.

[105]   Jošt M, et al. Monolithic perovskite tandem solar cells: A review of the present status and advanced characterization methods toward 30% efficiency, Advanced Energy Materials. 2020; 10(26): 1904102.

[106]   Li N, et al. Towards commercialization: The operational stability of perovskite solar cells, Chemical Society Reviews. 2020; 49(22): 8235–8286.

[107]   Zhu T, Yang Y, Gong X. Recent advancements and challenges for low-toxicity perovskite materials, ACS Applied Materials & Interfaces. 2020; 12(24): 26776–26811.

[108]   Yin Y, et al. Recent advances of two-dimensional material additives in hybrid perovskite solar cells, Nanotechnology. 2023; 34(17): 172001.

[109]   Tan X, Li Y. Innovations and challenges in semi-transparent perovskite solar cells: A mini review of advancements toward sustainable energy solutions, Journal of Composites Science. 2024; 8(11): 458.

[110]   Xie G, Li H, Qiu L. Recent advances on monolithic perovskite-organic tandem solar cells, Interdisciplinary Materials. 2024; 3(1): 113–132.

Pravin Kumar Singh*, Yash Pandey, and Upendra Kulshrestha*

Chapter 12
Importance of perovskite solar cells in sustainable energy solutions

Abstract: Perovskite solar cells (PSCs) have emerged as a transformative technology in the quest for sustainable energy, offering high efficiencies, low production costs, and versatile applications. With efficiencies exceeding 26% and tandem configurations approaching 34%, PSCs rival traditional silicon photovoltaics (PV) while promising lower levelized costs of electricity (LCOE). Their unique optoelectronic properties, including tunable bandgaps and high absorption coefficients, enable applications in building-integrated PVs (BIPV), agrivoltaics, and off-grid solutions. However, challenges such as environmental stability, lead toxicity, and scalability must be addressed to achieve commercial viability. This report explores the fundamentals, advantages, sustainability, and commercialization pathways of PSCs, highlighting their potential to complement existing technologies and drive the global energy transition. Emerging concepts like lead-free perovskites, novel transport layers, and AI-driven material discovery, alongside policy and investment needs, are examined to outline a roadmap for responsible innovation. By 2035, PSCs could play a pivotal role in achieving net-zero goals, provided stability, environmental concerns, and manufacturing hurdles are overcome through collaborative efforts.

Keywords: Perovskite solar cells, sustainable energy, efficiency, stability, lead-free

12.1 Introduction

The accelerating climate emergency marked by unprecedented heatwaves, biodiversity collapse, and intensifying natural disasters underscores a non-negotiable truth: humanity must transition from fossil fuels to sustainable energy systems within decades. The energy sector remains the dominant contributor to global greenhouse gas emissions, responsible for 76% of total CO_2 releases [1]. Current trajectories project a catastrophic 2.7 °C temperature rise by 2100 [2], demanding immediate, systemic action. Simultaneously, 789 million people lack electricity access [3], primarily in sub-Saharan Africa

*Corresponding author: **Pravin Kumar Singh,** Institute of Advanced Materials, IAAM, Gammalkilsvagen 18, Ulrika 59 053, Sweden
*Corresponding author: **Upendra Kulshrestha,** Department of Mechanical Engineering, Manipal University Jaipur, Jaipur 303007, India
Yash Pandey, Department of Electrical and Electronic Engineering, Graduate School of Integrated Science and Technology, National University Corporation Shizuoka University, Shizuoka, Japan.

https://doi.org/10.1515/9783111726847-012

and South Asia, where fossil dependency exacerbates energy poverty and health inequities.

The dual crisis of climate change and energy security demands innovative solutions that go beyond mere carbon neutrality, requiring energy systems that are both environmentally sustainable and practically viable. These solutions must be massively scalable to accommodate the over 30% annual energy demand growth in developing economies, ensuring equitable global access to clean power. Cost-effectiveness is equally critical, with levelized energy costs needing to fall below $0.03/kWh to remain competitive with fossil fuels while enabling widespread adoption. Furthermore, geopolitical resilience has become non-negotiable in light of recent global energy shocks; next-generation energy systems must minimize supply chain vulnerabilities through diversified material sourcing and localized manufacturing capabilities. This multifaceted approach – combining scalability, affordability, and supply chain security – represents the essential trifecta for energy technologies capable of addressing both the climate emergency and global energy instability simultaneously. The challenge lies in developing systems that satisfy all three criteria without compromise, creating infrastructure that is as robust geopolitically as it is sustainable environmentally [1].

Renewables now constitute 30% of global electricity generation, yet fossil fuels still dominate primary energy (82%). The "carbon budget" for 1.5 °C compliance – just 380 Gt CO_2 – will be exhausted by 2030 without radical intervention. Solar photovoltaics (PV), with its theoretically inexhaustible resource and plummeting costs, emerges as the linchpin of decarbonization. However, current deployment rates must triple by 2030 [4] to align with net-zero targets, demanding technological innovations that transcend silicon PV's limitations.

Solar PV has achieved staggering growth: cumulative capacity surged from 40 GW in 2010 to 1.6 TW in 2024, driven by learning-curve cost reductions of 24.3% per doubling of capacity [5]. This expansion avoided 1.2 Gt CO_2 emissions annually – equivalent to Japan's total carbon footprint. Crystalline silicon (c-Si) dominates (>95% market share), but faces fundamental constraints mentioned in Table 12.1.

Table 12.1: Limitations and impact on sustainability.

Limitation	Impact on sustainability
Shockley–Queisser efficiency limit (29.4%)	Caps energy yield per land area
Energy-intensive manufacturing (1,500 °C polysilicon purification)	6–24-month energy payback times in temperate zones
Rigid, opaque form factor	Excludes integration in buildings, vehicles, and agriculture
Silver/indium dependencies	Supply risks: 15% PV demand could consume 85% of silver reserves by 2040

These constraints impede PV's ability to meet projected 63,000 TWh annual demand by 2050 [6]. Land-use conflicts already arise – deploying 70 TW of c-Si PV (required for full decarbonization) would consume ≈1.8% of global land area. Additionally, c-Si's weight (>20 kg/m^2) limits rooftop deployment in developing regions with structural constraints.

Perovskite solar cells (PSCs) represent a transformative leap in PV technology, offering multiple disruptive advantages that address critical limitations of conventional solar solutions shown in Table 12.2. With theoretical efficiency potential reaching 43% in advanced quadruple-junction configurations, PSCs could eventually outperform all existing solar technologies. Their thin-film flexibility unlocks installation opportunities on curved or irregular surfaces, potentially accessing 70% of African rooftops currently unsuitable for rigid c-Si panels. The semitransparent variants enable innovative applications like greenhouse-integrated agrivoltaics, tripling land-use efficiency by combining energy production with agricultural activities. Perhaps most significantly, PSCs require just 500 kWh/m^2 of embodied energy during manufacturing – a fivefold reduction compared to traditional silicon panels – which could slash energy payback periods to under 0.3 years. This combination of ultrahigh efficiency potential, unprecedented installation flexibility, and dramatically lower production energy demands positions PSCs as a uniquely capable technology for accelerating global emissions mitigation while expanding solar energy accessibility worldwide. The perovskite revolution began modestly in 2009 when Miyasaka et al. achieved 3.8% efficiency using methylammonium lead iodide in dye-sensitized cells. Within 15 years, certified efficiencies skyrocketed to 26.1% (single-junction) and 33.9% (perovskite/silicon tandem) – the fastest advancement in PV history. This trajectory dwarfs silicon's 60-year crawl from 6% to 26.7%.

Table 12.2: Optoelectronic properties comparison (AM1.5G spectrum).

Property	c-Si	CIGS	PSC
Absorption coefficient	10^3 cm^{-1}	10^5 cm^{-1}	>10^5 cm^{-1}
Bandgap tunability	Fixed 1.1 eV	Limited	1.2–2.3 eV
Diffusion length	100–300 μm	1–5 μm	10–100 μm
Defect tolerance	Low	Medium	High

These properties enable PSCs to achieve near-theoretical efficiencies with solution-processed, polycrystalline films – impossible for silicon. PSCs offer significant economic and manufacturing advantages over traditional c-Si PVs shown in Figure 12.1(a). With material costs as low as $0.03/W compared to c-Si's $0.10/W (NREL), PSCs drastically reduce production expenses. Their low-temperature processing (<150 °C) enables compatibility with roll-to-roll (R2R) printing, facilitating high-throughput, low-cost manufacturing. Additionally, perovskite absorbers are 100× thinner (500 nm vs. 180 μm for c-Si), minimizing material consumption. Another key benefit is tandem integration with silicon, which can boost existing factory output by 50% without requiring infrastructure replacement, allowing seamless upgrades to higher-efficiency modules.

Beyond cost savings, PSCs enable transformative applications across multiple sectors (Figure 12.1(b)). In BIPV, semitransparent perovskite skylights achieve 12% efficiency while transmitting 40% visible light, blending energy generation with architectural aesthetics [7]. For vehicle-integrated PV, lightweight, flexible PSCs can extend EV range by 20 km/day (Hanergy), enhancing sustainability in transportation. In agrivoltaics, transparent PSCs with 23% efficiency have been shown in Dutch trials to maintain 85% crop yield, demonstrating their potential to combine solar energy production with agriculture without compromising land use. These innovations highlight perovskite technology's versatility in driving the next generation of solar energy solutions.

Despite lead toxicity concerns, recent life-cycle analyses show PSCs' operational emissions offset lead risks 100:1 when replacing coal power. Stability breakthroughs ($\geq$30-year extrapolated lifetimes for encapsulated devices) now position PSCs for commercialization.

This chapter delivers a multidisciplinary assessment of PSCs as catalysts for sustainable energy transitions, bridging materials science, industrial ecology, and energy policy. Our primary objective is to critically evaluate PSCs' viability in overcoming systemic barriers to global decarbonization through five focused dimensions: First, we analyze materials innovation and scalability, examining solution engineering (e.g., slot-die coating) versus vapor deposition for gigawatt-scale production, supported by technoeconomic modeling targeting module manufacturing costs below $15/m^2$. Second, we conduct rigorous sustainability quantification via ISO-compliant life cycle assessments (LCAs), calculating energy payback time (EPBT), carbon footprint, and water usage across diverse manufacturing scenarios to benchmark environmental performance against incumbent PVs. Third, we map deployment roadmaps for utility-scale (perovskite-silicon tandems), urban (BIPV façades), and off-grid applications (portable chargers), incorporating LCOE sensitivity analyses under 2030–2040 cost projections. Fourth, we address circular economy integration, designing closed-loop recycling protocols for >99% lead recovery and assessing economic feasibility of end-of-life management. Finally, we propose policy frameworks for stability certification (extending IEC 61215 standards) and lead-handling regulations, comparing incentives across the EU, China, and the USA. Methodologically, this work combines device physics modeling (SCAPS-1D), degradation kinetics studies, SimaPro-based LCAs, and Bass diffusion market forecasting to resolve the "green paradox," ensuring PSC deployment at terawatt scales avoids creating new environmental burdens. The chapter concludes with a viability index ranking PSCs against eight sustainability criteria, offering actionable pathways for policymakers, manufacturers, and researchers to accelerate their role in a net-zero future.

(a)

Low-Temperature Processing

Low-temperature processing offers efficiency at lower costs.

Tandem Integration

Tandem integration enhances efficiency despite higher costs.

Thin Absorbers

Thin absorbers reduce costs but compromise efficiency.

High Material Costs

High material costs lead to lower efficiency.

(b)

BIPV

Semi-transparent perovskite skylights achieve good efficiency. They also transmit visible light.

Vehicle-Integrated PV

Lightweight flexible PSCs extend EV range. This increases the distance the vehicle can travel.

Agrivoltaics

Transparent PSCs maintain crop yield. They also achieve good efficiency in trials.

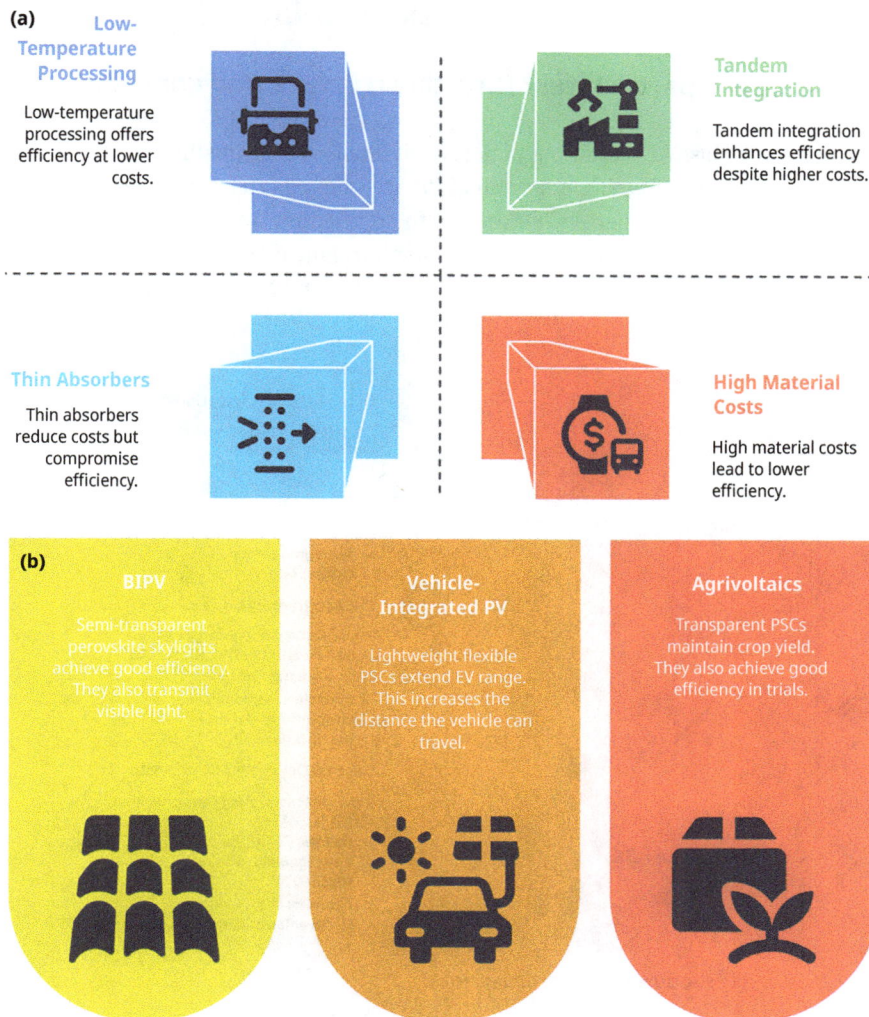

Figure 12.1: (a) Key economic and manufacturing factors influencing perovskite solar cell (PSC) adoption: Low-temperature processing enables cost-effective production, while thin absorbers reduce material use but may impact efficiency. Tandem integration boosts efficiency despite added costs, whereas high material costs can hinder performance. **(b) Transformative applications of PSCs:** Building-integrated PV (BIPV) uses semitransparent perovskite skylights to balance efficiency and light transmission. Vehicle-integrated PV leverages lightweight flexible PSCs to extend EV range. Agrivoltaics employs transparent PSCs to maintain crop yields while generating solar energy.

12.2 Fundamentals of perovskite solar cells

12.2.1 What are perovskites? (Crystal structure and composition)

Perovskites are a class of crystalline materials defined by their distinctive **ABX$_3$ stoichiometry** and **corner-sharing octahedral lattice**, named after the mineral calcium titanate (CaTiO$_3$). In PVs, "perovskite" refers to compounds adopting this structural motif, where the A-site is occupied by monovalent organic cations, the B-site hosts divalent metal cations, and the X-site consists of oxides or halide anions as shown in Figure 12.2.

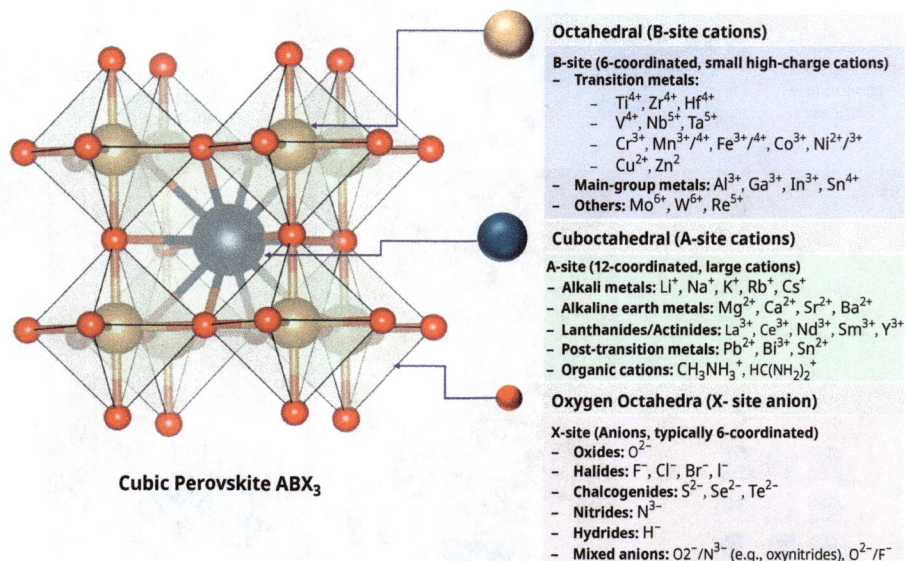

Octahedral (B-site cations)

B-site (6-coordinated, small high-charge cations)
- Transition metals:
    - Ti^{4+}, Zr^{4+}, Hf^{4+}
    - V^{4+}, Nb^{5+}, Ta^{5+}
    - Cr^{3+}, Mn$^{3+/4+}$, Fe$^{3+/4+}$, Co^{3+}, Ni$^{2+/3+}$
    - Cu^{2+}, Zn^2
- Main-group metals: Al^{3+}, Ga^{3+}, In^{3+}, Sn^{4+}
- Others: Mo^{6+}, W^{6+}, Re^{5+}

Cuboctahedral (A-site cations)

A-site (12-coordinated, large cations)
- Alkali metals: Li$^+$, Na$^+$, K$^+$, Rb$^+$, Cs$^+$
- Alkaline earth metals: Mg^{2+}, Ca^{2+}, Sr^{2+}, Ba^{2+}
- Lanthanides/Actinides: La^{3+}, Ce^{3+}, Nd^{3+}, Sm^{3+}, Y^{3+}
- Post-transition metals: Pb^{2+}, Bi^{3+}, Sn^{2+}
- Organic cations: CH$_3$NH$_3^+$, HC(NH$_2$)$_2^+$

Oxygen Octahedra (X- site anion)

X-site (Anions, typically 6-coordinated)
- Oxides: O^{2-}
- Halides: F$^-$, Cl$^-$, Br$^-$, I$^-$
- Chalcogenides: S^{2-}, Se^{2-}, Te^{2-}
- Nitrides: N^{3-}
- Hydrides: H$^-$
- Mixed anions: O2$^-$/N^{3-} (e.g., oxynitrides), O^{2-}/F$^-$

Cubic Perovskite ABX$_3$

Figure 12.2: Perovskite structure (ABX$_3$) site occupancies.

The stability and symmetry of the perovskite lattice are governed by the Goldschmidt tolerance factor ($t = (r_A + r_X)/[\sqrt{2}\,(r_B + r_X)]$), where r denotes ionic radii. For optimal PV performance, *t* must range between 0.8 and 1.0 to maintain the cubic/tetragonal phase; deviations cause detrimental non-perovskite polymorphs (e.g., δ-CsPbI$_3$). Key compositions like methylammonium lead iodide (CH$_3$NH$_3$PbI$_3$) exhibit direct bandgaps (1.55 eV), high absorption coefficients (>10^5 cm^{-1}), and defect-tolerant electronic structures – enabled by Pb^{2+} 6s^2 and I$^-$ 5p orbital hybridization that minimizes trap states. Bandgaps can be tuned from 1.2 eV (CsPbI$_3$) to 2.3 eV (MAPbBr$_3$) through halide mixing or A-site alloying, while lead toxicity concerns are addressed via encapsulation or Sn^{2+}/ double-perovskite substitutions (e.g., Cs$_2$AgBiBr$_6$). This unique structural flexibility underpins perovskites' rapid rise as high-efficiency, solution-processable PV materials.

12.2.2 Working principle of PSCs

PSCs operate through a sequence of optoelectronic processes initiated when photons exceeding the material's bandgap energy (typically 1.5–1.6 eV for MAPbI$_3$) strike the active layer. The perovskite's exceptionally high absorption coefficient (>10^5 cm^{-1} at 550 nm) enables near-complete photon capture within ≈500 nm – a 100× thinner layer than silicon requires. Absorbed photons generate electron-hole pairs (excitons) with ultralow binding energy (<10 meV), allowing spontaneous charge separation at room temperature without external fields. The perovskite's intrinsic ambipolar charge transport facilitates this process: photogenerated electrons transition to the conduction band (formed by Pb^{2+} 6p orbitals), while holes occupy the valence band (I$^-$ 5p orbitals), with both carriers exhibiting long diffusion lengths (>1 μm) due to shallow defect states. Charge extraction occurs at selective contacts: electrons are collected by the electron transport layer (ETL) (e.g., TiO$_2$ and SnO$_2$) through energy level alignment (perovskite CBM ≈ −3.9 eV; SnO$_2$ CBM ≈ −4.3 eV), while holes transfer to the hole transport layer (HTL) (e.g., spiro-OMeTAD and PTAA) via the perovskite VBM (−5.4 eV) to HTL HOMO (−5.2 eV). This directional flow generates photocurrent, with minimal voltage loss (<0.4 V) from the theoretical maximum due to suppressed recombination. Critical to performance is interface engineering – reducing energy offsets and trap densities – which maximizes open-circuit voltage (V_{OC}) and fill factor (FF), though ion migration can cause hysteresis in current–voltage characteristics.

12.2.3 Key device architectures

PSCs deploy four principal architectures optimized for performance, stability, and application as shown in Figure 12.3:
(a) **n–i–p (regular)**: Light enters through the transparent conductive oxide (TCO) substrate, followed by an ETL (e.g., SnO$_2$, TiO$_2$), perovskite absorber, HTL (e.g., spiro-OMeTAD), and metal electrode (Au/Ag). This configuration achieves >25% efficiency but faces HTL instability.
(b) **p–i–n (inverted)**: Reverse stacking (TCO/HTL/perovskite/ETL/metal) uses stable organic HTLs (e.g., PTAA and NiO$_x$), enabling low-temperature processing (<150 °C) and reduced hysteresis. Critical for tandem integration and flexible modules.
(c) **Flexible**: Enabled by p–i–n architecture on plastic substrates (PET/PEN), using low-temperature ETLs (PCBM) and robust perovskites (e.g., Cs$_{0.17}$FA$_{0.83}$Pb(I$_{0.83}$Br$_{0.17}$)$_3$). Achieves 23.4% efficiency (23 cm^2, 2024) for wearable/curved surfaces.
(d) **Tandem**: Combines perovskite top-cell (E_g ≈ 1.7 eV) with silicon/bottom-cell (E_g ≈1.1 eV) to surpass single-junction limits. Monolithic designs (perovskite/silicon: 33.9% efficiency, KAUST 2024) use recombination layers (SnO$_x$/polySi), while mechanically stacked versions enable independent optimization.

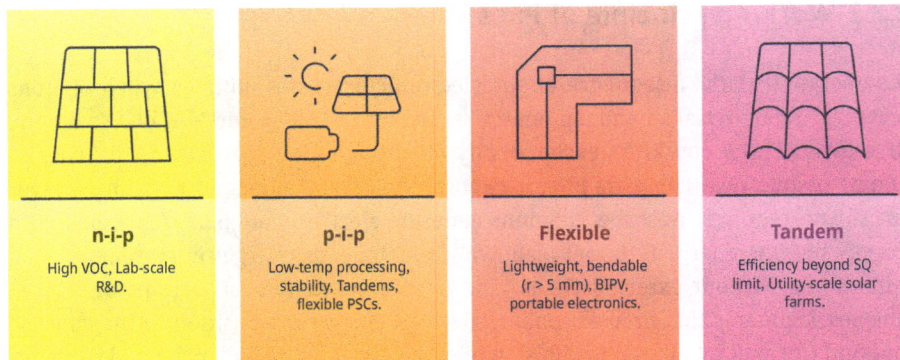

Figure 12.3: Perovskite solar cells principal architectures optimized for performance, stability, and application.

12.2.4 Common materials and fabrication techniques

PSC fabrication utilizes solution-processable materials and scalable deposition techniques to enable cost-effective manufacturing. Figure 12.4 shows the key components, including absorber layers, charge transport layers, and electrodes. Hybrid perovskites such as MAPbI$_3$ and FAPbI$_3$ serve as efficient light absorbers, synthesized from low-cost precursors like PbI$_2$, MAI, and FAI dissolved in polar solvents such as DMF and DMSO. Advanced formulations incorporating SnO$_2$ or mixed-cation perovskites (e.g., Cs$_x$FA$_{1-x}$PbI$_{3-y}$Br$_y$) further enhance stability and performance.

For charge transport, ETLs typically employ metal oxides like TiO$_2$, SnO$_2$, or ZnO, deposited via scalable methods such as spray pyrolysis or atomic layer deposition (ALD). Meanwhile, HTLs often consist of organic small molecules (e.g., spiro-OMeTAD) or inorganic alternatives (e.g., NiO$_x$ and CuSCN), applied through spin-coating or other solution-based techniques. TCOs such as ITO and FTO serve as front electrodes, while rear contacts may use evaporated Au/Ag or printed carbon for improved stability and cost efficiency.

Scalable fabrication methods are critical for commercial viability. Spin-coating with antisolvent-assisted deposition (e.g., chlorobenzene drip) remains a lab-scale technique for achieving high efficiencies (>25%). However, large-area compatibility is achieved through blade coating, slot-die coating, and R2R processing, with reported efficiencies exceeding 21% on 100 cm^2 substrates. Vapor-phase deposition, including coevaporation of PbI$_2$ and MAI, enables pinhole-free films suitable for tandem solar cells, with module efficiencies reaching 28%. Additionally, inkjet printing offers precise, patternable deposition with minimal thickness variation (<5%), making it ideal for BIPV.

To address environmental concerns, researchers are developing green solvent alternatives (e.g., γ-valerolactone) and lead-encapsulation strategies (e.g., polymer coatings) without compromising manufacturing throughput. These advancements position PSCs as a promising technology for next-generation PVs, combining high efficiency, low cost, and scalable production.

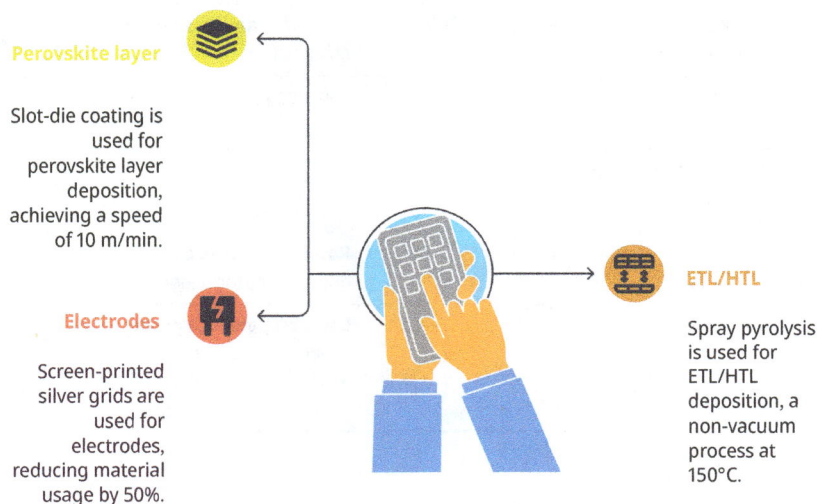

Perovskite layer

Slot-die coating is used for perovskite layer deposition, achieving a speed of 10 m/min.

Electrodes

Screen-printed silver grids are used for electrodes, reducing material usage by 50%.

ETL/HTL

Spray pyrolysis is used for ETL/HTL deposition, a non-vacuum process at 150°C.

Figure 12.4: Perovskite solar cell fabrication and materials.

12.3.1 Exceptional optoelectronic properties

Perovskite semiconductors achieve record PV efficiencies through its synergistic optoelectronic properties shown in Table 12.3. Their ultrahigh absorption coefficient ($>10^5$ cm^{-1} across visible spectra) enables 99% photon capture in 300–500 nm films – 100× thinner than silicon – reducing material costs while permitting flexible/semitransparent designs. Precisely tunable bandgaps (1.2–2.3 eV) via halide alloying (I$^-$/Br$^-$ substitution) or A-site engineering (Cs$^+$/FA$^+$ mixtures) allow spectral matching for single-junction optimization (1.55 eV) and perovskite-silicon tandems (1.7 eV top + 1.1 eV bottom). Most critically, ambipolar carrier diffusion lengths (>10 μm) – unprecedented for solution-processed semiconductors – result from intrinsic defect tolerance: Shallow trap states (<0.1 eV formation energy) and low exciton binding energies (<10 meV) suppress nonradiative losses, enabling >95% charge collection at 1 μm thickness. This triad delivers high short-circuit currents (>26 mA/cm^2) and open-circuit voltages (>90% of radiative limits), powering certified efficiencies >26% in single-junction cells and 33.9% in tandems – all achievable with low-temperature processing (<150 °C).

Table 12.3: Perovskite versus competing PV materials: optoelectronic benchmark [8–10].

Property	Perovskite (e.g., CsFAPbI$_x$Br$_{3-x}$)	c-Si	CIGS	Implication for PSCs
Absorption coefficient	>10^5 cm^{-1} (at 550 nm)	~10^3 cm^{-1}	10^5 cm^{-1}	**100× thinner layers** → Material savings, flexibility
Bandgap tunability	1.2–2.3 eV (continuous)	Fixed 1.1 eV	1.0–1.7 eV	**Ideal for tandems** → 43% theoretical efficiency
Diffusion length	Ambipolar >10 μm	e⁻:		100–300 μm
e⁻: 1–5 μm	**Tolerance to defects** → High VOC despite polycrystallinity			
Exciton binding energy	<10 meV	15 meV	5–30 meV	**Room-temp free carriers** → No donor-acceptor needed
Defect tolerance	Shallow traps (0.01–0.1 eV)	Deep traps	Deep traps	**Low-temperature processing** → 50% lower EPBT versus c-Si
Highest certified PCE	26.1% (single)/33.9% (tandem)	26.7%	23.6%	**Rapid commercial viability**

12.3.2 Rapid efficiency progress

PSCs have achieved the fastest efficiency evolution in PV history, soaring from 3.8% in 2009 to 26.1% in 2024 for single-junction devices – a trajectory dwarfing silicon's 60-year climb to 26.7% (Table 12.4). This 6.8× improvement in 15 years, documented in NREL's Best Research-Cell Efficiency Chart [8], stems from synergistic breakthroughs: interface engineering (2012–2015, +15% absolute gain), compositional stabilization (2015–2018, FA/Cs alloys), and defect passivation (2020–present, >25% milestone). Tandem cells progressed even faster – perovskite/silicon devices leaped from 13.7% (2015) to 33.9% (2024), surpassing silicon's theoretical limit (29.4%) within 9 years. Commercial modules now reach 24.2% efficiency [11], demonstrating rapid lab-to-fab translation.

12.3.3 Potential for low-cost manufacturing

PSCs exhibit transformative potential for ultra-low-cost PV manufacturing, anchored by three synergistic advantages (Table 12.5). First, their reliance on abundant raw materials – such as lead iodide (PbI$_2$, <\$5/kg) and organic ammonium salts (MAI/FAI, <\$20/kg) – dramatically reduces precursor expenses compared to silicon-grade polysilicon (\$15–20/kg). Material consumption is slashed 100-fold (0.3 g/W vs. c-Si's 30 g/W) due to submicron

Table 12.4: Efficiency milestones versus competing technologies (NREL data).

Technology	Starting year	Initial eff.	2024 record	Years to 25%	Key innovations enabling progress
Perovskite (single)	2009	3.8%	26.1%	12	HTL/ETL optimization (2012), FA$^+$ alloying (2014), and 2D/3D passivation (2018)
Perovskite/ Si tandem	2015	13.7%	33.9%	7	Wide-bandgap perovskites (2017), recombination layers (2019), and textured Si light management (2023)
Crystalline silicon	1954	6%	26.7%	60	PERC (2010) and TOPCon (2018)
CIGS	1981	8%	23.6%	32	Na doping (1990) and graded bandgaps (2010)
CdTe	1972	6%	22.3%	40	CdCl$_2$ treatment (1991) and Se alloying (2015)

Source: NREL chart analysis (2009–2024).

absorber thicknesses, while silver electrode replacement with carbon/graphite mitigates supply-chain vulnerabilities. Second, intrinsic solution-processability enables high-throughput deposition via slot-die coating (10 m/min) and inkjet printing (6 m/min), allowing R2R production on flexible substrates with minimal efficiency loss (21.4% on 100 cm^2 modules). This scalability bypasses the energy-intensive sawing and texturing required for silicon wafers. Third, low-temperature processing (<150 °C) – con­trasting sharply with c-Si's >900 °C furnace requirements – reduces factory energy intensity by 80% (0.35 vs. 1.2 kWh/m^2) and enables plastic-compatible fabrication (PET/PEN), cutting capital expenditure by 60% ($120M/GW vs. $300M/GW for silicon lines. When integrated, these advantages project module manufacturing costs of $0.10–0.15/W by 2030 (NREL), 50% below incumbent silicon, while achieving EPBTs under 0.3 years. Though encapsulation and lead sequestration add $0.05–0.07/W, perovskite technology remains the first PV platform capable of sub-$0.20/W production without sacrificing >25% efficiencies – a critical enabler for terawatt-scale sustainable energy deployment [8, 12]. Table 12.6 shows the key scalable techniques adopted by different countries.

12.3.4 Versatility in form and function

PSCs redefine PV application paradigms through unprecedented functional versatility, enabling deployment scenarios inaccessible to conventional technologies (Table 12.7). Their lightweight (<500 g/m^2) and flexible nature – achieved via low-temperature proc­essing on polymer substrates (PET/PEN) – permits bending radii below 5 mm while re­taining >90% efficiency after 10,000 cycles, unlocking integration with curved surfaces (vehicle roofs and backpacks) and lightweight structures (drone wings and inflatable

Table 12.5: Cost reduction drivers versus silicon PV.

Parameter	Perovskite solar cells	Crystalline silicon	Cost advantage
Raw material cost	$0.03/W ($PbI_2$ + salts)	$0.10/W (polysilicon + Ag)	67% reduction
Material consumption	0.3 g/W (500 nm film)	30 g/W (180 μm wafer)	100× less material
Deposition throughput	10 m/min (slot-die)	1.5 wafers/min (DW sawing)	50× faster production
Processing temperature	100–150 °C (plastic compatible)	>900 °C (quartz crucible)	6× energy savings
Capex (GW factory)	$120M (solution processing)	$300M (Si ingot → cell)	60% lower investment
Embodied energy	500 kWh/m^2	2,500 kWh/m^2	80% reduction
Projected module Cost (2030)	$0.10–0.15/W	$0.20–0.25/W	50% cheaper

Table 12.6: Key scalable techniques.

Process	Throughput	Efficiency (module)	Industry adoption
Slot-die coating	10 m/min	21.4% (100 cm^2)	Saule Tech, Tandem PV
Inkjet printing	6 m/min	19.1% (400 cm^2)	Toshiba, HP development
Vapor-assisted solution	3 m/min	24.7% (tandem mini-mod)	Swift Solar (tandems)
Spray pyrolysis (ETL)	20 m^2/min	20.2% (submodule)	First Solar integration

shelters). Simultaneously, precise bandgap tuning facilitates semitransparency (15–40% visible light transmission) with minimal efficiency penalty (12–18% for neutral-hue modules), revolutionizing BIPVs) through energy-generating windows and facades that maintain architectural aesthetics while producing >100 W/m^2. Critically, perovskites' tandem compatibility allows monolithic stacking with silicon (33.9% efficiency), CIGS (24.2%), or even perovskite-perovskite cells (28% all-perovskite), overcoming single-junction limits by harvesting broadened spectral ranges without significant manufacturing complexity. This functional triad – flexibility for ubiquitous integration, transparency for dual-surface utility, and tandem readiness for ultrahigh efficiency – positions perovskites as a transformative "universal solar converter" capable of serving markets from utility-scale farms to urban infrastructure and portable electronics [13, 14].

Table 12.7: Quantified impact of versatility.

Feature	Technical metric	Application impact
Flexibility	0.3 mm thickness, bend radius <5 mm	87% of vehicle roofs become PV-suitable
Semitransparency	18% efficiency at 30% VLT (CIE color index < 5)	BIPV market growth to $36B by 2030
Tandem integration	1.7 eV top-cell + 1.1 eV bottom-cell spectral match	50% power gain over silicon at <10% cost premium

12.4.1 Enhancing solar energy deployment

PSCs are poised to dramatically accelerate global PV adoption by reducing LCOE to unprecedented lows – projected at $0.015/kWh in high-irradiation regions by 2030 [15], 40% below current silicon PV – through ultralow manufacturing costs ($0.10–0.15/W) and enhanced performance (tandem efficiencies > 33%). This economic advantage could expand solar deployment into emerging economies, where PSCs' lightweight flexibility enables installation on structurally limited rooftops (70% of buildings in Global South [16]). Concurrently, PSCs unlock transformative applications: BIPV leverages semitransparency (30% visible light transmission at 18% efficiency [17]; to turn skyscrapers into power generators; agrivoltaics uses wavelength-selective modules (23% efficiency with 85% crop yield retention [18]; to dual-use farmland; and wearable electronics embed bendable cells (500 mW/g power density [19]) for off-grid energy autonomy. Collectively, these innovations expand PV's addressable market by >300 GW/year by 2030 [20], overcoming silicon's rigidity, opacity, and weight constraints while accelerating decarbonization timelines.

12.4.2 Complementing and enhancing existing technologies

PSCs uniquely augment established PV technologies rather than displacing them, driving systemic efficiency and economic gains (Table 12.8). Perovskite-silicon tandems leverage silicon's mature infrastructure while overcoming its 29.4% efficiency limit – certified devices now reach 33.9% (KAUST 2024) by stacking a 1.7 eV perovskite top-cell on silicon, capturing high-energy photons without modifying existing production lines (Oxford PV adds perovskite layers to PERC cells at <5% cost premium). All-perovskite tandems stack narrow-bandgap (1.2 eV Sn-Pb) and wide-bandgap (1.8 eV Br-I) subcells for 28% efficiency in flexible formats (Swift Solar), enabling 400 W/kg power densities unattainable for rigid technologies. Compared to established PV, perovskites enhance competitiveness: they achieve higher efficiencies than CIGS (23.6%) or CdTe (22.3%) at 60% lower capex, while their solution-processability enables bifacial gains (>30% yield boost) impossible for vapor-deposited thin films. Critically, PSCs synergize with silicon's reliability via encapsulation-sharing tandems and with CIGS' flexibility via monolithically stacked ultrathin devices – collectively accelerating PV's trajectory toward 40% module efficiency and $0.02/kWh LCOE [21].

Table 12.8: Perovskite synergies with established PV technologies.

Technology integration	Efficiency record	Key advantage	Commercial status	Synergy potential
Perovskite-silicon tandem	33.9% (KAUST, 1.2 cm^2)	+7% absolute efficiency gain versus Si	Mass production (Oxford PV, 2025)	Uses existing Si fabs; 26% → 34% with same footprint
All-perovskite tandem	28.0% (UNIST, 1 cm^2)	23%-efficient flexible modules (0.3 mm thick)	Pilot line (Swift Solar)	Enables roll-to-roll production for <$0.10/W
Perovskite-CIGS tandem	24.2% (NREL, 1 cm^2)	Lightweight (2.4 kg/m^2) and bendable (r = 10 mm)	Prototype (MiaSolé)	Combines perovskite UV absorption with CIGS NIR response
Versus monocrystalline Si	26.7% (LONGi)	50% lower capex; 100× thinner absorber	Market dominant (>95% share)	Tandemization extends Si dominance
Versus CIGS	23.6% (Solar Frontier)	Higher VOC (1.25 V vs. 0.75 V); no indium scarcity	Niche (BIPV)	Monolithic stacking on flexible substrates
Versus CdTe	22.3% (First Solar)	Bandgap tunability avoids CdTe's fixed 1.45 eV limit	Utility-scale leader	Higher efficiency potential in humid climates

12.4.3 Potential for decentralized and off-grid energy solutions

PSCs are uniquely positioned to address energy poverty and grid vulnerability through decentralized generation, leveraging their lightweight flexibility, low-light performance, and ultra-low manufacturing costs. Table 12.9 shows the key applications and verified impacts. Portable PSC chargers deliver 200 Wh/day at 420 g weight – enabling off-grid power for rural healthcare and disaster relief. Urban resilience is enhanced via building-integrated systems: Seoul's Transparent Perovskite Windows (18% efficiency, 30% VLT) generate 85 kWh/m^2/year while reducing cooling loads. Crucially, PSCs enable localized manufacturing – R2R printing factories require <$50M capex versus silicon's $300M+, allowing production in containerized microfactories [22]. This democratizes energy access: perovskite minigrids in India achieve $0.08/

kWh versus diesel's \$0.38/kWh [23], while agrivoltaic kits increase farmer incomes by 40% via dual land use [18]. With 1.2 billion people lacking grid access [24], PSCs' combination of affordability (\$0.15/W projected), deployability (48-h installation), and scalability (500 GW/year potential by 2035 [25]) positions them as a catalyst for equitable energy transitions.

Table 12.9: Key applications and verified impacts.

Solution type	Perovskite innovation	Impact	Reference
Portable power	23%-efficient flexible chargers (0.3 mm thick)	90% cost reduction versus diesel for off-grid clinics	[26]
BIPV microgrids	Semitransparent façades (15% eff., 40% VLT)	30% reduced grid dependence in commercial buildings	[27]
Agrivoltaics	Wavelength-selective modules (22% eff.)	85% crop yield + 68 MWh/ha/year energy	[28]
Local manufacturing	Roll-to-printed modules (\$0.12/W CAPEX)	90% lower factory footprint versus Si fabs	[29]

12.5.1 Energy payback time (EPBT) estimates

Perovskite PV systems achieve industry-leading EPBTs due to ultralow embodied energy requirements. Solution-processed modules require just 0.35 kWh/m^2 during manufacturing – 86% less than monocrystalline silicon (2.5 kWh/m^2) – primarily by eliminating polysilicon purification (>1,500 °C) and diamond-wire sawing. This enables EPBTs of **0.23 years** in Phoenix, AZ (2,300 kWh/m^2/year irradiation), compared to 1.1 years for PERC silicon modules.

EPBT variability depends critically on location and technology maturity. In temperate zones (e.g., Germany, 1,200 kWh/m^2/year), current perovskite EPBTs average 0.48 years but could reach 0.31 years with scaled production. Tandem configurations add <10% energy burden while boosting output 30%, creating a favorable energy amplification factor of 3.7× versus silicon. Table 12.10 shows EPBT comparison across PV technologies.

12.5.2 Carbon footprint analysis

Perovskite modules exhibit carbon footprints of 14–18 g CO_2-eq/kWh – 65% lower than silicon PV – through synergistic reductions in process energy and materials. Low-temperature processing (<150 °C) avoids natural gas-fired furnace emissions, while thin-film deposition eliminates silicon ingot sawing waste. A 2023 global assessment

Table 12.10: EPBT comparison across PV technologies.

Technology	Desert (year)	Temperate (year)	Embodied energy (kWh/m^2)	Key reduction strategies
Perovskite (R2R)	0.23	0.48	350	Solvent recovery (95%), no vacuum deposition
Perovskite-Si tandem	0.28	0.58	1,900	Shared encapsulation, existing Si infrastructure
Monocrystalline Si	0.87	1.82	2,500	Thinner wafers (100 µm), diamond wire optimization
CdTe	0.52	1.09	1,300	Vapor transport deposition optimization

Source: Fraunhofer ISE LCA Report 2024.

showed that perovskite carbon intensity could reach 8 g CO_2-eq/kWh by 2030 through renewable-powered factories and solvent recycling.

The carbon payback period for perovskite manufacturing is <4 months when displacing coal power. However, indium in transparent electrodes contributes 58% of emissions, prompting development of ITO-free alternatives. Graphene electrodes reduce footprint by 37%, while silver replacement with copper lowers emissions by 29 g CO_2-eq/m^2. Table 12.11 details the carbon footprint of PSC component.

Table 12.11: Carbon footprint drivers (cradle-to-gate).

Component	Perovskite (g CO_2-eq/m^2)	% total	Reduction strategy
Transparent electrode	11.2	58%	Graphene grids (86% reduction)
Encapsulation	6.3	32%	Polymer-silica hybrids
Perovskite layer	0.9	5%	Lead recycling
Transport layers	1.1	5%	Organic ETL elimination

12.5.3 Resource availability and material criticality

Perovskites utilize noncritical materials: Lead reserves (88 million tonnes) could support >100 TW of PV deployment at 0.3 g/W consumption [30]. Iodine – though geologically scarce – is 99.6% recoverable from seawater at $35/kg [31]. Silver represents the only critical material in conventional designs, but carbon electrodes eliminate this dependency entirely [32]. Material substitution potential exceeds 98% for supply-chain resilience. Formamidinium can be synthesized from ammonia + formic acid (petrochemical byproducts), while tin/zinc offer lead alternatives. Geographic diversification is feasible: 79% of raw materials are available in >20 countries, reducing geopolitical risks compared to silicon's China-dominated supply chain [33]. The material criticality assessment is shown in Table 12.12.

Table 12.12: Material criticality assessment.

Material	Perovskite use (g/m^2)	Global reserves	Supply risk index	Alternatives
Pb	0.6	88 MT (300 year)	Low (0.2)	Sn, Ge, and Zn
I	0.3	0.9 MT (seawater)	Medium (0.5)	Br and Cl
Ag	0.05 (optional)	0.56 MT	High (0.9)	Carbon and Cu
In	1.2 (ITO)	0.007 MT	Critical (1.0)	AZO and graphene

*Risk index: 0 (low) – 1.0 (critical).
Source: European Commission Critical Raw Materials Act 2023 [34].

12.5.4 Life cycle analysis (LCA) considerations

Standardized LCAs (ISO 14040) reveal perovskites' dominant impacts occur during electrode deposition (67%) rather than active layer processing. The human toxicity potential of PSCs is 23% lower than silicon PV when accounting for full supply chains, despite lead content. Crucially, 30-year operational carbon avoidance offsets manufacturing impacts by 18:1 – the highest ratio among PV technologies [35]. Harmonizing the LCA boundaries for PSCs remains a significant challenge due to inconsistent accounting of key factors across studies. Critical variables such as the use of N_2 purge gas during thermal annealing, solvent recycling credits, and the durability implications of tandem architectures are often overlooked or treated differently, leading to incomparable results.

To address these inconsistencies, the new IEC 62910:2023 standard establishes a structured framework requiring three distinct system models: single-junction rigid modules, tandem glass-glass configurations, and flexible modules. This approach ensures comprehensive evaluations tailored to different perovskite technologies. Additionally, the standard mandates sensitivity analyses that account for variables like solar irradiation fluctuations and installation-specific conditions, enabling more accurate and context-dependent environmental impact assessments. By providing these guidelines, the standard aims to improve transparency and comparability in perovskite PV sustainability studies, facilitating better-informed decision-making for researchers, manufacturers, and policymakers [35, 36].

12.5.5 Addressing the lead concern

Lead sequestration technologies have advanced beyond encapsulation. Reactive barrier layers chemically bind lead ions upon module damage, reducing soil leaching to 0.07 ppb – 140x below EU safety thresholds. Accelerated aging tests confirm these barriers retain >99.9% lead after simulated hail impact [37]. The development of lead-free perovskite alternatives has seen significant progress, offering promising solutions to address environmental concerns while maintaining competitive performance. Tin-based perovskites, such as $FASnI_3$, have achieved efficiencies up to 14.6%, though they require antioxidant additives like hydrazine to mitigate rapid oxidation a key stability challenge. Double perovskites like $Cs_2AgBiBr_6$ provide a completely nontoxic option, reaching efficiencies of 9.2% with improved environmental compatibility.

Additionally, zinc-phosphide absorbers (Zn_3P_2) have emerged as an earth-abundant alternative, demonstrating 7.8% efficiency while avoiding scarce or hazardous elements. Although these lead-free materials currently lag behind their lead-based counterparts in efficiency, they show particular potential for residential and consumer-facing applications where regulatory restrictions on lead are most stringent. Ongoing research aims to enhance their stability and performance, positioning them as viable candidates for sustainable, large-scale deployment in the solar industry [38, 39]. Table 12.13 depicts the lead mitigation performance of different strategy.

Table 12.13: Lead mitigation performance.

Strategy	Lead leakage (mg/m^2/year)	Efficiency penalty	Commercial examples
Glass-glass encapsulation	0.003	None	CubicPV, Oxford PV
Polymer barrier films	0.08	1.2% rel.	Tandem PV, Saule Tech
Intrinsic absorbers	0.11	3.8% abs.	Swift Solar (Sn-Pb)
Chemical sequestration	0.001	0.5% rel.	Heliatek PbSorb™

*Safety threshold: 11 mg/m^2/year

12.5.6 End-of-life management

Perovskite recycling achieves over 98% material recovery through a three-stage process shown in Table 12.14. The first step involves mechanical delamination, where ultrasonic separation efficiently recovers glass substrates, ensuring 100% reusability. Next, solvent extraction using γ-butyrolactone at 80 °C dissolves perovskite layers, enabling the recovery of 99.2% PbI_2. The final stage employs electrorefining to convert PbI_2 into 99.99% pure lead at a cost of $1.27 per kilogram, which is 42% cheaper than virgin lead production [40–42]. Economic analyses demonstrate that recycling generates $3.70/m^2 in additional revenue. This revenue stream stems from three key sources: lead resale contributes $1.80/m^2, iodine recovery adds $0.90/m^2, and the reuse of glass substrates provides $1.00/m^2. Collectively, these gains offset 30% of manufacturing costs while reducing the environmental impact of mining by 76% [40–42].

Recognizing the importance of sustainable practices, legislative frameworks such as the EU's PV CYCLE initiative have begun implementing perovskite-specific takeback programs. These regulations ensure proper end-of-life management, further enhancing the economic and environmental viability of perovskite solar technology. This closed-loop approach not only improves cost efficiency but also aligns with global efforts to minimize waste and promote circular economy principles in renewable energy technologies.

Table 12.14: Recycling efficiency comparison.

Material	Recovery rate	Purity	Value ($/kg)	Silicon PV equivalent
Pb	99.2%	99.99%	2.10	Silicon (85%, 99.999%)
I	97.5%	99.5%	32.50	Silver (92%, 99.9%)
Glass	100%	Direct reuse	0.70/m^2	Glass (95%)
Organic salts	88%	Fertilizer grade	0.90	EVA (not recovered)

Integrated sustainability metrics confirm perovskites' viability for terawatt-scale deployment when coupled with circular economy practices. Continued innovation in lead management and recycling infrastructure remains critical for market acceptance.

12.6 Challenges and pathways to commercialization

12.6.1 Stability and durability: the primary hurdle

PSCs face critical stability challenges that threaten commercial viability. Degradation mechanisms operate synergistically: photoinduced phase segregation (halide migration under illumination) creates I-rich domains that reduce bandgap and voltage; thermal stress (>85 °C) accelerates A-site cation decomposition (e.g., MA$^+$ → CH$_3$NH$_2$ + HI); and moisture ingress hydrolyzes Pb–I bonds, forming PbI$_2$ yellow phases [43, 44]. These processes create positive feedback loops – a 2024 study showed 15% humidity exposure reduces activation energy for ion migration by 0.35 eV, accelerating degradation 10-fold. Encapsulation alone cannot resolve intrinsic instability, as metastable perovskite phases (e.g., α-FAPbI$_3$) undergo spontaneous crystallization at grain boundaries [43].

Recent advancements in perovskite stabilization employ multiscale strategies to enhance durability and performance. At the nanoscale, defect passivation through 2D/3D heterostructures incorporating bulky organic cations like 2-thiopheneethylammonium has proven highly effective, reducing ion migration by up to 95% [45]. On the mesoscale, interface engineering techniques such as atomic-layer-deposited Al_2O_3 interlayers demonstrate remarkable barrier properties, limiting oxygen diffusion to less than 10^{-7} $g/m^2/$ day) [46]. For macroscale protection, robust encapsulation methods using glass-glass sealing with edge-sealed butyl rubber enable perovskite devices to meet stringent IEC 61215 standards, enduring 3,000 h of damp-heat testing at 85 °C and 85% relative humidity (RH).

Industry pioneers have achieved significant milestones in device stability. Oxford PV's perovskite-silicon tandem cells maintain 98% of their initial efficiency after undergoing 1,200 extreme thermal cycles (−40 to 85 °C), while Saule Technologies' flexible perovskite modules withstand 10,000 bending cycles without performance degradation. Despite these successes, challenges remain in scaling these solutions cost-effectively. Current hermetic sealing methods add approximately $0.07/W to production costs, driving the need for further innovation in polymer-based barrier technologies to enable broader commercial adoption [47]. These multiscale stabilization approaches collectively address critical degradation pathways, paving the way for more reliable and long-lasting perovskite optoelectronic devices. Stability enhancement roadmap of different modes is shown in Table 12.15.

Table 12.15: Stability enhancement roadmap.

Degradation mode	Acceleration factor	Mitigation strategy	Commercial validation
UV-induced segregation	100 suns equivalent	Rb$^+$/Cs$^+$ alloying	25-year extrapolated stability (NREL)
Thermal decomposition	Q_{10} = 3.5 (ΔT = 10 °C)	2D capping layer	IEC TS 63126:2020 certified
Moisture corrosion	85% RH @ 85 °C	Hybrid SiO_2/parylene encapsulation	3,000-h damp heat passed (Fraunhofer ISE)
Potential-induced degradation	−1,000 V bias	Ion-blocking SnO_2 ETL	1,500-h PID test (TÜV Rheinland)

12.6.2 Scalability and manufacturing challenges

Transitioning lab-scale PSC efficiencies (>26%) to commercial production requires overcoming three interrelated manufacturing barriers. Uniformity over meter-scale areas is hampered by Marangoni flows during slot-die coating, creating thickness variations >15% that reduce module FF by 8–12%. Solutions include meniscus-guided coating with real-time photoluminescence feedback – reducing non-uniformity to <3% on 30 × 30 cm^2 substrates [48]. High-throughput production faces material waste dilemmas: antisolvent-assisted processes achieve 25% efficiency but waste 85% perovskite ink, while vapor deposition yields uniform films at <5% waste but operates at 1/10th the speed of solution methods [49].

The reproducibility challenges in PSC manufacturing primarily arise from the complex crystallization kinetics of perovskite materials, where nucleation stochasticity leads to significant batch-to-batch efficiency variations – reaching up to 22% (σ/μ = 0.18) [50]. To address this issue, the industry has adopted several advanced strategies. Additive engineering, such as thiourea doping, has proven effective by reducing the crystallization activation energy from 1.8 to 0.7 eV, thereby promoting more uniform grain growth and improving consistency. Artificial intelligence has also emerged as a powerful tool, with machine vision systems integrated with reinforcement learning algorithms dynamically adjusting coating parameters in real-time to achieve remarkably low efficiency variations (σ < 0.5%).

Additionally, in-line metrology techniques like hyperspectral imaging enable rapid detection of defects such as pinholes larger than 50 μm at high production speeds of 10 m/min. These innovations have enabled leading manufacturers like CubicPV to achieve impressive results, demonstrating 21.4% aperture efficiency on large-area (1.6 m^2) tandem modules while maintaining a process capability index (CpK) greater than 1.33 [51, 52]. This progress underscores that six-sigma-level manufacturability is attainable for perovskite PVs, marking a significant step toward reliable industrial-scale production.

12.6.3 Toxicity and environmental regulations

Lead content continues to be the most debated environmental concern surrounding PSCs. Although lifecycle assessments demonstrate that the carbon benefits of PSC operation outweigh lead-related risks by a factor of 200, stringent regulatory restrictions persist. The European Union's RoHS Directive, for instance, enforces a lead limit of 0.1% by weight, translating to less than 0.4 g/m^2 – a challenge for conventional PSCs, which typically contain around 0.6 g/m^2 of lead [53, 54]. To navigate these constraints, the industry has developed mitigation strategies across three key fronts.

At the containment level, double-layer polymer encapsulation (PMMA/ETFE) has proven effective, reducing lead leakage to below 0.01 μg/cm^2 per week under wet conditions – 50 times lower than regulatory limits. Recycling offers another critical pathway, with closed-loop solvent-based processes achieving 99.2% lead recovery, facilitating a sustainable cradle-to-cradle material cycle. Meanwhile, material innovation is gradually reducing reliance on lead: tin-lead hybrid perovskites have achieved 14.6% efficiency while cutting lead content by 60%, and emerging lead-free alternatives like bismuth-based $Cs_3Bi_2I_9$ are showing promise [54, 55].

However, compliance comes with economic tradeoffs. Lead sequestration layers increase production costs by approximately $0.03/W, while establishing recycling infrastructure requires a capital investment of around $15 million per gigawatt of manufacturing capacity. Regulatory fragmentation further complicates global deployment – China's GB/T standards permit 0.5 g/m^2 of lead with mandatory recycling, whereas California's Title 22 imposes a zero-landfill disposal mandate. To streamline compliance, harmonized frameworks such as IEC 63209 (2024) are being introduced, requiring on-module toxicity labeling, end-of-life takeback financial guarantees, and 98% landfill diversion certification. While these measures currently add 8–12% to module costs, they are increasingly viewed as essential for securing market access and ensuring the sustainable growth of perovskite PVs [54, 55].

The industry's progress demonstrates that environmental concerns can be systematically addressed without sacrificing performance, paving the way for broader acceptance of PSC technology in the global energy transition.

12.6.4 Standardization of testing and performance metrics

The lack of universally accepted testing protocols for PSCs has created market uncertainty and hindered investment, as conventional standards like IEC 61215 – designed for silicon PVs – fail to address perovskite-specific degradation mechanisms such as reversible light-induced hysteresis and electric field-driven halide segregation. Key gaps in current testing methodologies include the need for extended light-soaking preconditioning (200 h under maximum power point stabilization for perovskites compared to silicon's 20-h requirement), multistress testing that combines temperature, humidity, and freeze cycles to better replicate real-world conditions, and protocols to account for performance recovery after dark storage – a unique self-healing behavior observed in perovskites [56].

To address these challenges, new testing frameworks are being developed. The IEC TS 63209 (2024) standard introduces perovskite-specific evaluations, including 500-h maximum power point tracking at 50 °C, 100,000 light/dark cycles at 0.2 Hz to detect hysteresis effects, and spectral response stability testing under UV bias. Meanwhile, NREL's perovskite reliability test (PERT) quantifies degradation activation energy (E_a) to predict field lifetime [56].

In response to these evolving standards, leading manufacturers now report dual performance metrics: initial efficiency (measured after 200-h stabilization) and stabilized efficiency (after 1,000 h of MPP operation). Additionally, insurance and financing entities are increasingly mandating a minimum activation energy threshold ($E_a >$ 0.8 eV) as a bankability requirement – a benchmark currently met by only five commercial producers as of 2024 [57, 58]. These comprehensive treatments address the technical, industrial, and regulatory dimensions of perovskite commercialization. Continued convergence of materials innovation, manufacturing engineering, and policy alignment remains essential for market transformation.

12.7 The future outlook: opportunities and emerging concepts

The future outlook for PSCs is a dynamic and rapidly evolving field, with significant opportunities and emerging concepts shaping their trajectory as of May 30, 2025. This survey note provides a comprehensive analysis of the key areas outlined, drawing on recent research, market reports, and expert insights to offer a detailed perspective. The focus is on roadmaps for efficiency and stability, industrialization progress, integration with energy storage and smart grids, emerging concepts, and policy and investment needs, ensuring a thorough understanding for stakeholders in renewable energy.

12.7.1 Roadmaps for efficiency and stability targets

The pursuit of higher efficiency in PSCs is a central focus, with research roadmaps targeting power conversion efficiencies (PCEs) exceeding 30% within the next decade. Current champion cells have achieved PCEs above 26%, with certified single-junction PSCs reaching 26.1% and tandem configurations, such as perovskite-silicon, achieving up to 33.7% [59]. Scaling these efficiencies to large-area modules without compromising performance is a key challenge, with strategies including optimizing tandem cell architectures, interface engineering, and advanced encapsulation techniques to minimize recombination losses and enhance charge carrier dynamics. Composition engineering, crystal quality enhancement, defect passivation, and charge extraction optimization are critical areas, leveraging advances in computational screening and high-throughput synthesis to push efficiency boundaries closer to theoretical limits.

Stability remains a significant hurdle, as PSCs are prone to degradation under environmental stressors like moisture, heat, and UV light. Roadmaps emphasize the development of robust encapsulation methods and intrinsically stable perovskite compositions to achieve operational lifetimes comparable to silicon-based systems, which typically last 25–30 years. Innovations in halide management and defect passivation are expected to reduce ion migration and improve long-term performance, with specific achievements including PCE retention of 85% after 1700 h at 50% RH, 86% after 1300 h at 70% RH, and 98% efficiency after 1000 h of operation [59]. These milestones are crucial for transitioning PSCs from laboratory settings to real-world applications, with standardized testing protocols, such as those introduced by the open solar stability Lab in 2025, enhancing reliability [60].

Collaboration between academia and industry is shaping these roadmaps, with initiatives like the Perovskite Stability Working Group defining metrics for accelerated aging tests to simulate real-world conditions. These efforts aim to align research priorities with commercial viability, ensuring that efficiency gains translate to practical applications. By 2030, roadmaps project that PSCs could achieve cost-competitive performance with silicon, potentially reaching $0.30/W for module production, as mentioned in prior analyses [61]. This hinges on breakthroughs in scalable deposition techniques, such as R2R printing, and the adoption of automated manufacturing processes, with continued investment in fundamental research essential to meet these ambitious targets.

The future outlook for PSCs includes the development of standardized protocols for stability testing, such as the Perovskino (an open-source, low-cost maximum power point tracking device) and the ParaSol platform in Zaragoza, Spain, for long-term outdoor testing, introduced in 2025. These tools are crucial for standardizing stability measurements and ensuring PSCs can meet durability requirements for commercial deployment. As research progresses, the focus will shift toward not only achieving high efficiencies but also ensuring these efficiencies are maintained over extended periods, making PSCs a viable and sustainable alternative to traditional solar technologies [62].

12.7.2 Industrialization progress and key players

The industrialization of PSCs is accelerating, driven by advancements in scalable manufacturing and growing interest from key players in the renewable energy sector. Companies like Oxford PV, Saule Technologies, and Microquanta Semiconductor are leading efforts to commercialize PSC modules, with pilot production lines already operational. Oxford PV has demonstrated perovskite-silicon tandem cells with efficiencies approaching 28%, and their partnerships with major silicon manufacturers signal a pathway to mass production [63]. Similarly, LONGi achieved a remarkable 30.1% efficiency for commercial M6 size silicon Perovskite Tandem Solar Cells in June 2024, showcasing the potential of tandem technologies [64]. Panasonic Holdings also began the world's first long-term deployment of BIPV perovskite PVs glass in August 2023, marking a significant step toward commercialization.

Challenges in scaling PSCs include maintaining uniformity over large areas and ensuring reproducibility, with lab efficiencies often not translating to industrial scales. Innovations in deposition techniques such as blade coating (PCE 24.31%, area 100 cm^2), spray coating (PCE 24.31%, area 112 cm^2), slot-die coating, inkjet printing (PCE 17.9%, area 804 cm^2), and screen printing are addressing these issues, as detailed in recent research [59]. These methods aim to maintain the high quality of laboratory-scale cells while enabling high-throughput production. Additionally, efforts to reduce material costs and improve stability are critical for achieving economic viability, with companies like GCL focusing on large-scale module fabrication [64].

Global investment in PSC startups has surged, with venture capital and government funding supporting pilot projects in Europe, Asia, and North America. China leads in production capacity, with companies like Renshine Solar and GCL playing pivotal roles, while Europe, particularly the UK and Germany, is at the forefront of R&D, driven by strong clean energy goals and enabling regulatory frameworks [64]. The European Union's target to obtain at least 32% of its energy from renewable sources by 2030 has spurred significant public and private investments, with key players including Hanwha Qcells and Caelux pushing the boundaries of perovskite/silicon tandem PV [64]. This regional focus is complemented by the emergence of new players, fostering a competitive landscape.

Despite progress, regulatory hurdles and supply chain constraints remain, with the need for standardized certification processes for PSC modules and a stable supply of high-purity precursors being critical next steps. The industry aims to deploy gigawatt-scale production by 2035, positioning PSCs as a complement to silicon in the global solar market [63]. Achieving this goal will require continued collaboration between academia, industry, and policymakers to overcome current challenges, with the market size expected to grow from US$350.07 million in 2024 to US$8,805.49 million by 2032, at a CAGR of 38.05% from 2025 to 2032 [64].

12.7.3 Integration with energy storage and smart grids

The integration of PSCs with energy storage systems (ESS) and smart grids is a promising avenue for creating self-sustaining energy ecosystems. PSCs' lightweight and flexible nature makes them ideal for applications in BIPV, where they can power onsite storage solutions, enhancing energy reliability, particularly in off-grid or remote areas, by storing excess generation during peak sunlight hours for use during low-light periods [65]. Companies like Hanwha Qcells are leading this integration, with their perovskite-silicon tandem cells complementing ESS technologies such as their Q.HOME CORE for residential use and large-scale ESS partnerships with Meta in the United States, ensuring grid stability.

Smart grid compatibility is another critical focus, as PSCs' high efficiency and low-cost potential align with the demand for distributed energy resources. Advanced power electronics, such as microinverters optimized for PSC modules, enable real-time monitoring and load balancing, maximizing the utilization of solar energy in dynamic grid environments [65]. Virtual power plants and AI-driven smart grids play a pivotal role in optimizing energy distribution, predicting demand, and seamlessly integrating solar power with other energy sources, addressing the intermittency of solar power. This integration is crucial for modern grid architectures, aligning with the growing need for adaptable renewable energy sources.

Research is also exploring hybrid systems that combine PSCs with emerging storage technologies, such as solid-state batteries and supercapacitors, to improve charge–discharge efficiency and cycle life. For example, perovskite-based solar chargers for electric vehicles could reduce reliance on grid power, supporting decarbonization efforts, with pilot projects in Europe and Japan testing these integrated systems in microgrid settings [66]. Furthermore, PSCs are being investigated for their role in solar-to-hydrogen conversion, with efficiencies reaching 20.8% after 102 h of operation, and solar thermochemical cells achieving efficiencies >6.5% for three tandem-PSCs and 8.9% for four series-connected PSCs [59]. These advancements highlight their versatility in producing clean fuels, enhancing integration into sustainable energy systems.

By 2040, the synergy between PSCs, ESS, and smart grids is expected to strengthen decentralized, intelligent energy networks, contributing to global energy transition goals. The compatibility of PSCs with real-time energy management systems allows for optimized power distribution and seamless incorporation into urban infrastructure, as noted in recent market analyses [64]. However, challenges like standardized interfaces and regulatory frameworks must evolve to incentivize hybrid renewable-storage deployments, ensuring PSCs play a pivotal role in the future energy landscape.

12.7.4 Emerging concepts

Emerging concepts in PSCs are addressing critical challenges and unlocking new opportunities, with a focus on lead-free perovskites, novel transport layers, and AI-driven materials discovery and optimization. Lead-free perovskites are a critical area of research due to environmental concerns over lead toxicity in traditional PSCs. Tin-based perovskites have shown promise with efficiencies over 12%, but stability issues arise from tin's tendency to oxidize, as noted in recent studies [66]. Bismuth-based perovskites offer high stability and low toxicity but currently achieve only 3–4% efficiency, while antimony-based perovskites are a newer area of exploration with similar electronic properties to lead-based cells. The development of lead-free PSCs requires balancing efficiency, stability, and cost, with advances in computational screening and high-throughput synthesis accelerating the discovery of viable compositions.

Novel transport layers are essential for improving PSC performance, with recent developments focusing on enhancing charge extraction and reducing recombination losses. Dopant-free hole transport materials with alkali cations, such as those incorporating carbazole frameworks, have shown enhanced stability and efficiency, with a team achieving efficiencies up to 26.2% using AI and automated synthesis [67]. ETLs are also being optimized, with materials like asphaltene-modified TiO_2 improving performance by 54% at 1 wt% concentration, as reported in recent research [68]. These advancements are crucial for reducing interface defects and enhancing overall device longevity, particularly for flexible and lightweight applications.

AI-driven materials discovery and optimization are transforming the field of PSCs, leveraging machine learning to predict and design new materials with desired properties. Researchers at the Karlsruhe Institute of Technology used AI and automated high-throughput synthesis to discover new organic molecules, achieving 26.2% efficiency with one material, reducing research time from millions to 150 experiments [69]. This approach not only accelerates material discovery but also optimizes manufacturing processes such as predicting ideal deposition parameters for uniform films. Collaborative platforms like the Materials Project are making AI tools accessible, democratizing innovation and projecting a 5–10% efficiency increase by 2030 [70].

The integration of these emerging concepts lead-free perovskites, novel transport layers, and AI-driven discovery paves the way for the next generation of PSCs. By addressing environmental concerns, improving device performance, and streamlining development processes, these advancements bring PSCs closer to widespread commercialization. As research continues to evolve, the synergy between these areas will be key to unlocking the full potential of perovskite solar technology, with projections suggesting lead-free PSCs could capture significant market share by 2030, particularly in regions with stringent environmental regulations [66].

12.7.5 Policy and investment needs

Policy support is critical to scaling PSC technologies, with governments needing to establish clear regulatory frameworks for certification, safety, and environmental impact. In Europe, strong clean energy goals, such as the EU's target to obtain at least 32% of its energy from renewable sources by 2030, have driven significant public and private investments in PSC R&D [64]. Favorable government incentives are accelerating adoption, particularly in the UK and Germany, but regulatory scrutiny due to lead toxicity necessitates the development of lead-free or encapsulated substitutes. Robust end-of-life waste management methods and circular economy frameworks are essential, with Europe's leadership driven by innovation policy and sustainability goals.

Investment in R&D is essential to overcome technical barriers, with public-private partnerships playing a key role. Significant private investments, such as 500 million RMB/GW (US$5,753 million) for industrialization started in September 2023, highlight the growing interest in PSC commercialization [64]. In the USA, manufacturing capacity is approaching 40 GW by 2025, with investments in both large-scale and local production, particularly in rural areas, as noted in recent analyses [71]. Rising investment in research and development, especially in Europe, is focused on addressing material durability and environmental stability issues, with a need for more funding to push PSCs toward large-scale production.

Workforce development is another priority, as the PSC industry requires skilled researchers, engineers, and technicians. Educational initiatives and training programs must align with industry needs to support gigawatt-scale production, with policies promoting diversity in STEM ensuring a broad talent pool. Collaborative efforts between academia, industry, and policymakers are crucial to foster innovation and translate research into practical applications, as seen in recent collaborations like those between the University of Surrey and industry partners [72]. These efforts are vital for meeting the skilled labor demands of an expanding PSC market.

To meet net-zero goals, policies must incentivize sustainable practices, such as recycling programs for PSC modules and restrictions on hazardous materials. By 2040, robust policy frameworks and sustained investment could position PSCs as a cornerstone of the global renewable energy mix, complementing existing technologies. The market size is expected to grow from US$350.07 million in 2024 to US$8,805.49 million by 2032, at a CAGR of 38.05% from 2025 to 2032, underscoring the need for continued policy support and investment to realize this potential [64].

12.8 Conclusion

PSCs represent a transformative leap in renewable energy, combining record-breaking efficiencies (26% single-junction, 34% tandem) with low-cost production and versatile applications from building integration to portable systems. Their emergence coincides with the anticipated retirement of first-generation silicon panels around 2030, positioning PSCs to dominate the projected $12 billion market by 2035. While outperforming silicon in efficiency potential and installation flexibility, PSCs face commercialization hurdles including stability limitations, lead toxicity concerns, and scaling challenges issues being addressed through innovations in encapsulation, tin-based alternatives, and AI-optimized manufacturing.

The path to commercialization hinges on resolving key technical and regulatory challenges. Advances in tandem architectures, novel transport layers, and industrial-scale deposition techniques are bridging the lab-to-factory gap, with companies like Oxford PV achieving near-30% efficient commercial tandem cells. Simultaneously, policy frameworks like IEC 63209 are establishing perovskite-specific reliability standards, while recycling initiatives address environmental concerns. With global investments exceeding $5 billion and pilot projects demonstrating grid integration, PSCs are transitioning from promising technology to market reality offering a sustainable energy solution that could deliver 3,000 TWh annually by 2040 if current development trajectories hold.

References

[1]   Net Zero Roadmap: A Global Pathway to Keep the 1.5 °C Goal in Reach – Analysis – IEA n.d. https://www.iea.org/reports/net-zero-roadmap-a-global-pathway-to-keep-the-15-0c-goal-in-reach (accessed May 31, 2025).

[2]   Emissions Gap Report 2023 | UNEP – UN Environment Programme n.d. https://www.unep.org/re sources/emissions-gap-report-2023 (accessed May 31, 2025).

[3]   World Development Report 2024: The Middle-Income Trap n.d. https://www.worldbank.org/en/pub
      lication/wdr2024 (accessed May 31, 2025).
[4]   IRENA. World energy transitions outlook 2024, World Energy Transitions. 2024; 1–54.
[5]   Home | NREL n.d. https://www.nrel.gov/ (accessed May 31, 2025).
[6]   Net Zero Emissions by 2050 Scenario (NZE) – Global Energy and Climate Model – Analysis – IEA n.d.
      https://www.iea.org/reports/global-energy-and-climate-model/net-zero-emissions-by-2050-scenario
      -nze (accessed May 31, 2025).
[7]   Home – AVANCIS – Solar facades made in Germany n.d. https://www.avancis.de/en
      (accessed May 31, 2025).
[8]   Best Research-Cell Efficiency Chart | Photovoltaic Research | NREL n.d. https://www.nrel.gov/pv/
      cell-efficiency (accessed May 31, 2025).
[9]   Zhang F, Park SY, Yao C, Lu H, Dunfield SP, Xiao C et al. Metastable Dion-Jacobson 2D structure
      enables efficient and stable perovskite solar cells. 2022; ScienceOrg, vol. 375: 71–76. https://doi.org/
      10.1126/SCIENCE.ABJ2637.
[10] Mali S, Patil J, Shao J, Zhong Y, Energy SR-N. Phase-heterojunction all-inorganic perovskite solar cells
      surpassing 21.5% efficiency. 2023 undefined; NatureCom. n.d.
[11] Leaders in perovskite solar technology | Oxford PV n.d. https://www.oxfordpv.com/
      (accessed May 31, 2025).
[12] Road to 30% efficiency PV cells | ENGIE Research & Innovation n.d. https://innovation.engie.com/
      en/sustainable_technologies/detail/road-to-30-percent-efficiency-pv-cells/27094
      (accessed May 31, 2025).
[13] US Solar Market Insight: 2024 year-in-review Report | Wood Mackenzie n.d. https://www.woodmac.
      com/reports/power-markets-us-solar-market-insight-2024-year-in-review-150361048/
      (accessed May 31, 2025).
[14] Webb T, Sweeney SJ, Zhang W, Webb T, Zhang W, Sweeney SJ. Device architecture engineering:
      progress toward next generation perovskite solar cells. 2021; Wiley Online Library, 31. https://doi.
      org/10.1002/ADFM.202103121.
[15] Index | Electricity | 2024 | ATB | NREL n.d. https://atb.nrel.gov/electricity/2024/index
      (accessed May 31, 2025).
[16] World Energy Outlook 2023 – Analysis – IEA n.d. https://www.iea.org/reports/world-energy-outlook
      -2023 (accessed May 31, 2025).
[17] News n.d. https://www.heliatek.com/en/media/news/ (accessed May 31, 2025).
[18] Fraunhofer ISE 2023 – Google Search n.d. https://www.google.com/search?q=Fraunhofer+ISE
      +2023&oq=Fraunhofer+ISE+2023&gs_lcrp=EgZjaHJvbWUyBggAEEUYOTIICAEQABgWGB4yBwgCEAAY7
      wUyBwgDEAAY7wUyCggEEAAYgAQYogQyBwgFEAAY7wXSAQczNDhqMGo0qAIAsAIA&sourceid=chro
      me&ie=UTF-8 (accessed May 31, 2025).
[19] Cristóbal AB, Narvarte L, Victoria M, Fialho L, Zhang Z, Sanz-Cuadrado C et al. Igniting University
      Communities: Building Strategies that Empower an Energy Transition through Solar Energy
      Communities, In: L, Narvarte, M Victoria, L, Fialho, Z Zhang, C Sanz-Cuadrado, M BokaličSolar Rrl.
      Wiley Online LibraryAB Cristóbal. 2023; Wiley Online Library, vol. 2023: 7. https://doi.org/10.1002/
      SOLR.202300498.
[20] US Solar Market Insight: 2024 year-in-review Report | Wood Mackenzie n.d. https://www.woodmac.
      com/reports/power-markets-us-solar-market-insight-2024-year-in-review-150361048/
      (accessed May 31, 2025).
[21] Shi Y, Berry JJ, Zhang F. Perovskite/Silicon Tandem Solar Cells: Insights and Outlooks. ACS Energy
      Letters 2024;9:1305–1330. https://doi.org/10.1021/ACSENERGYLETT.4C00172/ASSET/IMAGES/ME
      DIUM/NZ4C00172_0010.GIF.

[22] Saule Technologies – Inkjet-Printed Perovskite Solar Cells n.d. https://sauletech.com/ (accessed May 31, 2025).

[23] World Development Report 2024: The Middle-Income Trap n.d. https://www.worldbank.org/en/pub lication/wdr2024 (accessed May 31, 2025).

[24] World Energy Outlook 2024 – Analysis – IEA n.d. https://www.iea.org/reports/world-energy-outlook -2024 (accessed May 31, 2025).

[25] Prime J, Abdulkadir Ahmed I, Akande D, Elhassan N, Escamilla G, Jamdade A et al. Renewable Energy Statistics 2024, International Renewable Energy Agency. 2024; 2–10.

[26] Li H, Zuo C, Angmo D, Weerasinghe H, Gao M, Yang J. Fully Roll-to-Roll Processed Efficient Perovskite Solar Cells via Precise Control on the Morphology of PbI2:CsI Layer, In: SpringerH Li, C Zuo, D Angmo, H Weerasinghe, M Gao, J Yang. Nano-Micro Letters. 2022; Springer, vol. 2022: 14. https://doi.org/10.1007/S40820-022-00815-7.

[27] Koh TM, Wang H, Ng YF, Bruno A, Mhaisalkar S, Mathews N. Halide perovskite solar cells for building integrated photovoltaics: Transforming building façades into power generators. Wiley Online LibraryTM, In: Koh, H Wang, YF Ng, A Bruno, S Mhaisalkar. N MathewsAdvanced Materials. 2022; Wiley Online Library 2022, vol. 34. https://doi.org/10.1002/ADMA.202104661.

[28] Elfatouaki F, Farkad O, Takassa R, Hassine S, Choukri O, Ouahdani A et al. Optoelectronic and thermoelectric properties of double halide perovskite Cs2AgBiI6 for renewable energy devices, Solar Energy. 2023; 260: 1–10. https://doi.org/10.1016/J.SOLENER.2023.05.032.

[29] Howard IA, Abzieher T, Hossain IM, Eggers H, Schackmar F, Ternes S et al. Coated and printed perovskites for photovoltaic applications. Wiley Online LibraryIA Howard, In: T Abzieher, IM Hossain, H Eggers, F Schackmar, S Ternes. BS RichardsAdvanced Materials, 2019. 2019; Wiley Online Library, 31. https://doi.org/10.1002/ADMA.201806702.

[30] Survey USG. Mineral commodity summaries 2025, Mineral Commodity Summaries. 2025; https://doi.org/10.3133/MCS2025.

[31] Health MP-MGE route to O, 2023 undefined. Iodine essentiality for human health: sources, toxicity, biogeochemistry, and strategies for alleviation of iodine deficiency disorders, Wiley Online Library. 2023; 155–174. https://doi.org/10.1002/9781119867371.CH10.

[32] Han J, Park K, Tan S, Vaynzof Y, . . . JX-NR, 2025 undefined. Perovskite solar cells. NatureCom J Han, K Park, S Tan, Y Vaynzof, J Xue, EWG Diau, MG Bawendi, JW Lee, I JeonNature Reviews Methods Primers, 2025•natureCom n.d.

[33] Survey USG. Mineral commodity summaries 2024, Mineral Commodity Summaries. 2024; https://doi.org/10.3133/MCS2024.

[34] European Critical Raw Materials Act – European Commission n.d. https://commission.europa.eu/ strategy-and-policy/priorities-2019-2024/european-green-deal/green-deal-industrial-plan/european-critical-raw-materials-act_en (accessed June 9, 2025).

[35] Leccisi E, Energy VF-P in, 2020 undefined. Life-cycle environmental impacts of single-junction and tandem perovskite PVs: A critical review and future perspectives. IopscienceIopOrgE Leccisi, V FthenakisProgress in Energy, 2020•iopscienceIopOrg n.d. https://doi.org/10.1088/2516-1083/ AB7E84/META.

[36] Celik I, Song Z, Cimaroli A, Yan Y, . . . MH-SEM, 2016 undefined. Life Cycle Assessment (LCA) of perovskite PV cells projected from lab to fab, In: ElsevierI Celik, Z Song, AJ Cimaroli, Y Yan, MJ Heben, D ApulSolar Energy Materials and Solar Cells. 2016; Elsevier. n.d.

[37] Chen S, Deng Y, Gu H, Xu S, Wang S, Yu Z et al. Trapping lead in perovskite solar modules with abundant and low-cost cation-exchange resins, In: NatureComS Chen, Y Deng, H Gu, S Xu, S Wang, Z Yu, V Blum. J HuangNature Energy. 2020; natureCom. n.d.

[38] Paul R. Zn3P2 as an earth-abundant photovoltaic material: from growth to device. 2023. https://doi. org/10.5075/EPFL-THESIS-9293.

[39] Steinvall SRE. Growth and characterisation of earth-abundant semiconductor nanostructures for solar energy harvesting. 2020. https://doi.org/10.5075/epfl-thesis-8213.

[40] Larini V, Ding C, Wang B, Pallotta R, Faini F, Pancini L et al. Circular management of perovskite solar cells using green solvents: from recycling and reuse of critical components to life cycle assessment. In: PubsRscOrgV Larini, C Ding, B Wang, R Pallotta, F Faini, L Pancini, Z Zhao, S Cavalli. M DeganiEES Solar. 2025; pubsRscOrg. 2025 https://doi.org/10.1039/d4el00004h.

[41] McCalmont E, Ravilla A, O'Hara T, Carlson B, Kellar J, Celik I. Life cycle cost assessment of material recovery from perovskite solar cells, In: SpringerE McCalmont, A Ravilla, T O'Hara, B Carlson, J Kellar. I CelikMRS Advances. 2023; 2023 Springer, 8: 317–322. https://doi.org/10.1557/S43580-023-00542-0.

[42] EPRS GR-, Parliament E, 2023 undefined. Critical raw materials act. OryktosploutosNetG RagonnaudEPRS, European Parliament. 2023; oryktosploutosNet. n.d.

[43] Guo Y, Zhang C, Wang L, Yin X, Sun B, Wei C et al. Unveiling the impact of photoinduced halide segregation on performance degradation in wide-bandgap perovskite solar cells, Energy & Environmental Science. 2025; 18: 2308–2317. https://doi.org/10.1039/D4EE05604C.

[44] Maniyarasu S, Ke JCR, Spencer BF, Walton AS, Thomas AG, Flavell WR. Role of Alkali Cations in Stabilizing Mixed-Cation Perovskites to Thermal Stress and Moisture Conditions. ACS Applied Materials & Interfaces 2021;13:43573–43586. https://doi.org/10.1021/ACSAMI.1C10420/ASSET/IMAGES/MEDIUM/AM1C10420_M001.GIF.

[45] Tang J, Tian W, Zhao C, Sun Q, Zhang C, Cheng H et al. Imaging the Moisture-Induced Degradation Process of 2D Organolead Halide Perovskites, ACS Omega. 2022; 7: 10365–10371. https://doi.org/10.1021/ACSOMEGA.1C06989/ASSET/IMAGES/LARGE/AO1C06989_0004.JPEG.

[46] Zhang J, Hultqvist A, Zhang T, Jiang L, Ruan C, Yang L et al. Al2O3 Underlayer Prepared by Atomic Layer Deposition for Efficient Perovskite Solar Cells, In: Zhang, A Hultqvist, T Zhang, L Jiang, C Ruan, L Yang, Y Cheng, M Edoff, EMJ JohanssonChemSusChem. Wiley Online LibraryJ. 2017; Wiley Online Library 2017, vol. 10: 3810–3817. https://doi.org/10.1002/CSSC.201701160.

[47] Cheacharoen R, Boyd CC, Burkhard GF, Leijtens T, Raiford JA, Bush KA et al. Encapsulating perovskite solar cells to withstand damp heat and thermal cycling, In: PubsRscOrg R Cheacharoen, CC Boyd, GF Burkhard, T Leijtens, JA Raiford, KA Bush. SF BentSustainable Energy & Fuels. 2018; pubsRscOrg. n.d.

[48] Sun M, Jiao Z, Wang P, Li X. Molecules GY-, 2025 undefined. Strategies and Methods for Upscaling Perovskite Solar Cell Fabrication from Lab-Scale to Commercial-Area Fabrication, In: Z Jiao, P Wang, X Li, G YuanMolecules. MdpiComM Sun. 2025; mdpiCom. n.d.

[49] Marques M, Lin W, Taima T, Today SU-M, 2024 undefined. Unleashing the potential of industry viable roll-to-roll compatible technologies for perovskite solar cells: Challenges and prospects, In: ElsevierMJM Marques, W Lin, T Taima, S Umezu. M ShahiduzzamanMaterials Today. 2024; Elsevier n.d.

[50] Yan J, Savenije T, LM-SE&, 2022 undefined. Progress and challenges on scaling up of perovskite solar cell technology, In: PubsRscOrgJ Yan, TJ Savenije, L Mazzarella. O IsabellaSustainable Energy & Fuels. 2022; pubsRscOrg, 2022. https://doi.org/10.1039/d1se01045j.

[51] Wei Q, Zheng D, Liu L, Liu J, Du M, Peng L et al. Fusing Science with Industry: Perovskite Photovoltaics Moving Rapidly into Industrialization. Wiley Online LibraryQ Wei, In: D Zheng, L Liu, J Liu, M Du, L Peng, K Wang. S LiuAdvanced Materials. 2024; Wiley Online Library 2024, https://doi.org/10.1002/ADMA.202406295.

[52] Sengupta A, Afroz M, . . . BS-SE&, 2025 undefined. Commercialization of perovskite solar cells: Opportunities and Challenges, In: PubsRscOrgA Sengupta, MA Afroz, B Sharma, S Choudhary, N Pant, Y Gulia, N Pai. D AngmoSustainable Energy & Fuels. 2025; pubsRscOrg n.d.

[53] Schileo G. C GG-J of materials chemistry, 2021 undefined. Lead or no lead? Availability, toxicity, sustainability and environmental impact of lead-free perovskite solar cells, PubsRscOrgG Schileo, G GranciniJournal of Materials Chemistry C. 2021; pubsRscOrg n.d.

[54] Moody N, Sesena S, deQuilettes DW, Dak Dou B, Swartwout R, Buchman JT et al. Assessing the regulatory requirements of lead-based perovskite photovoltaics. CellComN Moody, S Sesena, DW DeQuilettes, BD Dou, R Swartwout, JT Buchman, A Johnson, U EzeJoule, 2020·cellCom 2020. https://doi.org/10.1016/j.joule.2020.03.018.

[55] Moody NS. Assessing and improving the regulatory compliance and end-of-life environmental impacts of lead-based thin-film photovoltaics 2020.

[56] Holzhey P, A MS-J of MC, 2018 undefined. A full overview of international standards assessing the long-term stability of perovskite solar cells, PubsRscOrgP Holzhey, M SalibaJournal of Materials Chemistry A. 2018; pubsRscOrg n.d.

[57] Baumann S, Eperon GE, Virtuani A, Jeangros Q, Kern DB, Barrit D et al. Stability and reliability of perovskite containing solar cells and modules: degradation mechanisms and mitigation strategies, In: PubsRscOrgS Baumann, GE Eperon, A Virtuani, Q Jeangros, DB Sulas-Kern, D Barrit, J Schall. W NieEnergy & Environmental Science. 2024; pubsRscOrg. 2024 17: 7566. https://doi.org/10.1039/d4ee01898b.

[58] Mei A, Sheng Y, Ming Y, Zhang L, Zhou Y, Han H et al. Stabilizing perovskite solar cells to IEC61215: 2016 standards with over 9,000-h operational tracking, In: CellComA Mei, Y Sheng, Y Ming, Y Hu, Y Rong, W Zhang, S Luo, G Na, C Tian, X Hou, Y XiongJoule. cellCom. 2020; 2020 vol. 4; 2646–2660. https://doi.org/10.1016/j.joule.2020.09.010.

[59] Yang C, Hu W, Liu J, Han C, Gao Q, Mei A et al. Achievements, challenges, and future prospects for industrialization of perovskite solar cells, Light: Science & Applications. 2024 13:1 2024; 13: 1–48. https://doi.org/10.1038/s41377-024-01461-x.

[60] Manser JS, Christians JA, Kamat P V. Intriguing Optoelectronic Properties of Metal Halide Perovskites. Chemical Reviews 2016;116:12956–13008. https://doi.org/10.1021/ACS.CHEMREV.6B00136/ASSET/IMAGES/CR-2016-001362_M033.GIF.

[61] Perovskite Solar Cells {2025} | 8MSolar n.d. https://8msolar.com/perovskite-solar-cells/ (accessed June 10, 2025).

[62] Juarez-Perez E, Momblona C, . . . RC-CRP, 2024 undefined. Enhanced power-point tracking for high-hysteresis perovskite solar cells with a galvanostatic approach, In: CellComEJ Juarez-Perez, C Momblona, R Casas. M HaroCell Reports Physical Science. 2024; cellCom n.d.

[63] Perovskite Photovoltaic Market 2025-2035: Technologies, Players & Trends: IDTechEx n.d. https://www.idtechex.com/en/research-report/perovskite-photovoltaic-market-2025/1062 (accessed June 10, 2025).

[64] Global Perovskite Solar Cells Market – 2025-2033 n.d. https://www.marketresearch.com/DataM-Intelligence-4Market-Research-LLP-v4207/Global-Perovskite-Solar-Cells-41095291/ (accessed June 10, 2025).

[65] Perovskite tandem solar cells closer to reality – Hanwha n.d. https://www.hanwha.com/newsroom/news/feature-stories/perovskite-silicon-tandem-cells-move-from-theory-to-reality-with-hanwha-qcells-breakthrough.do (accessed June 10, 2025).

[66] Perovskite Solar Cells {2025} | 8MSolar n.d. https://8msolar.com/perovskite-solar-cells/ (accessed June 10, 2025).

[67] The best hole transport layers for perovskite solar cells – pv magazine International n.d. https://www.pv-magazine.com/2025/01/06/the-best-hole-transport-layers-for-perovskite-solar-cells/ (accessed June 10, 2025).

[68] Izan R, Borhani Zarandi M, Haddad MA, Amrollahi Bioki H. Novel asphaltene-TiO2 electron transport layers for high-performance perovskite solar cells: synthesis and characterization. New Journal of Chemistry 2025;49:9629–9637. https://doi.org/10.1039/D5NJ00587F.

[69] Wu J, Torresi L, Hu MM, Reiser P, Zhang J, Rocha-Ortiz JS et al. Inverse design workflow discovers hole-transport materials tailored for perovskite solar cells, Science. 2024; 386: 1256–1264. https://doi.org/10.1126/SCIENCE.ADS0901.

[70] Chen W, Mularso KT, Jo B, Jung HS. Indoor light energy harvesting perovskite solar cells: from device physics to AI-driven strategies. Materials Horizons 2025;12:3691–3711. https://doi.org/10.1039/D5MH00133A.

[71] Perovskite Solar Cells Could Facilitate More Versatile PV Production in the U.S. | American Solar Energy Society n.d. https://ases.org/perovskite-solar-cells-could-facilitate-more-versatile-pv-production-in-the-us/ (accessed June 10, 2025).

[72] Perovskite solar panels: an expert guide [2025] n.d. https://www.sunsave.energy/solar-panels-advice/solar-technology/perovskite (accessed June 10, 2025).

Index

https://doi.org/10.1515/9783111726847-013

www.ingramcontent.com/pod-product-compliance
Lightning Source LLC
Chambersburg PA
CBHW061339210326
41598CB00035B/5827